Matthew W. King

Political Ecology of Mangroves in Southern Honduras

Matthew W. King

Political Ecology of Mangroves in Southern Honduras

Emergence and Evolution of Environmental Conflict in the Gulf of Fonseca 1973 - 2006

VDM Verlag Dr. Müller

Impressum/Imprint (nur für Deutschland/ only for Germany)

Bibliografische Information der Deutschen Nationalbibliothek: Die Deutsche Nationalbibliothek verzeichnet diese Publikation in der Deutschen Nationalbibliografie; detaillierte bibliografische Daten sind im Internet über http://dnb.d-nb.de abrufbar.

Alle in diesem Buch genannten Marken und Produktnamen unterliegen warenzeichen-, marken- oder patentrechtlichem Schutz bzw. sind Warenzeichen oder eingetragene Warenzeichen der jeweiligen Inhaber. Die Wiedergabe von Marken, Produktnamen, Gebrauchsnamen, Handelsnamen, Warenbezeichnungen u.s.w. in diesem Werk berechtigt auch ohne besondere Kennzeichnung nicht zu der Annahme, dass solche Namen im Sinne der Warenzeichen- und Markenschutzgesetzgebung als frei zu betrachten wären und daher von jedermann benutzt werden dürften.

Coverbild: www.purestockx.com

Verlag: VDM Verlag Dr. Müller Aktiengesellschaft & Co. KG
Dudweiler Landstr. 99, 66123 Saarbrücken, Deutschland
Telefon +49 681 9100-698, Telefax +49 681 9100-988, Email: info@vdm-verlag.de

Herstellung in Deutschland:
Schaltungsdienst Lange o.H.G., Berlin
Books on Demand GmbH, Norderstedt
Reha GmbH, Saarbrücken
Amazon Distribution GmbH, Leipzig
ISBN: 978-3-639-20264-9

Imprint (only for USA, GB)

Bibliographic information published by the Deutsche Nationalbibliothek: The Deutsche Nationalbibliothek lists this publication in the Deutsche Nationalbibliografie; detailed bibliographic data are available in the Internet at http://dnb.d-nb.de .

Any brand names and product names mentioned in this book are subject to trademark, brand or patent protection and are trademarks or registered trademarks of their respective holders. The use of brand names, product names, common names, trade names, product descriptions etc. even without a particular marking in this works is in no way to be construed to mean that such names may be regarded as unrestricted in respect of trademark and brand protection legislation and could thus be used by anyone.

Cover image: www.purestockx.com

Publisher:
VDM Verlag Dr. Müller Aktiengesellschaft & Co. KG
Dudweiler Landstr. 99, 66123 Saarbrücken, Germany
Phone +49 681 9100-698, Fax +49 681 9100-988, Email: info@vdm-publishing.com

Printed in the U.S.A.
Printed in the U.K. by (see last page)
ISBN: 978-3-639-20264-9

Political ecology of mangroves in southern Honduras:

the emergence and evolution of environmental conflict in the Gulf of Fonseca
1973–2006

Matthew Wilburn King
Pembroke College
June 2008

Declaration

This dissertation is the result of my own work and includes nothing that is the outcome of work done in collaboration except where specifically indicated in the text. All Spanish translations are my own unless otherwise noted. The contents are my responsibility and do not reflect the views of the US Government, except where cited.

This dissertation is 75,509 words in length.

Signature:

Contents

Dedication...i

Acknowledgements...ii

Mangrove Wetlands...iv

Abstract...v

Abbreviations...vi

List of Figures...xii

List of Tables...xiii

List of Maps..xiv

Table A.1: Chronology of Events History and Ecology........................xv

Table A.2: List of Honduran Presidents 1973–2006..........................xix

Table A.3: Average Annual Exchange Rate (Honduran Lempiras against US
    Dollars) from 1980 to 2007...xx

Figure A.1: Jorge Varela, Executive Director of CODDEFFAGOLF....................xxi

Part I: Conflict, Theory and Geography

Chapter 1: The Global Aquaculture and Environment Debate............................1

Chapter 2: Poststructuralist Political Ecology Framework...................16

Chapter 3: Geographic Overview..34

Part II: The Emergence of Neoliberalism and Aquaculture 1973–1988

Chapter 4: The Emergence of Aquaculture, Southern Honduras: 1973–1988....57

Chapter 5: State Discourse and the Legal Construction of the Mangrove
    Wetlands..75

Chapter 6: Actions of the State, USAID, and International Financial
    Institutions for Aquaculture Development......................................90

Chapter 7: The Honduran Environmental Movement, Local Actors and
    Alternative Discourses on the Mangrove Wetlands........................113

Part III: Neoliberal Futures and the Rise of Contestation 1988-1998

Chapter 8: The Rise of Contestation to Shrimp Aquaculture............................136

Chapter 9: The Formation of Transnational Environmental Political Networks..149

Chapter 10: Expansion of the Fisheries and Aquaculture Sectors.....................162

Chapter 11: Neoliberal and Anti-Neoliberal Coalitions...............................179

Part IV: Post Hurricane Mitch: 1998–2006
      Towards Neoliberal Environmental Governance

Chapter 12: Reconstruction and the Convergence of Neoliberal and
      Environmental Discourse...205
Chapter 13: Violence and Protected Areas..222
Chapter 14: Conclusion...245

Appendices
Appendix 1   Research Design and Methodology...256
Appendix 2   Socioeconomic Questionnaire and List of Respondents...........265
Appendix 3   Estimates of Mangrove Cover in the Gulf of Fonseca.............291
Appendix 4   Shareholders: Latin American Agribusiness Development
              Corporation S.A..294
Appendix 5   Choluteca Declaration 16 October 1996................................296
Appendix 6   2006 Aquaculture Law...300
Appendix 7   World Wildlife Fund for Nature's Position Statement on
              Aquaculture...302
Appendix 8   Honduran Laws Consulted..305
Appendix 9   Interview list 307

Bibliography...324

Dedication

. . . where epiphany and serendipity beckon the soul.

Perseverance Trail
Ebner Falls,
Mt. Roberts
Juneau, Alaska
April 2001

58.312ºN   134.376ºW

Acknowledgements

I am indebted to a number of people who have helped in a variety of ways during the research and writing of this thesis.

I wish to begin by acknowledging my wife, Lina Maria Barrera. Without her, the dissertation would have been my 'tomb,' instead it became a chrysalis from which I was able to emerge. She endured the ups-and-downs of my emotions: excitement, self-doubt, frustration, and happiness over an extensive period of time. As the English winter etiolated my soul, she was my strength and light where there was none. I shall be grateful in perpetuity for her unbelievable ability to give when I had nothing to share, to remain empathetic while I toiled selfishly in solitude, and to revitalize my spirit when I thought it might dissipate into an existential vacuum. Fortunately, I have the entirety of my life to give that which I have received, along with enduring love for her elegant and intelligent style!

I am extraordinarily grateful for the support of my supervisor, Dr. Tim Bayliss-Smith for providing me with the opportunity to complete this work within the Department of Geography at Cambridge. Gratitude is extended to my internal and external examiners, Dr. Sarah Radcliffe (Senior Lecturer in Latin American Geography and Fellow, New Hall Cambridge) and Dr. Tim Forsyth (Senior Lecturer Environment and Development, Development Studies Institute, London School of Economics and Political Science).

I would not have been able to accomplish this research without the support of the National Oceanic and Atmospheric Administration (NOAA), in particular NOAA Research and the Office of International Activities (Mr. Rene Eppi, Dr. Ron Baird, Dr. Jim Murray, Dr. Leon Cammen, Captain Craig McLean, and Dr. Richard Spinrad were supportive in numerous and varied ways. I would like to extend special thanks to Ms. Janelle Bruce and Ms. Jill Hepp, both were a tremendous help navigating the ships of bureaucracy.

My initial work in the Central American region would not have been possible without the encouragement of Mr. David Alarid, US Diplomat, Central American Regional Environmental Hub Office, US Embassy, Costa Rica. Our initial encounter at the CCAD/CARICOM conference in Belize City, Belize led to a fruitful relationship and, eventually, a US Department of State Oceans, Environment, and International Scientific Affairs Initiative grant. The grant set me on a path that permitted me to develop an intimate knowledge of the Central American region, in particular the Gulf of Fonseca.

The Pan-American Agricultural School, University of Zamorano, Honduras assisted in numerous ways. Dr. Daniel Meyer was extraordinarily supportive, encouraging, and helpful throughout the course of this research. He is truly one of the most humble gentlemen that I have encountered in my life. Dr. Meyer's generosity equals that of any Honduran's. I would also like to thank his wife, Suyapa, for opening up their home whilst I was visiting or conducting my fieldwork. I am particularly indebted to Dr. Arie Sanders and his employees in the Department of Socioeconomics, Environment, and Development. Mr. Alfredo Kaegi's and Ms. Maria Cecilia Peña Paz's research assistance made my life easier than I could have imagined from the outset. I would also like to thank the numerous individuals in Honduras and throughout the Central American region that were supportive and willing to provide information as I worked to complete the objectives of this thesis.

Given that I am rarely constrained by a lack of gratuitous sentiments, I would also like to thank the following for their helpful comments, discussions, and development of the final product: my wife (Lina), Dr. Sarah Beth Wilkerson, Mr. Steven Hunt, Dr. Denise Stanley, Dr. Sarah Gammage, Dr. Anja Nygren, Dr. Steve Cohn, Dr. James Tobey, Dr. Barbara Bodenhorn, Dr. Loraine Gelsthorpe, Mr. Stephen Olsen, Dr. Dominique Gautier, Mr. J. Romero, Ms. Laura Sosa, Mr. Jeremy Bristow, Ms. Agnes Saborio Coze, Ms. Elsa Zamora, Mr. Jorge Varela, Mr. Edas Muñoz and numerous employees in the US and Honduran governments along with all those that provided me with their time to discuss coastal resource issues in Central America and the Gulf of Fonseca.

I would also like to thank Terry and Sharon Wilburn (my parents), Lance and Michael Wilburn (my brothers), and a seemingly endless number of others that made the completion of this thesis possible, directly and indirectly. Finally, I would like to extend a load of gratitude to Sir Rupert Debonair the Second! Corky, you'll always hold a place in my heart. Pug and Sprinkles, well . . .

I would like to thank the following for their financial support: US Department of State, US Department of Commerce's National Oceanic and Atmospheric Administration, University of Zamorano (Pan-American Agricultural School); Cambridge Philosophical Society; Lundgren Research Award; Worts Bartle Frere Travel Scholarship; Searl Fund, Franklin and Marshall Trust Fund, Pembroke College; Society for Latin American Studies (UK); The Grovesnor-Shilling Bursary in Land Economy, University of Cambridge; Cambridge European Trust, University of Cambridge; Philip Lake II Fund and the William Vaughn Lewis Fund, Department of Geography; and last, but not least, The Free Press Society, Cambridge (mostly in the form of cocktails).

Mangrove Wetlands

Southern Honduras, Gulf of Fonseca
Photo: Courtesy of CODDEFFAGOLF

Abstract

Political Ecology of Mangroves in southern Honduras: The Emergence and Evolution of Environmental Conflict in the Gulf of Fonseca 1973 - 2006
Matthew Wilburn King

My thesis thoroughly analyzed a conflict over access to and control over the mangrove wetlands of the Gulf of Fonseca, southern Honduras between 1973 and 2006. The conflict is related to changes in the local labour market and the environment due to the promotion of shrimp aquaculture in the coastal zone. The research employed a poststructuralist political ecology framework, combining an actor-network approach to discourse analysis with event history and ecology. The objective was to analyze the discourses and actions that have framed the conflict socio-historically and to identify each actor's associated political and economic interests in relation to the physical geographic space in dispute. The result was the development of an analytical tool that led to a more cogent apperception of the historical nature of the environmental conflict explicated. This approach was used to adduce the discourses, interests and actions of the social actors involved in the conflict, as well as their postulations associated with specific socio-economic and ecological outcomes, within the context of crucial events in the struggle over discursive and physical geographic space.

The results affirm that this environmental conflict emerged due to two distinct but inter-related social imaginaries. Both neoliberalism and environmentalism have increasingly become powerful ideologies, dominating the debates related to 'environment and development' in Honduras. Both of these ideologies have been pursued through discursive means and actions that ultimately have affected the lives of local people and the mangrove ecosystem of the Gulf. My argument is that the conflict was manifested as the result of ordinary people resisting the impacts consequent on the Honduran government's adoption of policies rooted in neoliberal ideology and perpetuated through their discourses and actions. Powerful social actors affiliated with a number of institutions were backing these policies. Due to the hegemonic tendency of neoliberal ideology, both discursive and physical spaces related to the coastal wetlands of southern Honduras have been affected. Local resistance to the Honduran government and the shrimp industry's actions led civil society actors to re-imagine these spaces within the public realm, leading to nascent forms of environmental governance within the context of democratic and capitalist transformations in Honduras, reshaping rural coastal livelihoods and the environment.

Key words: political ecology, Honduras, shrimp aquaculture, Central America, environmental governance

Abbreviations

ACC	Aquaculture Certification Council
AFE	Administración Forestal del Estado [State Forest Administration]
AFE–COHDEFOR	Administración Forestal del Estado–Corporación Hondureña de Desarrollo Forestal [Honduran Corporation for Forestry Development]
AHE	Asociación Hondureña de la Ecología [Honduran Ecological Association]
ALIDES	Alliance for Sustainable Development
ALCAHUMO	Asociación Local de Caballeros Humanos Montañeses [Local Association of Mountain Gentlemen]
AMADHO	Asociación de Madereros de Honduras [Loggers Association of Honduras]
ANDAH	Asociación Nacional de Aquacultores Hondureños [National Aquaculture Association of Honduras]
ANDI	Asociación Nacional de Industrialistas [National Association of Industrialists]
ANEXHON	Asociación Nacional de Exportadores Hondureños [National Association of Honduran Exporters]
ARD	Associates in Rural Development
ARMASUR	Asociación de Recolectores de Mariscos del Sur [Association of Southern Shellfish Harvesters]
ASOMAR	Asociación de Marisqueros del Sur [Association of Shellfishermen of the South]
BANADESA	Banco Nacional de Desarrollo Agrícola [National Bank for Agricultural Development]
BANAFOM	Banco Nacional de Fomento [National Bank for Development]
BANASUPRO	Suplidora Nacional de Productos Básicos [National Supplier of Basic Products]
BAP	Best Aquaculture Practices
BCH	Banco Central de Honduras [Central Bank of Honduras]
BCIE	Banco Centroamericano para Integración Económica [Central American Bank for Economic Integration]
CA	Central America
CAFTA	Central American Free Trade Agreement
CARE	CARE International
CARICOM	Caribbean Economic Community
CBERA	(US) Caribbean Basin Economic Recovery Act
CBI	(US) Caribbean Basin Initiative Act
CCAD	Comisión Centroamericana de Ambiente y Desarrollo [Central American Commission on Environment and Development]

CDI	Centro de Desarrollo Industrial [Center for Industrial Development]
CIA	(US) Central Intelligence Agency
CIFAR	Industria Acuicultura de Honduras [Aquaculture Industry of Honduras]
CITES	Convention on the International Trade of Endangered Species of Wild Flora and Fauna
CODDEFFAGOLF	Comité para la Defensa y Desarrollo de la Flora y la Fauna del Golfo de Fonseca [Committee for the Defense and Development of the Flora and Fauna of the Gulf of Fonseca]
CODESUR	Comisión de Desarrollo de la Región Sur [Commission for the Development of the Southern Region]
COHEP	Consejo Hondureño de la Empresa Privada [Honduran Council for Private Business]
CONAACUIH	Consejo Nacional de Acuicultura de Honduras [National Council for Honduran Aquaculture]
CONAFEXI	Consejo Nacional para Promover Exportación e Inversión [National Council to Promote Exports and Investment]
CONAMA	Consejo Nacional para el Medio Ambiente y el Desarrollo [National Commission for Environment and Development]
CONAPH	Consejo Nacional de Areas Protegidas de Honduras [National Comission of Honduran Protected Areas]
CONCAUSA	Central American-United States Joint Accord
CONGESA	Consultores en Gestión Ambiental [Consultants for Environmental Implementation]
CONSUPLANE	Secretaria Técnica del Consejo Superior de Planificación Económica [Technical Secretariat for the Superior Council on Economic Planning]
CRISUR	Conjunto Centroamérica USA [Central American USA Agreement]
CRSP	Collaborative Research Support Program
CVC	Comisión para la Vigilancia, Verificación y Control de los Recursos Natrales del Golfo de Fonseca [Commission for the Vigilance, Verification, and Control of the Natural Resources of the GOF]
DANIDA	Danish Agency for International Development
DCP	Departamento de Caza y Pesca [Department of Fish and Game]
DGA	Dirección General del Medio Ambiente [Director General of the Environment]
DIFOCOOP	Cooperativa Regional de Pescadores del Golfo de Fonseca [Regional Fishermmen's Cooperative of the Gulf of Fonseca]
DIGEPESCA	Dirección General de Pesca y Acuacultura [General Directorate of Fisheries and Aquaculture]
DNI	Dirección Nacional de Investigaciones [National Directorate of Investigations]
DOC	(US) Department of Commerce
DOD	(US) Department of Defense

EAP	Escuela Agrícola Panamericana [Panamerican Agricultural School/University of Zamorano]
ECA	Environmental Cooperation Agreement
ECLAC	Economic Commission for Latin America and the Caribbean
EDF	Environmental Defense Fund
EEC	European Economic Community
EEP	Environmental Education Programme
EEZ	Exclusive Economic Zone
EIE	Environmental Impact Evaluation
EIS	Environmental Impact Studies
EJF	Environmental Justice Foundation
ENA	Escuela Nacional de Agricultura [National Agricultural School]
ENGO	Environmental Non-Governmental Organisation
ENP	Empresa Nacional Portuario [National Port Council]
ESA	Endangered Species Act
ESF	Economic Support Funds
ESNACIFOR	Escuela Nacional de Ciencias Forestales [National School of Forest Science]
FAO	Food and Agricultural Organisation
FECOPRUH	Federación de Colegios Profesionales Universitarios de Honduras [Federation of Professional Colleges and Universities of Honduras]
FEDAMBIENTE	Federación de Organizaciones Ambientales e No Gubernamentales Hondureñas [Federation of Honduran Non-Governmental Environmental Organisations]
FENAGH	Federación Nacional de Agricultores y Ganaderos de Honduras [Honduran National Federation of Agriculturalists and Ranchers]
FEPROEXAAH/FPX	*See* FPX
FHIA	Fundación Hondureña de Investigación Agrícola [Honduran Foundation for Agricultural Research]
FHIS	Fundación Hondureña de Inversiones Sociales [Honduran Foundation for Social Investment]
FIDE	Fundación para la Investigación y el Desarrollo Empresarial [Foundation for the Research and Development of Business]
FOE	Friends of the Earth
FPX	Federación de Agroexportadores de Honduras [Federation of Agro-Exporters of Honduras, also known as Honduran Federation of Agricultural and Agro-Industrial Producers and Exporters (FEPROEXAAH/FPX)]
FSP	Fuerzas Seguridades del Publico [Public Security Forces]
FTA	Free Trade Agreement
FTAA	Free Trade Agreement of the Americas

FUNDAHEH	Fundación Nacional de Desarrollo de Honduras [National Development Foundation of Honduras]
GAA	Global Aquaculture Alliance
GDP	Gross Domestic Product
GEMAH	Gerentes y Empresarios Asociados de Honduras Associated Managers and Entrepreneurs of Honduras
GGMSB	Grupo Granjas Marinas San Bernardo
GOF	Gulf of Fonseca
IAF	Inter-American Foundation
ICJ	International Court of Justice
ICZM	Integrated Coastal Zone Management
IDB	Inter-American Development Bank
IGO	Inter-Governmental Organisations
IFC	International Finance Corporation
IFI	International Financial Institution
IHDECOOP	Instituto Hondureño de Cooperativas [Honduran Institute for Cooperatives]
IHT	Instituto Hondureño de Turismo [Honduran Institute of Tourism]
IICA	Inter-American Institute for Cooperation on Agriculture
IMF	International Monetary Fund
INA	Instituto Nacional Agrario [National Agrarian Institute]
INFOP	Instituto Nacional de Formación Profesional [National Institute for Professional Development]
ISA-net	Industrial Shrimp Action Network
ISI	Import Substituting Industrialisation
ISO	International Standardisation Organisation
ITTO	International Tropical Timber Organisation
IUCN	International Union for the Conservation of Nature and Natural Resources (aka World Conservation Union)
LAAD	Latin American Agribusiness Development
LBT	Latinoamericana de Bosques Tropicales [Latin American Network for Tropical Forests]
LMDSA	La Ley para la Modernización y Desarrollo del Sector Agrícola [Law for the Modernisation and Development of the Agricultural Sector]
LME	Large Marine Ecosystem
MAP	Mangrove Action Project
MBC	Meso-American Biological Corridor (formerly Paseo Pantera)
NAFTA	North America Free Trade Agreement
NDP	National Development Plan
NGO	Non-Governmental Organisation
NOAA	(US) National Oceanic and Atmospheric Administration
NRDC	Natural Resources Defense Council

NSM	New Social Movement
NTAX	Non-traditional Agricultural Exports
OAS	Organisation of American States
OECD	Organisation for Economic Cooperation and Development
OIMIT	Organización Internacional de las Maderas Tropicales
OLDEPESCA	Organización Latinoamericana de Desarrollo Pesquero [Latin American Organisation for Fisheries Development]
PACOH	President of Conservation Professors
PDO	Private Development Organisation
PFZ	Protected Forest Zone
PINU	Partido Innovación y Unidad Party of Innovation and Unity
PLH	Partido Liberal de Honduras [Liberal Party of Honduras]
PNH	Partido Nacional de Honduras [National Party of Honduras]
PRADEPESCA	Proyecto de Apoyo al Desarrollo de la Pesca en el Istmo Centroamericano [Project to Support the Development of the Fishing Industry on the Central American Isthmus]
PROARCA	Programa Regional Ambiental para Centro América [Regional Environmental Programme for Central America]
PROGOLFO	Proyecto del Ordenamiento del Ecosistema del Golfo de Fonseca [Conservation Project for the Ecosystems of the Gulf of Fonseca]
PROMANGLE	Manejo y Conservación de los Manglares del Golfo de Fonseca, Honduras [Conservation and Management of the mangroves in the Gulf of Fonseca, Honduras]
RAMSAR	Convention on Wetlands of International Importance
REDES	Red para el Desarrollo Sostenible de Centroamérica [Network for the Sustainable Development of Central America]
RENARE	Dirección General para Recursos Naturales Renovables [General Directorate for Renewable Natural Resources]
SAG	Secretaría de Agricultura y Ganadería [Secretariat for Agriculture and Cattle Ranching]
SAP	Structural Adjustment Programme
SECPLAN	Unidad de Planificación Regional de la Secretaria de Coordinación, Presupuesto, y Planificación [Secretariat of Planning, Coordination and Budget]
SECTUR	Secretaria de Cultura y Turismo [Secretariat for Culture and Tourism]
SEDA	Secretaría de Desarrollo y Ambiente [Secretariat for Environment and Development]
SERNA	Secretarías de Recursos Naturales y del Medio Ambiente (formerly SEDA) [Secretariat for Natural Resources and Environment]
SFI	Sea Farms International
SICA	Sistema de Integración de Centroamérica [System for Regional Integration]
SINAPH	Sistema Nacional de Areas Protegidas de Honduras [National System of Protected Areas]

SINEIA	Reglamento Sistema Nacional de Evaluación de Impacto Ambiental [National System for the Regulation and Evaluation of Environmental Impacts]
SRN	Secretaria de Recursos Naturales [Secretariat of Natural Resources]
TED	Turtle-Excluding Device
TNC	The Nature Conservancy
TOR	Terms of Reference
TRD	Tropical Resources and Development Institute
UMA	Municipal Environmental Unit
UN	United Nations
UNCED	UN Conference on Environment and Development
UNCSD	United Nations Commission for Sustainable Development
UNAH	Universidad Nacional Autonomous de Honduras [National Autonomous University of Honduras]
UNDP	United Nations Development Program
USAID	United States Agency for International Development
WCS	Wildlife Conservation Society
WWF	World Wildlife Fund
WTO	World Trade Organisation

List of Figures

Figure A.1: Jorge Varela, Executive Director of CODDEFFAGOLF

Figure 1.1: 'Honduras Peeled King Prawns'

Figure 2.1: Sunset over a shrimp pond in the Gulf of Fonseca

Figure 3.1 Salt Flats

Figure 3.2: Depiction of the location from which remittances originate

Figure 3.3 'Forced Return'

Figure 3.4a: Mangrove wetlands

Figure 3.4b: Mangrove wetlands

Figure 3.5: Variation in mangrove forest types

Figure 3.6: COHDEFOR/PROMANGLE reforestation project in the Gulf of Fonseca

Figure 5.1: Development of land concessions for shrimp operations in Southern Honduras (1985–2000)

Figure 6.1: Satellite image of GGMSB's operations in 1985

Figure 6.2: GGMSB's main operations in San Bernardo

Figure 6.3: GGMSB processing facilities in San Lorenzo

Figure 7.1: Salt production

Figure 7.2: Woman cooking with mangrove wood

Figure 7.3: Restaurant constructed with red mangrove wood

Figure 7.4: Example of mangrove clearance for shrimp pond construction

Figure 7.5: Shrimp lifecycle *penaeidos*

Figure 7.6: CODDEFFALGOLF's Junta Directiva

Figure 8.1: CODDEFFALGOLF headquarters in San Lorenzo, Southern Honduras

Figure 8.2: Diminished access to traditional fishing grounds

Figure 10.1: Satellite image of GGMSB's operations in 1989

Figure 10.2: Local fisherman

Figure 11.1: The Moby Dick sailing into the Gulf of Fonseca

Figure 11.2: Satellite image of GGMSB's operations in 1997

Figure 11.3: Protests against shrimp farm expansion

Figure 11.4: Mangrove Wetlands near Punta Condega

Figure 12.1: Shrimp production in Honduras, 1985–2003

Figure 12.2: GGMSB's operations in 2003 in San Barnardo and other shrimp operations in the region

Figure 13.1: Perceptions of the household respondents in relation to how violent they perceived the conflicts surrounding the mangrove wetlands

Figure 13.2: Fisherman/protester being arrested

List of Tables

Table A.1:   Chronology of Events History and Ecology

Table A.2:   List of Honduran Presidents 1973–2006

Table A.3:   Average Annual Exchange Rate (Honduran Lempiras against US Dollars) from 1980 to 2007

Table 2.1:   Coastal communities selected for the socio-economic analysis

Table 3.1:   Characteristics of the Department of Choluteca and Valle, Southern Honduras

Table 3.2:   Species of mangrove found in the Gulf of Fonseca

Table 3.3:   Species of mangrove and annual inundations

Table 3.4:   Direct use of the areas of the mangrove forests in the Gulf of Fonseca

Table 3.5:   The rate of change in coverage and corresponding area 1989. 1995. 1998

Table 3.6:   Loss in coverage of mangrove forest in the Gulf of Fonseca

Table 3.7:   Distribution of mangrove loss in hectares due to shrimp farming and salt production

Table 4.1:   Shrimp farms: concessions issued or in process of adjudication between 1984 and 1988

Table 7.1:   Amount of shrimp exported, by weight and value: 1986-1988

Table 7.2:   Lands conceded for the development and production of shrimp in Southern Honduras 1985–1996

Table 11.1:   Lost coverage of mangrove forest in the Gulf of Fonseca

Table 12.1:   Hurricane Mitch: number of deaths, missing and injured as of 14 November 1998

Table 12.2:   Distribution of lands of ANDAH affiliates in operation in 1997, prior to Hurricane Mitch

List of Maps

Map 3.1:   Central America
Map 3.2:   Honduras
Map 3.3:   The Gulf of Fonseca
Map 3.4:   Mangrove wetlands of the Gulf of Fonseca
Map 8.1:   The Gulf of Fonseca with San Bernardo and Punta Condega
Map 13.1:   Protected areas in the south
Map 13.2:   Protected areas in the Departments of Valle and Choluteca
Map 13.3:   Geographic division of the mangrove forest zones in Southern Honduras
Map 13.4:   Protected areas and general division of forestry administration

TABLE A.1: CHRONOLOGY OF EVENTS HISTORY AND ECOLOGY

Event	Year
Fisheries Law	1959
Sea Farms Incorporated, Delaware, USA	1966
FAO Completes a Study on Potential Land Use in S. Honduras	1968
Shrimp Experimental Laboratories Established in Florida, USA	1972
Government Issues Concessions for First Shrimp Pond Built in S. Honduras	1973
Forestry Law Passed	1972
COHDEFOR Law Passed	1974
First Newspaper Article on NTAX published	1975
Chinese and Taiwanese Technical Assistance Sent to Honduras	1975
Government Takes Action to Promote Aquaculture Development Further	1979
Government Creates DIFOCOOP (Fishermen's Cooperative)	1980
Decree 968 of 1980 Passed: Transfer of Coastal Land Use Rights	1980
US begins supporting Contras against the Sandinistas in Nicaragua	1980
Honduran Ecological Association (AHE) Created	1981
UN Law of the Sea	1982
Caribbean Basin Initiative Act Passed by US Congress	1983
USAID Initiates Cooperative Support Research Programme (CRSP) in Honduras	1983
Inter-American Development Bank (IDB) Develops Fisheries Sector	1984
First Newspaper Article Associated with Mangrove Degradation Published	1984
FPX, FHIA, and FDIE Established	1984
IDB Recognises Lack of Coastal Management Plan in Honduras	1984
Salt Producers and Artisan Fishermen Trained in Aquaculture	1984
Temporary Import Tax Exemption Decree 37 Passed	1984
International Finance Corporation (IFC) Invests in Aquaculture in the S. Honduras	1984
INA Continues to Promote Expansion of Aquaculture	1986
AHE Establishes wider-campaign to protect the GOF	1986
ANDAH Established by the Aquaculture Industry	1986
USAID Invests Further to Support Growth in Aquaculture Industry	1986
USAID PTR Programme Initiated	1988
CODDEFFAGOLF Created by Artisan Fishermen and Jorge Varela, GOF	1988
First Forest Management Meeting in Honduras	1989
Central American Commission for Environment and Development (CCAD) Created	1989
Enterprise for the Americas Initiative Launched by President Bush Sr.	1989
Territorial Boundaries Increasingly Contested Nicaragua, Honduras, and El Salvador	1990
OLDEPESCA Supports Growth of the Industry Further	1990
Municipalities Law Passed	1990
Presidential Decree 041-90 Passed	1990

Tegucigalpa Protocol: Charter of the Organisation of Central American States Signed	1991
General Directorate of Fisheries and Aquaculture (DIGEPESCA) Created	1991
First Central American Aquaculture Conference Held	1991
First Suspension of the Capture of Wild Shrimp Postlarvaes	1991
First General Environmental Law Passed by Honduran National Congress	1992
Presidential Decree 1118-92 Passed Reserves and Protected Areas	1992
International Court of Justice (ICJ) Rules on Territorial Dispute between HN & ES	1992
ANDAH Establishes its First Sustainable Development Policy & New Articles of Assoc.	1992
First UN Conference on Environment and Development (UNCED) in Brasil	1992
Aquaculture Incorporated into University Curriculum	1993
Ministry of Natural Resources Announces Restrictions on NTAX	1993
COHDEFOR Elaborates Regulations on Protected Areas	1993
National System of Protected Areas (SINAPH) Created	1993
Standards for the Certification of Shrimp Aquaculture Operations Developed	1993
First Suspension of Land Concessions	1993
WWF Awards CODDEFFAGOLF J. Paul Getty Award	1993
ANDAH Water Quality Programme Commences	1993
First Co-Management Regime Established for the GOF (CVC)	1994
Alliance for Sustainable Development (ALIDES) Created	1994
CONCAUSA Signed between Central American Governments and USA	1994
Greenpeace Sails Moby Dick into the GOF to Protest Shrimp Farming	1994
3,000 Hondurans Protest the Industry, Organised by CODDEFFAGOLF/Greenpeace	1994
Greenpeace Announces its International Campaign to Protect the GOF	1994
PRADPESCA Promotes Aquaculture Projects in the GOF	1994
Local Agenda 21 for Honduras Created National Policy on the Environment	1994
Taura Virus Plagues Shrimp Farms	1994
Conference on Sustainability Sponsored by ANDAH to Educate Producers	1994
Regional Meeting Convened to Discuss Sustainable Development of the GOF	1994
Paseo Pantera Launched and Later Becomes Meso-American Biological Corridor	1994
ACTRIGOLFO Created: Trinational Civil Association (Transboundary Resource Mngt.)	1995
INA Signs Agreement w/ SINAPH (No titles to lands in Protected Areas)	1995
Transboundary Natural Resource Disputes Intensify between Nic, HN, El Salv.	1995
ANDAH Develops Code of Conduct for Aquaculture Practices	1995
Global Boycott on Shrimp Campaign Initiated by Greenpeace	1995
Government Moratorium on Shrimp Pond Development Exec. Decree 5-96	1995
USAID Launches PROARCA-Costas to address Marine Resource Issues in the GOF	1995
Tourism Development Plan for the South is Established	1996
5 Year Strategic Forestry Plan is Developed by COHDEFOR	1996
UN Shrimp Tribunal Held in NY, NY	1996
Choluteca Declaration is Signed	1996
CRIMASA, Finca Sur, and El Faro Sign Agreements to Give Access to Artisan Fishermen	1996

World Conservation Congress Convened Central American Marine Issues Addressed	1996
COHDEFOR Completes Estimate of Mangrove Forest Loss	1996
First Meso-American Congress of Protected Areas	1997
Industrial Shrimp Action Network (ISA-net) Established in CA, USA	1997
Honduran Congress Passes Decree No. 105-97 Moratorium on Shrimp Farming	1997
Global Aquaculture Alliance (GAA) Established by the Commercial Industry	1997
Grupo Deli Signs Agreement with CODDEFFAGOLF to Permit Access to Fishermen	1997
National University Develops First Management Plans for the Bay of Chismuyo	1997
Hurricane Mitch Devastates Honduras	1998
First Buoys Installed to Demarcate Territorial Boundaries in the Sea, GOF	1998
Honduran National Environmental Plan Completed	1998
WWF Releases Position Statement on Aquaculture	1998
Honduran Structural Reform Accelerates	1998
Conservation Project for the Coastal Ecoystem of the GOF Initiated (PROGOLFO)	1999
Honduran Congress Creates Protected Areas in the GOF, RAMSAR Site 1,000, and GOF becomes a part of the Meso-American Biological Corridor	1999
Honduras Signs Enhanced Structural Adjustment Package with the IMF	1999
World Bank and the IMF Declare Honduras Eligible for Debt Relief	1999
International Finance Corporation Invests $6 million into Aquaculture Industry	1999
International Tropical Timber Organisation (ITTO) Establishes PROMANGLE w/ COHDEFOR	1999
Jorge Varela Awarded the Goldman Environmental Prize	1999
Regional Workshop Held in Costa Rica on Legal Rules and Policies on Aquaculture	1999
Tribunal of Conscience Convened to address Wetland Degradation in the GOF	1999
Funds Released for Hurricane Mitch Reconstruction Efforts	1999
IMF and World Bank Provide $100 million for Honduras to Restructure	2000
Government and CODDEFFAGOLF Agree to a Truce on Shrimp Farm Expansion	2000
ANDAH and CODDEFFAGOLF work together to prevent illegal shrimp pond construction	2000
Demarcation of Territorial Boundaries in the GOF Continues	2000
Reformation on the Fisheries Law Begins	2000
Environmental Justice Foundation (EJF) Formed and Links with CODDEFFAGOLF	2000
Volvamos al Sur Campaign Initiated 'Return to the South'	2001
Industry Suffers Huge Financial Loss	2001
World Summit on Sustainable Development (WSSD) held in South Africa	2002
Government Restrictions Placed on Shrimp Farming	2002
US Department of State Gives OESI Grant to NOAA/OAR/IA to assess feasibility of adapting the US Sea Grant model of applied research, extension, and education for marine and coastal resource management in the GOF	2003
President Maduro Issues Statement on Protecting the Environment	2003
Smash and Grab is Published by EJF Attacks Shrimp Industry	2003
Central American Free Trade Agreement (CAFTA) Negotiations Launched	2003
NGOs Immediately Raise Concerns in Regard to Environment and Labour Issues related to CAFTA	2003

President Ricardo Maduro Gives National Environmental Award to GGM	2003
Meso-American Congress of Protected Areas Held	2003
White Water to Blue Water (WW2BW) Conference Held in Miami, FL	2004
Greenpeace, Mangrove Action Project, Red Manglar, & CODDEFFAGOLF launch global campaign against shrimp farming	2004
RAMSAR Guidelines Updated	2004
Ministry of Environment Begins Development of a Plan of Action for the GOF	2004
British Broadcasting Company (BBC) Releases Documentary on S. Honduras "Price of Prawns"	2004
Collaboration on New Fisheries and Aquaculture Law Commences	2004
CAFTA-DR Signed by US and Central American Countries	2005
Global Environment Facility/IDB Collaborate on GEF Grant for GOF	2006
Management Plans for the 1999 Protected Areas are completed	2007
GEF/IDB Integrated Coastal Zone Management Initiative for the GOF Commences	2007

TABLE A.2: LIST OF HONDURAN PRESIDENTS 1973–2006

President and Party	Dates in Office
Ramón Villeda Morales (PLH)	21 December 1957–3 October 1963
Oswaldo López Arellano	3 October 1963–7 June 1971
Ramón Ernesto Cruz Uclés	7 June 1971–4 December 1972
Oswaldo López Arellano (Head of State)	4 December 1972–22 April 1975
Juan Alberto Melgar Castro (Head of State)	22 April 1975–7 August 1978
Policarpo Paz García (Provisional)	7 August 1978–27 January 1982
Roberto Suazo Córdova (PLH)	27 January 1982–27 January 1986
José Azcona del Hoyo (PLH)	27 January 1986–27 January 1990
Rafael Leonardo Callejas (PNH)	27 January 1990–27 January 1994
Carlos Roberto Reina (PLH)	27 January 1994–27 January 1998
Carlos Roberto Flores Facussé (PLH)	27 January 1998–27 January 2002
Ricardo Maduro (PNH)	27 January 2002–27 January 2006
Manuel Zelaya (PLH)	27 January 2006–incumbent

PLH: Liberal Party of Honduras
PNH: National Party of Honduras

Retrieved from http://en.wikipedia.org/wiki/President_of_Honduras

TABLE A.3:   *AVERAGE ANNUAL EXCHANGE RATE (HONDURAN LEMPIRAS AGAINST US DOLLARS)* *FROM 1980 TO 2007*

Year	Exchange rate*	Source for this statistic
1980s	Fixed at an average of 2 Lempiras per US Dollar	Information located in: http://en.wikipedia.org/wiki/Honduran_lempira
1990	4.04	Information located in: Economic Survey of Latin America and the Caribbean 1998-1999, Table 1, Page 229. Published by the Economic Commission for Latin America and the Caribbean. Link: http://www.cepal.org/publicaciones/xml/8/4918/honduras_eng.pdf
1991	5.60	
1992	5.51	
1993	6.50	
1994	8.60	
1995	9.54	
1996	11.71	
1997	13.00	
1998	13.39	
1999	-------	Hurricane Mitch at the end of 1998 affected the government's ability to compile statistics for 1999
2000	15.05	Information located in: UK Forex Foreign Exchange Yearly Rate Calculater. Link: http://www.chartflow.com/ukforex/ averageRate.asp?period=yr&ccy1=USD&ccy2= HNL&days=3652&amount=1
2001	15.45	
2002	16.40	
2003	17.34	
2004	18.20	
2005	18.82	
2006	18.89	
2007	18.90	

*All figures represent the average nominal exchange rate for the year, rounded to 2 decimal places.

Mr. Jorge Varela: Executive Director of the Committee for the Defense and Development of the Flora and Fauna of the Gulf of Fonseca, Honduras. 2004

FIGURE A.1: JORGE VARELA, EXECUTIVE DIRECTOR OF CODDEFFAGOLF (CENTRE OF PHOTOGRAPH)
Source: CODDEFFAGOLF, 2004

Part I

Conflict, Theory and Geography

Chapter 1

The Global Aquaculture and Environment Debate

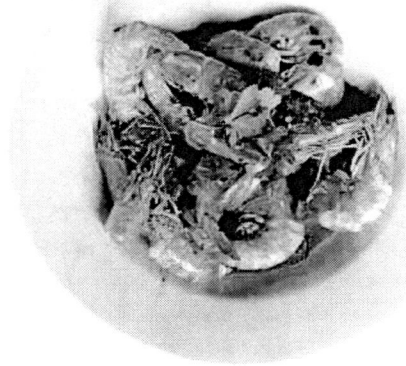

FIGURE 1.1: 'Honduran Peeled King Prawns, ready to eat from the tropical waters of Honduras, succulent and bursting with flavour! 280 grams only £5.99 – 0.62lb (Marks and Spencer packaging). *Source:* Environmental Justice Foundation (EJF) 2004

Introduction

It is a typical Saturday morning in Cambridge. King's Parade is milling with tourists and Market Square is full of students and locals strolling about. Located on one side of the square is Marks and Spencer, one of Britain's largest supermarket chains. If one were to walk into the seafood section one could find prawns or shrimp produced in Honduras via aquaculture.

On the packaging are the words "These large prawns are farmed in the warm waters of the Gulf of Fonseca, Honduras". While buying shrimp at Marks and Spencer, one of the largest buyers of Honduran King Prawns in the UK, the main concern of the average customer may well be whether or not to serve the shrimp with *cous cous* or salad, but some will wonder where the product originates, how it is produced, what types of conflicts surround the production of it, and even the socio-historical, political, economic, and environmental factors associated with the crustacean that they will soon wash down with a nice bottle of plonk from Oddbins. In Honduras, some people consider these questions important enough to fight, resist, and contest the corporations and the governments that have supported the production of this agro-export product for more than two decades. My thesis concerns those people in Honduras opposed to commercial shrimp farming and their role in the global aquaculture and environment debate.

The Global Aquaculture and Environment Debate

The global debate regarding aquaculture and the environment has emerged within the last two to three decades as countries have started using new economic technologies to produce a wide variety of molluscs, crustaceans, and fish in Africa, Latin America, Asia, and numerous developed countries. The resulting expansion of global aquaculture has become known as the "Blue Revolution" (*The New Internationalist*, August 1992; *The Economist*, 7 August, 2003). It can be attributed to a global decline in fisheries stocks, increased demand for marine species, and the efforts of primarily developing countries to identify ways to increase foreign exchange earnings. Although aquaculture (shrimp, fish, and mollusc farming) is practised in both developed and developing countries worldwide, it is most prominent in Asia which accounts for roughly 90 per cent of global aquaculture production; China alone contributes more than two-thirds of the total (Naylor et al, 2000: 1018). In terms of farmed crustaceans, cultured shrimp is the largest sector, and nearly 99 per cent of cultured shrimp are raised in the developing world, with nearly 72 per cent originating in Asia, while the rest comes from Latin America (Stonich and Bailey, 2000: 24). It is estimated that between 1987 and 1997, global production of farmed fish and shellfish more than doubled in weight and value, as did its contribution to world fish supplies (Naylor et al, 2000: 1017).

Shrimp farming has more recently become central to the global aquaculture-environment debate due to its proliferation throughout Asia, Africa, and Latin America over the last three decades. Shrimp farming has seen a 300 per cent increase in production between 1975 and 1985, and a 250 per cent increase between 1985 and 1995, making it the fastest-growing aquaculture product over the last three decades (Stonich and Bailey, 2000: 24). Overall, the expansion of the industry has been particularly rapid during the past decade: 30 per cent of the shrimp placed on the world market today comes from aquaculture (Boyd, 2001: 9). In 2003, 70 per cent of the shrimp produced in Honduras was exported to the US and the majority of the rest to Europe, a portion of which went to Marks and Spencer based in Britain. Coupled with

Spain, Britain accounted for $45 million of Honduras' total shrimp exports to Europe in the same year (*El Heraldo*, 17 July and 10 September, 2003).

The proliferation of the shrimp aquaculture industry throughout Latin America and worldwide can be directly attributed to the efforts of multilateral development banks, foreign governments, and private investors. These institutions have worked with countries to implement economic reforms that emphasise diversification of economic activities and the promotion of non-traditional agricultural exports (NTAX), such as aquaculture. Shrimp farming's rapid expansion has led to an increase in low level environmental conflicts as local people have contested the practices of the state and private interests. The corresponding politicisation of the social and environmental issues surrounding the industry pose particular challenges in relation to the management of marine and coastal resources (Stonich and Bailey, 2000: 24).

The conflicts surrounding shrimp farming in Southeast Asia, Africa, and Latin America are alike in some respects. In each case, the socio-political and environmental issues centre on the following discourses and claims: shrimp farming is associated with the destruction and subsequent loss of mangroves, poor water quality, salination of underground water supplies, displacement of local populations from coastal areas, enclosure of common pool resources, and loss of access to traditional fishing grounds (Boyd, 2001: 9). Some have even identified shrimp farming as the direct cause of violent conflict in a number of coastal areas (Stonich and Vandergeest, 1999: 261). I seek to thoroughly analyse these claims associated with the industry in southern Honduras through a case study of aquaculture development in the region of the Gulf of Fonseca (GOF) between 1973 and 2006.

My argument is that the conflict is the result of ordinary people resisting the impacts consequent on the Honduran government's adoption of policies rooted in neoliberal ideology and perpetuated through the discourses and actions of more powerful social actors affiliated with a number of institutions. I argue that the environmental conflict that has emerged in southern Honduras is about contested neoliberal futures. Neoliberal ideology, quite simply, is a set of doctrines or beliefs that can be linked to neoclassical economist Friedrich von Hayek, based at the University of Chicago in the 1960s. As an overarching concept, neoliberalism "refers to an intellectual and political movement that espouses economic liberalism [libertarian] as a means of promoting economic development and securing political liberty" (Portes, 1997: 230). Saad-Filho and Johnston argue that neoliberalism straddles a number of disciplines at different levels of complexity and is a particular organisation of capitalism rather than a specific mode of production (2005: 3). Although it is neither homogenously practiced nor a unified ideology, economic theory or political philosophy, it has been discussed analytically, as is accomplished in Chapter 4.

To assess the conflicts surrounding aquaculture that have emerged within the context of globalisation and capitalist transformation in Honduras, I have developed an analytical tool that links an actor-network approach to discourse analysis with event history and events ecology. Its application has revealed that local resistance to neoliberal change in southern Honduras has led to the creation of new social and political spaces as local actors have sought to redefine their future within the context of capitalist transformations. Furthermore, the analytical tool developed is unique in that it provided a means to assess the conflict discursively as actors have competed over physical geographic spaces.

The Geographic Context for the Research

FIGURE 1.1: SUNSET OVER A SHRIMP POND IN THE GULF OF FONSECA
Source: CODDEFFAGOLF

Due to its rich biodiversity, fertile soils, unique hydrology, and close proximity to the sea, the mangrove ecosystem of the Gulf of Fonseca (GOF) is the site of a variety of economic activities crucial to the southern Honduran economy. Several of these currently provide critical sources of export revenue and foreign exchange, particularly extensive agriculture and shrimp farming, which has recently become the third largest sector of the Honduran economy. The mangrove wetlands are also the site of numerous subsistence and small-scale economic activities such as salt production, collection of mangrove wood for fuel and construction, fishing and collection of crustaceans and molluscs for local consumption, and small-scale agriculture. Consequently, a number of local, national, and international social actors have varied political and economic interests in this region. Conflicts have thus arisen over access to and control of the natural resources as the Honduran government, supported by the US, International Financial Institutions (IFIs), and private investors, has embraced and enacted neoliberal ideology. The resulting policies have transformed the southern Honduran landscape for the production of shrimp over the last three decades, redefining the use of the GOF's wetlands. Consequently, the wetlands' commercial value has been exploited as investors, governments, and IFIs have capitalised the land for the commodification of shrimp. The ideas and actions underpinning the capitalist transformation of the coastal wetlands have generated conflicts within the artisan fishing community and between a segment of artisan fishermen, the government of Honduras, and the shrimp industry.

In particular, conflict between one group of the fishermen, represented by the Committee for the Defense and Development of the Flora and Fauna of the Gulf of Fonseca (CODDEFFAGOLF), and the shrimp industry, represented by the National Aquaculture Association of Honduras (ANDAH)[1], has become increasingly prominent since aquaculture commenced in the 1970s, and expanded throughout the 1980s and 1990s.

The conflicts surrounding aquaculture are shaped by a complex set of factors including tensions between local, national, and global political and economic interests. Furthermore, incongruent environment and development policies often fail to take into consideration local-level knowledge, customs, and traditions. I focus on these factors through a detailed local-level analysis of the social, political, and economic processes that have contributed to the emergence and evolution of environmental conflict in southern Honduras over the last three decades.

My aim is to place the discursive positions and actions of the actors, as well as their discursive claims associated with the social and ecological outcomes, within the context of key events in the struggle over physical space. My argument is that this conflict must not only be seen as a struggle over access to and control of a physical space but also a discursive conflict as narratives are a central means by which conflict is generated,

---

[1] When I refer to the industry I am generally referring to ANDAH. If I am referring to small-scale artisan shrimp farmers that are not members of ANDAH I will specify. ANDAH does not represent everyone who produces shrimp in the south, just the largest number of hectares under production.

sustained, mediated and historically represented at all levels of social organisation (Briggs, 1996: 3). Particular attention is placed on the discourses and actions underlying neoliberal reform in Honduras and those of the actors that have contested the reform objectives of the state and IFIs.

The discourses used by the governments and institutions aggrandising neoliberal ideas vary, and are diffused on a societal scale through various techniques of power and have more recently begun to dominate discursively within the economic and political spheres. Their dominance can be attributed to the institutional powers behind neoliberal ideas that translate ideas into policies, actions, and, subsequently, specific social and ecological outcomes. Powerful institutions at governmental and extra-governmental levels have been dominated by neoliberal ideologies for nearly two decades. However, resistance and contestation have arisen as the people affected by the policies and actions of these institutions have sought to mitigate the social, political, economic, and environmental impacts.

Neoliberalism has sought to reshape the state, civil society, and land use through the logic of the market and emphasises frequently-heard discourses and concepts such as free trade, privatisation, limited government intervention, and undistorted market prices. It is an ideological framework that empowers the market and private interests over the state and civil society and has underpinned the transformation of macro-economic and political structures worldwide. The process that has followed neoliberal economic reform is often referred to as globalisation, which some consider to be nefarious as it is often equated with the spread of global capitalism through government and military interventionism to secure the interests of transnational corporations (Robinson, 2003: 50-54). The proponents of neoliberalism, however, view it as a means towards economic growth, human development, and even freedom and liberty – totalising concepts used to legitimate its propagation.

The actors who translate neoliberal ideas into policy use differing techniques of power. I make this argument throughout and place particular emphasis on it in Part II by focusing on six techniques of power that the Honduran government, USAID, and IFIs, used to promote neoliberal reform in Honduras incrementally throughout the 1980s and more assertively in the 1990s. Although neoliberalism has hegemonic tendencies, it is not monolithic. The ideas that underpin it have generated innumerable counter-discourses within civil society; these have sought to redefine the future of neoliberalism by influencing both institutional policies and practices. Each is situated within a specific socio-historical context and is analysed as such.

In the case of Honduras, the social actors who have sought to change the narrative have opportunely acquired access to the discursive arena at a variety of scales within the social order, while creating new spaces within the public arena in which new alternatives have been imagined [discussed in Chapter 2]. Actors affiliated with the global environmental movement have successfully taken the local debates into the national, regional, and international arenas in order to reconsider how the mangrove wetlands of southern Honduras *ought* to be utilised, managed, and conserved. However, as attempts to put forth new conceptions of the physical space around the GOF have taken place, powerful actors have continued to advocate neoliberal approaches whilst co-opting the discourses of environmentalists. Furthermore, many local actors have been unable to participate in these debates due to a lack of access to the fora in which they take place: Davos, the G8, World Trade Organisation (WTO) meetings, World Bank annual meeting, the formal political arena, the media, and international conferences.

More specifically, rural coastal Hondurans are not only excluded from these fora, but are subject to a lack of state presence, institutional weakness, and relative isolation from the political apparatuses responsible for governing the marine and coastal resources on which they are dependent for sustenance. Those who have successfully found ways to articulate their interests within the public arena are often still denied access by more powerful actors to the physical space for which they compete. The physical and discursive marginalisation of these actors has resulted in some of the conditions that have motivated continuous forms of resistance against the state, industry, and IFIs that have been behind aquaculture development in southern Honduras over the last three decades.

The need for analysis

Johnston has suggested that empirical analyses of conflicts tend to focus on superficial aspects manifested as outcomes without appropriately developing a full understanding of the socio-historical contexts in which these conflicts have emerged and evolved (Johnston *et al.*, 2000: 106). Held likewise has argued that this form of analysis has resulted in cursory explanations of complicated processes of social change and leads to a lack of a sufficiently differentiated, concretely grounded, empirical analysis of the socio-historical trajectory of conflicts within the social sciences (Held, 1980: 389). Most approaches to more contemporary conflicts have also failed to consider the effects of neoliberalisation and globalisation (Demmers, 2005: 331) thus; I seek to complete a more thorough analysis by addressing each of these concerns.

I take these criticisms as cornerstones for the construction of this analysis by incorporating historical processes with a thorough assessment of the discursive strategies and actions that have taken place in relation to key events associated with the conflict. My aim is to examine the links between discourse and action in the social construction of environmental conflict in southern Honduras between 1973 and 2006 by utilising a poststructuralist political ecology framework (see Chapter 2). This is an important goal within the field of human geography, given that it is concerned with the social science aspects of how the world is discursively and physically arranged by people (Johnston *et al.*, 1986: 205-207).

Previous research in southern Honduras has significantly contributed to understanding current environment and development issues in the GOF but, more importantly, revealed the lack of knowledge associated with how conflicts surrounding aquaculture development have become known and metamorphosed within a specific socio-historical setting (Stonich, 1991; Vergne, *et al.*, 1993). Research focused on southern Honduras has generally situated analysis on a broader set of social, economic, environment, demographic, and political issues related to the region as a whole (Stonich, 1991; Vergne, *et al.*, 1993; Stanley, 1996). As a consequence, there is no single comprehensive study of the socio-historical context in which environmental conflict in the GOF has materialised and unfolded over time. This has complicated efforts to manage marine and coastal resource use in the region.

Some of this work has highlighted specific aspects of the conflicts surrounding the mangrove wetlands in the GOF. The work of Susan Stonich (1991), Philippe Vergne *et al.* (1993), Denise Stanley (1996, 1998, 1999a, 1999b), and Sarah Gammage (2000) are particularly notable since all are experts in their respective fields. Most non-academic publications pertaining to conflicts surrounding the mangrove wetlands in southern Honduras constitute the representation of the discursive positions of actors. As a result, international organisations, government agencies, NGOs, and universities interested in the management of marine and coastal resources in the GOF have inadequate knowledge of the socio-historical factors associated with the conflicts surrounding the wetlands. Furthermore, the lack of knowledge makes it exceedingly difficult to design appropriate institutional mechanisms for the governance of marine and coastal resources. Despite these biases, existing non-academic documents such as government reports or organisational pamphlets are invaluable when analysing social actors' discourses and actions related to specific events associated with the conflict. My thesis considers the multiple actors involved and attempts to ensure that local actors are not overlooked throughout the process of explanation.

Scope and research questions

The environmental conflict analysed in relation to disputes over the wetlands of the GOF is assessed as a struggle over physical geographic space that occurs through actions and discourse. I resist the temptation to reduce my analysis of social life to discourse, but at the same time I seek to demonstrate its importance in the lives of social actors by focusing on the link between discourse and action. My goal is to analyse discourse as an important variable in conflict while developing a further understanding of the power of discourse by assessing actions associated with various claims. The central enquiry centres upon understanding how this environmental conflict has been manifested, produced, and reproduced by a variety of social actors at micro and macro levels within the context of economic transformation led by neoliberal institutions and policies. As others have noted in other contexts, my data reveal the centrality of conflict to the constitution of social relations, institutions and ideologies, defined as a system of beliefs, meanings, and values (Briggs, 1996: 7-10).

In particular, here as elsewhere, environmental conflict in southern Honduras is the effect of neoliberal and environmental ideologies vying to influence political and socio-economic outcomes through discursive means and actions, ultimately affecting the lives of local people. To understand the conflict, my research combined an actor-network approach to discourse analysis with event history and events ecology (discussed in Chapter 2). The latter is referred to in this thesis as event history and ecology, which I combined in order to analyse the discourses and actions that have framed the conflict historically whilst also identifying each actor's associated political and economic interests in relation to the physical geographic space in dispute. The result was the development of an analytical tool that led to a more comprehensive understanding of the historical nature of the environmental conflict analysed.

My research had two primary goals. Firstly, to investigate how control and access to the physical geographic space of southern Honduras has been politicised, defined, and contested at the local, national, and international levels in order to apprehend the nature of the conflict. Secondly, to lay open the actors' explanation of the spatial and temporal changes within the GOF's mangrove ecosystem and the impacts on surrounding communities, and then compare them as a process of politicisation. The aim was to develop a more cogent theory regarding the relationship between the discourses of various social actors and the physical geographic space for which they compete and the means by which environmental conflict emerges and evolves. I assert that local actors are often not merely physically and socially marginalised from their material environment, but are also discursively and, subsequently, politically marginalised by more dominant social actors. In the context of neoliberal economic reform, this has serious implications for the conservation and development of the coastal wetlands and the strengthening of civil society; segments of capitalist society that lie outside of both the sphere of the state and production (Johnston, 2000: 85).

The core of my thesis seeks to unveil the discourses and actions that underlie neoliberal development processes in Honduras and the social and ecological outcomes that follow. Furthermore, my intent is to take into consideration heterogeneous local discourses situated within larger social, political, and economic processes in order to understand the negative aspects of conflicts associated with various development interventions. A number of questions will be addressed here in order to understand the discursive formations that influence actors to pursue specific actions and strategies, and understand these actions in relation to the larger discourses that frame the issues in the GOF. Throughout my thesis the following questions are addressed:

1. What are the direct and indirect political and economic interests, characteristics[2], and actions of actors (local people, the state, multi-lateral development institutions, non-governmental organisations, international governmental organisations, and private industry) in relation to the mangrove wetlands surrounding the GOF in southern Honduras?
2. What is the relationship between larger social, political, and economic factors and the interests and actions of these actors and what are the various

---

[2] Characteristics are used to denote the features and distinctive qualities of the social actors/institutions or organisations associated directly or indirectly with the conflict. Essentially, the thesis seeks to describe the key actors, their associated discourses, political and economic interests, and their actions which are considered to be features of any actor's character. The goal is to elucidate the relevant idiosyncrasies of the actors since these are key aspects of their identity. Both identity and interests are central to conflict analysis.

environment and development discourses that influence the interests and actions of these actors?

3. How do social and ecological changes, in relation to the mangroves, relate to the political and economic interests and actions of these actors?
4. What discourses have actors used to try to achieve their political and economic interests in relation to the physical geographic space of the GOF?

I draw on a poststructuralist political ecology framework that pulls together a number of interrelated theories to answer these questions. Political ecology integrates the concerns of ecology and a broadly defined political economy in a developing country context (Blaikie and Brookfield, 1987: 17). A further goal was to identify the discourses that underlie the aims and interests of the actors and how they come into conflict. Discourses are defined as "frameworks that embrace particular combinations of narratives, concepts, ideologies and signifying practices, each relevant to a particular realm of social action" (Peet and Watts, 1996: 14). The combination of approaches has resulted in the development of a new model for environmental conflict analysis.

Aims

The results of my research provide a solid empirical analysis of conflict that appreciates its vicissitudinary nature and the socio-historical context within which it has emerged and evolved. In turn, my thesis contributes towards a better academic understanding of the relationship between neoliberal transformations in Honduras and the environmental conflict that has arisen around the coastal wetlands between 1973 and 2006. My intent was to enhance the burgeoning field of political ecology by making an empirically grounded contribution towards understanding the role of discourse and action in processes of local-level socio-economic and environmental change in developing countries and the conflicts associated with those processes.

Structure of the thesis

In Chapter 2, I begin with a discussion of the ontological and epistemological issues associated with the theory and methodological approach. I provide an overview of the methods used and an in-depth discussion of the theoretical basis of the research. I argue that the poststructuralist emphasis on discourse analysis contributes towards a better understanding of the conflict in the GOF. In particular, I argue that a poststructuralist approach within a political ecological framework is highly appropriate for this type of research, resulting in an innovative approach to conflict analysis. While most previous scholarship has referred to the Committee for the Defense and Development of the Flora and Fauna of the Gulf of Fonseca (CODDEFFAGOLF) as a new social movement (NSM), in this chapter I argue that NSM theory is inadequate for analysing this environmental conflict although it remains useful for explicating the role of CODDEFFAGOLF as an actor central to the conflict.

Rather than include extensive descriptions of the methods used in the corpus of the text, I have decided to place a thorough account in Appendix I while still elucidating the methodological perspective used throughout the course of my research. This includes information on how I address the key questions, an overview of the research design, general considerations associated with the fieldwork, and how I conducted the socio-economic assessment, interviews, and archival research. It also includes relevant materials such as the survey questionnaire used for the collection of the socio-economic data.

In Chapter 3, I define further what is meant by physical geographic space throughout the course of the thesis. I briefly address the 'natural agency' of the environment through the lens of actor-network theory, and provide a thorough description of the setting in which the conflict has taken place by focusing on the ecology and physical geographic features of GOF. Whilst avoiding the trap of environmental determinism, I eschew a simplistic approach by acknowledging that the natural world constrains and influences how individuals and institutions perceive environmental change and, therefore, how they respond to them.

In Parts II, III, and IV, I begin by providing an overview of the macro-economic and political context of the periods covered, drawing particular attention to factors relevant to neoliberal reform, land use change and the environmental conflict in the south. Parts II, III, and IV of this thesis cover three different periods that define the socio-historical trajectory of the conflict.

In Part II, I focus on the emergence of aquaculture as a material manifestation of economic reform efforts that began in the 1970s and accelerated when neoliberal reforms were introduced in the 1980s. The period between 1973 and 1988 began with the construction of the first shrimp pond in southern Honduras, and ends when CODDEFFAGOLF was created to contest the future expansion of shrimp aquaculture in the south.

I demonstrate in Part II how the government and international institutions utilised the twin discourses of economic development and food security as justification for the development of shrimp aquaculture in southern Honduras between 1973 and 1988. These ideas led to decisions that resulted in the creation of new institutions to promote aquaculture, subsequently leading to land use changes in the south. By the end of the period, contestation emerged in response to the effects of the power exercised to promote shrimp aquaculture. Subsequently, contestation arose from within a segment of the local artisan fishing community resulting in the creation of an environmental movement, CODDEFFAGOLF, as a reaction to those who promoted this new NTAX crop. Resistance to the government and the industry's efforts to order the physical space in line with their interests for aquaculture expansion was then contested and negotiated as CODDEFFAGOLF sought to re-conceive of how the physical space of the GOF ought to be utilised, managed, and conserved.

In Part III, I examine the rise of resistance to aquaculture expansion as a process of contestation that sought to influence the trajectory of neoliberal transformations in southern Honduras. In Part III, I conclude the period in 1998 when Hurricane Mitch devastated the country, changing the nature of the issues associated with the conflict as extra-national factors played an increasingly important role.

Finally, in Part IV, I summarise the key findings in relation to the theoretical orientation used. I also discuss relevant issues associated with the conflict, post-Hurricane Mitch, between 1998 and 2006. I argue that this period should be seen as a transition towards neoliberal environmental governance that is the result of the convergence of two social imaginaries, neoliberalism and environmentalism, and is exemplified by the current efforts towards a market-based approach to marine and coastal resource management in the GOF. In the last chapter, I conclude the thesis and synthesise my ideas and their relevance to future research.

Summary

In this Chapter, I introduced the conflict surrounding shrimp aquaculture in southern Honduras as a material manifestation of the 'Blue Revolution'. I illustrated the extraordinary growth of aquaculture worldwide over the last three decades and identified the key actors behind the promotion of the industry in Latin America as well as those who have contested it. The key claims against the shrimp industry were highlighted. I argued that resistance to shrimp farming in southern Honduras was the result of a group of actors contesting the socio-economic and environmental impacts of policies and actions rooted in neoliberal ideology and perpetuated by a wide group of actors, including USAID, IFIs, and the Honduran government. In order to analyse the conflict, I developed an analytical tool within a poststructuralist political ecology framework that linked an actor-network approach to discourse analysis with event history and ecology.

An overview of the research site was provided in order to reveal the socio-economic and environmental context in which the conflict has taken place between 1973 and 2006 in conjunction with the factors associated with it. The aims of the research were established, several issues related to the analysis of conflict, and my contribution to the field of human geography. Since neoliberal ideology is central to the analysis, I explain it more thoroughly in Chapter 4, the actors and interests behind it, and the various techniques of power that have been used to translate neoliberal ideas into policy and action. Conversely, I also focus on the techniques of power used by civil society actors to contest neoliberal reform in southern Honduras.

I provided a brief overview of the literature that has been published on southern Honduras and justified the originality of my thesis. I also established my core arguments, research goals, and the key questions that I address throughout. The theoretical and methodological bases of my research are briefly explained and are thoroughly addressed in Chapter 2.

Chapter 2

Poststructuralist
Political Ecology Framework

Theoretical Orientation: A Poststructuralist Political Ecology Framework

In this thesis I have adopted a poststructuralist political ecology framework. Poststructuralism is a line of inquiry in 20th c. French Philosophy that is concerned with "the relations between human beings, the world, and the process of making and reproducing meanings" (Turner, 2006: 461). As such, it is concerned with the role of discourse in the production of knowledge and truth, both of which are conceived of as effects of power in poststructuralist literature. The poststructuralist turn in political ecology is intended to shift inquiry from framing environmental debates only in terms of their material attributes towards a focus on the struggles related to the social construction of knowledge and truth (Dove 2005: 231). Consequently, it is an ideal framework for analysing conflict because of its focus on discourse, power and the social construction of knowledge, all of which are key factors that influence the trajectory and outcomes in a conflict scenario.

The challenge is to develop an analytical tool that can unravel the inconsistencies between incongruent and conflicting versions of 'reality', while attempting to determine the central issues associated with a conflict by going behind the statements produced. Subsequently, an actor-network approach to discourse analysis is employed to assess the discourses of specific actors in order to understand the complexity of the conflict under consideration from their perspective. Methodologically, actor-network approaches have been used in political ecology to avoid interpretative closure and to resist the creation of binary oppositions while treating institutions, practices, and actors as materially heterogeneous and the source of 'agency', consistent with a poststructuralist framework (Turner, 2006: 4). The research combined this approach with event history and ecology in order to situate the analysis around key events that have influenced the socio-historical trajectory of the conflict.

Political ecology is a multidisciplinary field associated with anthropology, geography, international studies and development studies. The term 'political ecology' emerged in the 1970s as local and global environmental concerns became more prominent and politicised. Keil has identified "two major theoretical thrusts that have most influenced the formation of political ecology... political economy, with its insistence on the need to link the distribution of power with productive activity, and ecological analysis, with its broader vision of bio-environmental relationships" (1998: 13). Traditionally, political economy has focused on the connections between the fields of politics and economics to explain how political power influences economic outcomes and how economic factors affect political action (Crane and Amawi, 1997: 4). Through the extension of the tradition of political economy to the analysis of the role of the political and the economic in environmental change, political ecology has become a favoured approach to address environment and development issues in the developing world.

Blaikie and Brookfield's oft-cited definition of political ecology highlights its origins in political economy: "political ecology combines the concerns of ecology with a broadly defined political economy and together this encompasses the constantly shifting dialectic between society and land-based resources and also within classes and groups within society itself" (1987: 17). Political ecology's roots in political economy have led to a growing number of enquiries focused upon productive activities and the social relations that surround access to and control over natural resources – essential elements of political economy. Most early approaches to political ecology used an existing political-economic framework in an effort to elucidate the factors that contribute to environmental degradation and create models to assist with conservation, restoration, and sustainable alternatives (Paulson et al., 2003: 206).

Although poststructuralist approaches to political ecology vary, the analysis of the current and historical discursive positions of social actors in relation to each other and their natural environment plays a role in most analyses. A poststructuralist approach accepts that discourses are a part of social processes embedded in institutions and organisational practices, and are thus closely associated with both power and knowledge (Foucault 1967, 1973a, 1977b in Nuijten, 1992: 205). The approach "combines the structuralist[3] style of objective, technical, and even formal discourse

---

[3] It is important to note that *structuralism* within social theory, as used here, is not the same as structuralism as used in development theory. When referring to the former it is *italicized*. Here, structuralism refers to a line of thought in twentieth-century French philosophy. Gregory has defined it as "a set of principles and procedures originally derived from linguistics and linguistic philosophy, which involve moving 'beneath' the visible and conscious designs of active human subjects. The purpose is "to expose an essential logic which is supposed to bind these designs together in enduring and underlying structures that can be exposed through a series of purely intellectual operations" (Johnston et al., 2000: 797).

about the human world with a rejection of the structuralist claim that there is any deep or final truth that such discourse can uncover" (Gutting, 2001: 250).

More recently, scholars have emphasised the need to analyse the discursive elements associated with the politicisation of the natural environment. Particular attention here and in the literature more generally focuses on how political factors influence narratives used to explain environmental change (Peet and Watts, 1996: 263; Bryant and Bailey, 1997: 195; and Forsyth, 2003: 278). I emphasised more recent poststructuralist approaches and utilised an actor-network approach to discourse analysis to highlight the discursive elements associated with conflict over marine and coastal resources in the GOF. My theoretical framework drew on the 'liberation ecologies' approach articulated by Peet and Watts that:

> proposes studying the processes by which environmental imaginaries are formed, contested, and practiced in the course of specific trajectories of political-economic change while borrowing from poststructuralism a fascination with discourse and institutional power, yet remaining within that tradition of political ecology which sees imaginaries, discourses, and environmental practices as grounded in the social relations of production and their attendant struggles. (1996: 263)

Here I argue that the production and reproduction of specific discourses relating to a particular physical geographic space is an important variable in local resource struggles and, ultimately, it is a political process. Understanding the 'political' is particularly pertinent given that it "is no error, illusion, alienated consciousness, or ideology; it is truth itself" (Rabinow, 1984: 75). Furthermore, the struggle over who defines the 'truth' is essential to the analysis of environmental conflict; hence, the turn to discourse. In short, the struggle over who seeks to define the 'truth' is a political process and is central to the analysis of conflicts in order to determine who succeeds and why. This has great importance in conflict situations because those who control discourse can influence the terms of the debate, determine what is 'true', and affect the outcomes. It is for this reason that various social actors associated with the conflict in the GOF produce competing images of the 'reality' analysed, as depicted throughout the thesis. In turn, the 'agency' the actors have to exercise power regulates the extent discursive formations can emerge, be institutionalised, and dominate a given social context.

The intention here is to develop a more detailed understanding of the views of each actor, with competing interests over the coastal wetlands of the GOF, by assessing their discursive claims and how they contend to be seen as 'true' over others within a specific socio-historical context. Furthermore, the 'truths' concealed will be uncovered by focusing on how actors have put forth specific images of the social reality in which they have participated. The socially constructed images put forth by individual actors often suppress the underlying issues at stake in a conflict situation and, therefore, require elucidation. It is critical to understand that "representatives of each party strive to create impressions, images, and symbols supportive of their position" (Hall, 1972: 53). It is the positions and interests of these various actors that frame the conflict. Therefore, to illuminate such a conflict it is necessary to conduct an effective and comprehensive socio-historical analysis.

Within political ecology, discourse analysis is utilised as a tool to explicate the historical and cultural influences on evolving concepts of environmental change and degradation. Considered as linguistic and political forces in their own right, these images actively influence the social construction of knowledge in relation to the environment (Rocheleau, 1995; Leach and Mearns, 1996; Escobar, 1995, 1996, 1998; in Forsyth, 2003: 8). In other words, knowledge is historically and socially constructed through the interactions of people that have varying degrees of power to influence what counts as knowledge through negotiated struggles related to real, concrete worlds (Escobar, 1992b: 62). Three concepts are further elucidated, the 'political', 'truth', and power and their role in conflict scenarios in order to position them within a poststructuralist framework. The role of discourse analysis and the power embedded in the construction of meaning are also discussed as well as how various techniques of power are used to frame outcomes in conflict situations. My goal is to illuminate the difficulties of presenting an overall historical account of environmental conflict.

Power and Knowledge in the Social Construction of Environmental Conflict

To clarify the concept of power, it is best to turn to Foucault. He saw power as indistinguishable from knowledge and referred to them as a single concept pouvoir/savoir (power/knowledge) (Foucault, 1980). In these terms, power cannot be possessed because it is an active, resistive, or reactive force (Fox, 2000: 861). Power is relational and productive while diffuse and inherent in all social interactions, including discursive interactions (Braun and Castree, 1998: 19).

How is power an active, resistive, or a reactive force? How is it relational and productive? These questions are at the heart of assessing power, as is the need to "specify the practices, discourses, and representations, which operate to underpin a will to order and the will to resist" (Radcliffe, 2001a: 221). The exercise of power is relational because it involves speech and action between people. It is productive because the construction of new meanings or representations is the result of the relations between people as they speak and act. Furthermore, power is utilised to varying degrees to order the situation, meanings, or representations in the interests of a specific set of actors and sometimes to resist or react to the power exercised by other sets of actors. It was on these aspects of conflict in the GOF that I focused my analysis.

By analysing the discourses, actions and characteristics of the actors involved in the conflict, I sought to extrapolate the events relevant to the conflict from their perspectives. In sum, my research seeks to map the landscape of power in the GOF and the spatial character of discourse concerning its mangrove ecosystem between 1973 and 2006. The goal is to elucidate how various concepts associated with the socio-economic and environmental reality acted as political forces in their own right. Furthermore, the relationship between the discourses of social actors, the physical geographic space for which they compete and their role in the social construction of environmental conflict about those spaces is analysed.

The following section defines 'social actors,' and discusses the application of an actor-network approach to discourse analysis. To continue, I discuss actors' strategies, the relationship between discourse and action, and the importance of assessing conflict at various scales. The environmental movement that emerges at the end of the period covered in Part I raises questions around research related to new social movements (NSMs). My thesis does not advance NSM theory, but uses concepts from it to understand an environmental movement in a conflict scenario. Subsequently, I explain the limitations of NSM theory for the type of research and analysis I have chosen to conduct; examples are utilised as necessary and the positive contributions of NSM theory highlighted.

From Actor-Oriented to Actor-Network Approaches

Throughout the course of my thesis, the term 'actors' refers to a multiple set of people, groups and institutions involved in the conflict locally, nationally, and internationally.[4] By

---

[4] The actors are Honduran government agencies; inter-governmental organisations (IGOs); international organisations; entities representing foreign governments (e.g. USAID); and multi-

focusing on a wider set of actors, I reveal a multiple set of perspectives on the issues central to the conflict. Actor-oriented approaches were first utilised in the early-1970s in connection with the social movements that emerged during that period, and increased academic attention on the social change and conflict that those movements created (Long, 1992).

Actor-oriented approaches are a useful means of assessing the socio-historical nature of conflicts, especially when coupled with discourse analysis, since they seek to understand environmental conflicts and cooperation as an outcome of the discursive interaction and actions of different social actors.[5] However, by the 1980s, the early approaches were claimed to attribute too much 'agency' to individual actors while dismissing the collection of heterogeneous activities and networks within which people are embedded. In other words, actor-network theory (ANT) extended 'agency' beyond the individual in early actor-oriented approaches, arguing that an actor is only one constituent element in a wider network that is not only comprised of people, institutions, and their practices but also of the technologies and materials located in the network (Turner, 2006: 4). It is these "diverse 'actor-networks' that are the source of 'agency' in the world" (Johnston et al., 2000: 5). In sum, 'agency' is conferred through the relational aspects of each entity situated within any given network. Thus, by focusing on the social actors involved, as well as their attendant discourses and actions, an actor-network approach to discourse analysis is a theoretical and methodological avenue that enables the understanding of processes of social change and conflict (Long, 1992: 271).

Throughout the dissertation I use the terms 'actor-oriented' and 'actor-network approaches' interchangeably. My intention is for either to connote the more recent conceptualization of actors and 'agency'. I illustrate this broader conception of 'agency' in Chapter 3 when I discuss the 'natural agency' of the environment. My approach placed emphasis on the actors, not only in terms of what they have *done* but also in terms of the 'agency' embedded in what they have *said*. This is consistent with

---

lateral development institutions (e.g. World Bank and Inter-American Development Bank). Non-governmental organisations (NGOs); the Honduran environmental movement; private industry; academics; independent development experts; organisations or informal entities representing fishermen; shrimp farmers; salt producers; and local level people that utilise the mangrove wetlands for a variety of purposes.

[5] Researchers at the University of Manchester were the catalysts in the development of this approach; referred to as the 'Manchester School'. A number of fields, including sociology and anthropology have turned towards actor-oriented approaches. The theoretical and methodological significance of actor-oriented approaches was addressed, in probably the most comprehensive manner, in N. Long and A. Long (eds) (1992), *The Battlefields of Knowledge: The Interlocking of Theory and Practice in Social Research and Development*.

poststructuralist approaches in political ecology. Agrawal has argued that academic analysis of human-environment relations often lacks consideration for actors' direct experiences of environmental change and how new subject positions are created through the co-production of agency and structure (Agrawal, 2005). In order to address this gap, the role of the social actors was analysed by assessing what they have come to represent, through their discursive positions, in political struggles related to the natural environment (Forsyth, 2003: 140). Specific attention was directed towards the discourses and actions of state, civil society, and private industry actors. Focusing on their positions was critical in the analysis of this conflict since they are often incongruous and a frequent source of conflict.

Considering new social movements

New social movements are considered to differ from 'old' social movements, such as organised labour because they are more issue-specific, represent larger segments of society by breaking down class boundaries, use a number of different strategies and tactics, express both meaning and identity, and target issues that are complex and more than a zero-sum game (Johnston *et al.*, 2000: 759). Scholars have often relied on NSM theory to analyse how actors such as CODDEFFAGOLF, the ENGO organised by local artisan fishermen, seek to contest impacts on the wetlands. My primary focus was not to place specific attention on CODDEFFAGOLF as a 'new social movement' (NSM), but to utilise the NSM literature as a medium for understanding the role of this type of actor in an environmental conflict.

However, there were problems associated with using NSM theory when assessing this conflict. In some cases, NSM has regarded conflict as a struggle between a dominant actor, such as the state, and a subordinate actor, such as CODDEFFAGOLF, and sought explanation by highlighting the dichotomy between them. Dichotomising the conflict has led to insufficient forms of analysis because it has omitted consideration of the complex set of relations that exist between multiple actors in local resource struggles.

There are a number of well-acknowledged problems associated with researching NSMs. Foweraker in particular has identified two principal issues to consider when conducting research related to NSMs. First, social movement research has not focused on the relationship of movements, such as CODDEFFAGOLF, with other social actors in civil society. In this regard, my research had to understand CODDEFFAGOLF in relation to other social actors (1995: 21). Forsyth argues that environmental problems have to be analysed as socially constructed since the actors' co-construct meaning around specific problems through their interactions (2003: 16). In short, actors are not in a binary position to each other and an appropriate analysis should not view them as such.

Analysis required going behind the claims produced, the purpose for their production, and what sets of relations underlie them. An analysis of environmental conflict that focuses on the role of a single social movement, such as CODDEFFAGOLF, in opposition to another actor such as private industrial shrimp farmers, can tendentiously construct the division of a single social space into two opposing fields, geographically and discursively (Foweraker, 1995: 96). To avoid this common pitfall I focused on a wider set of claims around access to and control over the mangrove wetlands of the GOF.

Second, Foweraker argues that while social movements are often associated with marginalised people, they emerge because of a small subset of actors, which does not always represent the wider discourses of the community as a whole. Leach and Mearns have stated that the shared meanings or ideas that emerge among various actors in relation to physical geographic space and, or causal explanations of change associated with that space, can be viewed as 'received wisdom'. 'Received wisdom' "is, at the same time a product of the unintended and intended consequences of the actions of individual human agents, and a part of the [social] structure within which they act and which shapes future possibilities for action" (Long and Long, 1992 cited in Leach and Mearns, 1996: 9). Consequently, movements such as CODDEFFAGOLF can dominate discursive space like more powerful external actors, leading to the marginalisation of others when such movements become the source of explanation.

My proposition is to acknowledge that discourses deployed by social movements do not always represent or support the interests of all actors a movement purports to represent and, therefore, I could not rely on them to comprehend the complexity of this environmental conflict. In current political ecology, the 'liberation ecologies' approach suggests that when an analysis of environmental conflict emerges from a multiple set of voices, particularly marginalised voices, research can result in a more comprehensive assessment that appreciates the positions and interests of a wider set of actors (Forsyth, 2003: 159). Consequently, the triangulation of theory and methods assiduously avoided explaining the historical nature of the conflict by relying on 'received wisdom' generated by too narrow a group of actors.

Accordingly, the conflict over mangrove wetlands in the GOF is examined within its appropriate macro-economic and political context for each of the periods analysed: 1973-1988, 1988-1998, and 1998-2006. My research relied heavily on fieldwork (mentioned in this chapter and more extensively in Appendix I) to highlight the political and ecological realities, from the perspective of social actors at various scales, locally, nationally, and internationally. Discourse analysis was used to accomplish this goal.

Discourse Analysis

Discourse analysis is central to poststructuralism and its application within political ecology, since it highlights the co-construction of discourses and narratives regarding environmental change by various social actors (Forsyth, 2003: 9). A discourse constitutes knowledge through communication generating a distinctive way of understanding the world while competing in a given social context (Milton, 1996: 166). Discourses refer to systems of meaningful practices that form the identities of subjects, objects, and concrete systems of social relations and practices that are intrinsically political, always involve the exercise of power, and are contingent and historical constructions (Howarth et al., 2000: 3-4). Poststructuralist accounts reveal that discourse is often used by social actors to actuate a specific way of understanding and acting towards what is regarded as important, which can lead to conflict between actors when they pursue different interests and hold different positions in environmental struggles (Harvey, 1996: 77).

Here, discourse analysis[6] was utilised to create an historical account of the conflict by turning towards the narratives of the social actors involved. They were analysed comparatively in relation to specific actions and events associated with the socio-historical trajectory of the conflict in order to understand their relationship to specific actors' claims. How actors behave in relation to their perceptions of social reality is important when studying conflict over environmental resources as it "is typically a struggle over ideas as to what constitutes 'appropriate' environmental use and management" (Bryant and Bailey, 1997: 192). Conflicts over access to and control of natural resources arise due to the complexity and uncertainty of actors' claims to 'truth' in terms of how a space ought to be used, managed, and conserved. Claims can be

---

[6]The research relies significantly on discourse theory and analysis borrowing from the work of Foucault and Derrida's poststructuralist, also known as deconstructivist, approaches within the social sciences. Laclau and Mouffe are probably the most prolific writers in the field to date and their ideas apply to research within several disciplines in the social sciences. Howarth et al., Fairclough, and a number of others have also discussed the applications of this approach within various realms of the social sciences (Howarth, 2000).

divisive, and often become disputed as actors seek to achieve their political and economic interests in relation to the physical space disputed (Roe, 1994: 13-14). Actors' claims to 'truth' are often contested when the interests of others actors are affected or, are potentially affected, because of those claims. This process is inherently political.

Central to any discourse analysis is the concept of 'truth'. I do not intend to address the complex historical or philosophical arguments surrounding the nature of 'truth'.[7] Rather, the point is to illustrate that the concept is at stake in discursive struggles. Debates in philosophical circles, in religious forums, or in the back garden amongst friends illustrates that discursive struggles take place over the concept itself. As a result, it is central to the analysis of environmental conflict since claims to 'truth' or statements considered 'true' are, by nature, assertoric, sometimes put forth apodictically, and, consequently, conflictive in discursive struggles illustrating the complexity associated with developing an historical account of a conflict.

How something gets to count as true is particularly important for current research within political ecology, which is currently lacking a clear articulation of the political. Poststructuralist research views truths "as statements within socially produced discourses, rather than 'facts' about reality" (Peet and Watts, 1996: 228). The question of 'truth' or, perceptions of what is true, are loci of competing discourses associated with any social reality. Furthermore, it is through competing 'truths' associated with various discursive practices that the environment becomes politicised, and power exercised. Drawing on Foucault, Edkins states that "truth is linked in a circular relation with power: power and knowledge are mutually constitutive which means that the political question... is truth itself, and how something gets to count as true is a political process" (1999: 12).

The goal of a thorough analysis of environmental conflict is thus partially to understand "historically how effects of truth are produced within discourses which in themselves are neither true nor false", but have implications for society and the environment (Rabinow, 1984: 60). In other words, I focused on how 'truth' is created and constituted at various levels of the political process and how this process affected specific outcomes related to

---

[7] "A broad division can be drawn between theories of truth which regard truth as a property of representations (whether linguistic or mental) including sentences, statements, and beliefs and those which regard truth as a property of propositions. The latter are conceived as items represented or expressed in thought or speech; disputes between theorists of truth are sometimes confused by a failure to discern this division" (Lowe, 1995: 881-882). For more on the concept of truth see: Susan Haack, 1978.

disputes over physical geographic space. The advantage of poststructuralist approaches is that "by definition, a description of discourse allows for a multiplicity of truths (interpretations) that can only be revealed through the description of peoples' negotiations and accommodations as they are played out in an active context" (Long, 1992: 164).

My focus on discourse was to avoid directing too much attention on the material issues associated with environmental conflict, such as solely focusing on who gets what as an outcome of the conflict or, what physical changes have resulted from it. Such overemphasis can diminish the significance of the discursive issues central to explaining its historical nature. Nevertheless, discursive practices "establish the conditions of possibility for thought and action [and are] an active intervention in the social and physical realms" that become the material issues that researchers address (Duncan *et al.*, 2004: 85). Environmental conflict analysis therefore, must identify the underlying discourses of the actors' in relation to their strategies in order to deduce how discursive practices are an active intervention in the social and physical realms. My approach establishes the basis to arrive at a better understanding of the material issues at stake and recognises the full imbrication of the discursive and the material in poststructuralist research (Johnston *et al.*, 2000: 626). Stonich argued, "An overemphasis on constructivist discourse analysis may diminish the concern for the material issues that first provoked the emergence of political ecology" (1999: 24). In other words, discursive struggles should not overshadow the material aspects of a conflict. As a result, it was necessary to develop a more coherent analytical framework to understand the relationship between the material issues central to political ecology research as well as the discursive interventions that result in actions that affect a physical geographic space. In sum, my thesis took into account both the discursive and the material in order to avoid an analysis that placed too much attention on one or the other while further explaining the relationship between the two.

Assessing the manifestation, production, and reproduction of environmental discourses revealed the historical framings related to this environmental conflict and, subsequently, extrapolated the exigent issues associated with the conflict. However, a conflict analysis that fails to consider discourse in relation to the actions of the social actors involved would be inconclusive.

Discourse and Action

I argue that an analysis of environmental conflict taking an actor-network approach to discourse analysis must continually scrutinise the relationship between discourse and action, as they can be incongruous. The inconsistencies between actors' expressed intentions and their actions are a primary source of conflict as they are often contradictory. In other words, social actors often *say* one thing but *do* another, leading to conflicts between them.

Actors' discursive strategies change depending on how they interpret the discourses and conduct of others. Moreover, assessing such strategies requires recognising that they are not the result of effects of external factors but arise instead from how humans interpret and handle the actions which they are mutually constructing through social interactions (Blumer 1962: 183).[8] In this regard, actions are also discursive in nature as they connote, symbolically, certain meanings to the actors that perceive them. An assessment of the discursive reactions and the actions taken, if taken, allowed for the deduction of the interpretations of the actors. My analysis of this conflict, therefore, could neither rely exclusively on an analysis of discursive constructions nor of the actions associated with particular events. However, assessed conjointly they revealed indicative idiosyncrasies of the environmental conflict. My approach was fundamental to substantiate the power of discourse in the social construction of environmental conflict.

Event history and event ecology

My thesis argues that a third element necessary to conduct a socio-historical analysis of environmental conflict is the identification of key events associated with it, since events denote important instances associated with processes of social change. In other words, situated within social processes, events are defining moments that connote change that has resulted from specific actions. Thus, inherently bound up in discourse and action, events manifest from either. Actions are events therefore, like actions; events are also discursive in nature as they also signify, emblematically, certain meanings to the actors that perceive the events. Logically, construed as an event, each moment is a constituent element. However, events remain ambiguous until they acquire meaning within a specific socio-historical context, hermeneutically derived by turning towards the discourses and the actions of the social actors associated with a particular event. As actors derive meaning from specific events, the meaning derived influences their stance

---

[8]Symbolic interactionism is a valuable source of ideas about the ways in which groups of people or, sets of actors interact and collectively construct their social worlds, and the ways in which these worlds can come into conflict. See Herbert Blumer or George H. Mead

in relation to other actors and future events negotiated and contested within the public arena. Assessing and selecting the key events required the use of a combination of methodological approaches. My goal was to reconstruct the history of the conflict by focusing on the 'public' events of significance manifested through the discourses and actions of the social actors involved.[9]

In order to reconstruct an historical timeline of the conflict in southern Honduras, and identify the multiplicity of local, national, and international actors, the text of newspapers, national laws, official and unofficial documents and the text of open-ended and semi-structured interviews were analysed. Archival sources[10] collected through government, educational, non-profit, and international institutions in the region and in Washington D.C. were utilised (See Appendix I). The methods I borrowed from event history and ecology practitioners coupled with an actor-network approach to discourse analysis led to the development of a more complete framework for the socio-historical analysis of environmental conflict in the GOF between 1973 and 2006.

Events ecology, which is similar to event history, has been described as "the analytical methodology for evaluating causal links between historical events and integrating relevant biophysical and socio-economic information"; this approach is guided by "open-ended questions about why specific environmental changes of interest (events) have occurred" (Walters, 2003: 295). Forsyth has stated that as such events ecology "adopts a partly phenomenological attitude by seeking to understand 'events' as local changes of significance, rather than as 'facts' that can be incorporated into preexisting theories, or 'factors' that imply events have causal significance" (2003: 223).

Event history was linked with events ecology since its practitioners methodologically "compile databases from published accounts in the contemporary press on events that they consider worthy of notice" and proceed by creating a "temporal map of incidents through which the... activities and interactions [of various actors] can be traced" and

---

[9] For more on actions and events see Davidson (2001).

[10] The majority of the archival research conducted took place at the Hermeroteque (archives) at the Universidad Nacional Autonomous de Honduras (UNAH) which has the best archival collection of newspapers in the nation including *El Cronista*, *La Tribuna*, and *El Heraldo*. Newspaper research focused on different periods depending on the circulation rates and prominence of the newspapers at different times. For the period 1970-1980 *El Cronista* was searched, from 1980-1995 *La Tribuna*, and from 1995-present the focus was on *El Heraldo*, currently the most widely read. Archival research was also conducted to acquire the complete text of relevant laws through their original publication in *La Gazeta*, the official national record of laws. Once a law is printed in *La Gazeta* it officially goes into effect.

further assessed (Tarrow, 1996: 875). My approach isolated relevant events by assessing newspapers in conjunction with a broader set of official and unofficial documents (e.g. laws and project reports). The archival information also made a significant contribution towards the discursive analysis of the conflict. Assessing these documents historically helped to identify the changes in the actors' discourses and their role in the conflict, at least as conveyed through the discourses drawn from these various sources. It also enabled me to construct a diachronic narrative.

Focusing on particular events without collecting and assessing data that emphasised the actions of individuals as strategic agents rather than as occupants of particular statuses or structural locations would fail to place events within their appropriate social context. When situated in terms of everyday social practices, situations, or events, an analysis of the actors' discourses and actions are more meaningful (Long, 1992: 164). Employing these methods contributed towards a 'thick description' of the key events, the actors involved, the rules associated with their actions, and the social contexts in which they emerged (Berg, 2001: 165). In other words, the goal was to place the discursive positions and actions of the actors, as well as purported social and ecological changes, within the social context of key events. They were then analysed as significant points in the struggle over discursive and physical geographic space and the emergence and evolution of environmental conflict in the GOF between 1973 and 2006.

Methodological framework

I spent fourteen months in southern Honduras and the capital of Tegucigalpa while living in the Valle de Zamorano at the Pan-American Agricultural School (University of Zamorano). The research design and methods included:

- Initial Site Visit (Interviews)
- Literature Review
- Data Collection and Organisation
- Archival Research
- Identification of Historical Events
- Identification of Local, National, and International Actors
- Selection of 21 Coastal Communities to conduct Socio-economic Analyses
- Participant Observation
- See Appendix II Research Design and Methodology

The research relied on archival research to reconstruct an historical narrative of events surrounding the conflict between 1973 and 2006. Relevant newspaper articles elucidated the socio-historical context, the actors, their discursive positions, interests and actions in relation to the issues surrounding the conflict. Legal and policy documents determined how the legal framework in relation to marine and coastal resources has changed since 1973. The legal framework also assisted with identifying the characteristics and interests of the state as a key actor in the conflict. I conducted over 150 open-ended and 35 semi-structured interviews, completed with individuals involved in the conflict at local, national, and international levels (See Appendix I). Interviews with the actors served as the basis for determining the ways in which their ideas compete or are congruent with those of other actors associated with the issues surrounding the conflict.

TABLE 2.1: COASTAL COMMUNITIES SELECTED FOR THE SOCIO-ECONOMIC ANALYSIS

Communities	Municipality	Department
1.San Jeronimo	Namasigue	Choluteca
2.San Bernardo	Namasigue	Choluteca
3.Guameru	Namasigue	Choluteca
4.Costa Azul	Namasigue	Choluteca
5.Playa Negra	Namasigue	Choluteca
6.El Tulito	Choluteca	Choluteca
7.El Carrizo	Choluteca	Choluteca
8.Guapinol	Marcovia	Choluteca
9.Cedeño	Marcovia	Choluteca
10.Pueblo Nuevo	Marcovia	Choluteca
11.Punta Raton	Marcovia	Choluteca
12.Campamento	San Lorenzo	Valle
13.Coyolito	Amapala	Valle
14.Pintadellara	Amapala	Valle
15.Puerto Soto/Relleno	Amapala	Valle
16.Puerto Grande	Amapala	Valle
17.La Brea	Nacaome	Valle
18.Playa Grande	Nacaome	Valle
19.Las Playitas	Alianza	Valle
20.Valle Nuevo	Alianza	Valle
21.Los Guatales	Alianza	Valle

Finally, I trained 25 fourth year students in the Department of Development, Socio-economics, and Environment at the University of Zamorano to assist with the collection of socio-economic data through 368 household questionnaires in 21 coastal communities (see Table 2.1 above for list of communities) and analysed them (see Appendix II for Household Questionnaire). Although a significant amount of data was collected it was not the point of the research and, therefore, was not used extensively.

The lack of any data on southern Honduras compelled me to gather enough to understand the socio-economic characteristics of coastal communities in the south. Therefore, a stratified random sample was taken, entered into an Access database and analysed using the Statistical Package for the Social Sciences (SPSS). My aim was to be able depict the local 'reality' as needed throughout the thesis to make relevant points, particularly in Chapter 3 when I provide an overview of the case study site.

Summary

My theoretical orientation combined with my methods provided a powerful perspective from which to understand the discourses, actions, and events that have influenced the social construction of environmental conflict in southern Honduras. My research utilised a diachronic approach focusing attention on a specific phenomenon, conflicts surrounding mangrove forests in southern Honduras, and the ways in which discourses in relation to the actors' physical reality have emerged and evolved within the context of neoliberal transformation in the GOF, southern Honduras. Methodologically, the research accepted multiple realities that have defined the nature of the environmental conflict and started from the position that the way people act is based on their perceptions, and that their actions have consequences – thus subjective reality was no less real than an objectively defined and measured reality (Fetterman, 1989: 5).

My methodological intent was to avoid an essentialist, positivistic[11] or voluntaristic[12] approach towards environmental conflict. In this regard, my research drew upon a phenomenological[13] approach in the sense that it recognises the possibility of multiple realities. Fetterman argues that a perspective of this nature is fundamental towards examining environmental conflict since the documentation of different perceptions of any given 'reality' is the basis for understanding how and why social actors think and act in relation to their social and physical realities (Fetterman, 1989: 20). In this chapter, I have mentioned the various methods that I used to document multiple perspectives; these are discussed more extensively in the appendices.

---

[11] Positivism recognises only positive facts and observable phenomena, with the objective relations of these and the laws, which determine them, abandoning all inquiry into causes or ultimate origins (Little, Flower and Coulson, revised and edited by C. T. Onions (1985).

[12] Voluntarism is a philosophical doctrine that regards will over and against one's other mental faculties (Hare, Peter H.; The Oxford Companion to Philosophy, 1995: 902-903).

[13] Phenomenology is a "continental European philosophy which is founded on the importance of reflecting on the ways in which the world is made available for intellectual inquiry: this means that it pays particular attention to the active, creative function of language in making the world intelligible. One of phenomenology's main concerns is 'to disclose the world as it shows itself before scientific inquiry, as that which is pre-given and presupposed by the sciences'. As such, phenomenology provides a powerful critique of positivism" (Johnston et al., 2000: 579).

I chose to use the theoretical orientation discussed to assess the perceptions of the actors but grounded the analysis in the empirical data collected. In other words, I grounded my theoretical framework to illuminate perspectives on the ways in which human actions are motivated by the environment, or material reality, from an external, theoretical social scientific perspective (Fetterman, 1989: 22). The theoretical perspectives presented in this chapter placed the empirical data within current social scientific theoretical perspectives of 'reality', therefore, contributing to a more robust level of analysis whilst situating the research in a theoretical and comparative perspective. Chapter 3 turns towards the physical reality where the conflict has occurred.

Chapter 3

Geographic Overview

Introduction: The 'Natural Agency' of the Environment

In Chapter 3, I provide an overview of the Republic of Honduras, the region of the Gulf of Fonseca, and a description of the mangrove wetlands. I briefly address the *'natural agency'* of the environment, and provide a thorough description of the setting in which the conflict has taken place by focusing on the ecology and physical geographic features of the GOF. The concept of *'natural agency'* as conceived here, draws on the actor-network approach articulated by Sarah Whatmore. Whatmore's broader definition of 'agency' views "it as a relational effect generated by a network of heterogeneous, interacting components [of which the environment is but one] whose activity is constituted in the networks of which they form a part" (Whatmore, 1999: 28).

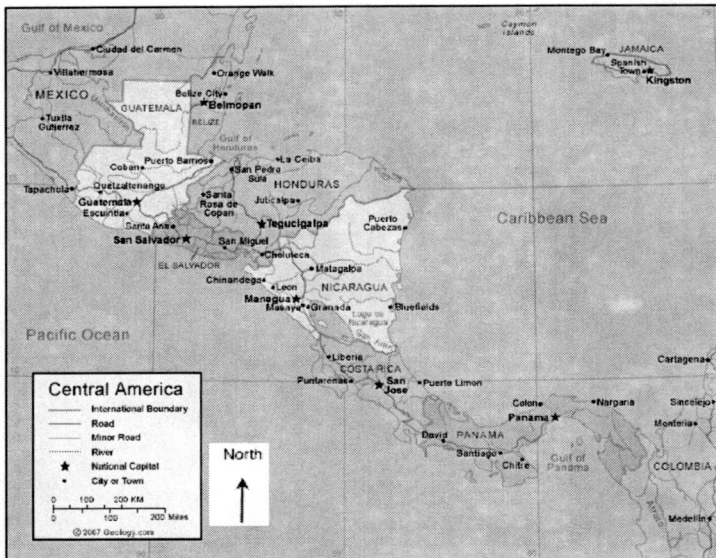

MAP 3.1: HONDURAS AND THE ENVIRONMENT OF THE GULF OF FONSECA
Source: http://geology.com/world/central-america-physical-map.shtml

Whilst avoiding the trap of environmental determinism, I periodically illustrate how the natural world influences the field of discourses, interests, and actions of social actors at varying scales. The purpose is to illuminate how the natural world conditions how we frame our understanding of it. In other words, the environment influences our perceptions and how we choose to act in relation to it. Naturally, there are certain constraints and limitations that arise; however, the environment does not determine how we act but influences our options in terms of how we might act in relation to it or value it.

When I speak of 'natural agency' here, I try to illustrate the concept through the ways in which humans perceive and interact with the natural environment of the GOF. The point is to demonstrate that human expressions or actions in relation to the environment are not homogenous. In turn, the cognitive disposition of any actor in relation to his/her environment is guided by certain value orientations. I acknowledge that different value orientations towards the natural world are influenced by the environment depending on an actor's knowledge and interests, the building blocks of their values towards it. I also accept that the process by which knowledge and interests are socially constructed in relation to the natural world is an iterative process. In other words, knowledge and interests do not come a priori to human engagement with the natural world.

Before proceeding, the concept of physical geographic space, as utilised in my thesis, is clarified and several points are made in regard to the 'natural agency' of the environment to illustrate further the importance of understanding the ecology of the region, illuminating how each has been conceived throughout the course of my research.

Throughout the course of my thesis, physical geographic space refers to the coastal wetlands of the GOF. More specifically, physical geographic space is intended to denote the material and social conditions of production situated around the wetlands referring to the natural resources located within this space on and in the ground, the wetlands, and the sea. Physical geographic space is influenced by natural processes and the people living and working in the region and affects both the material and social conditions of production (Dunford and Perrons, 1983: 70-71).

Traditionally, geographers have conceived of space as both natural space governed by natural laws, "studied by the historical sciences of nature including physical geography, and also as a natural and social space studied by human geography" (Dunford and Perrons, 1983: 68). The relationship between physical space and discourse associated with it is ontologically and epistemologically interdependent: physical geographic space influences how humans conceive, talk about, and act in relation to it. The physical space of southern Honduras thus has a certain agency that is 'situated at the intersection of different environmental discourses' which are central to the conflict because they affect what people do in relation to their environment and how they can do it (Johnston et al., 2000: 351).

I argue that physical geographic space plays an important role in terms of its capacity to influence the field of discourses, interests, and actions available to any given set of human or institutional actors in relation to a space in dispute. In other words, the environment mediates human relations and thus is the basis of the social construction of any 'environmental reality'. Further, I argue that the natural world influences the amalgamation of human expressions in relation to the environment and how they emanate within the social and public realms through both discourse and action, as they are politically co-constructed. The knowledge any given set of actors has in relation to a physical space also influences the degree of power that they possess to perform different types of actions in relation to that space.

The natural environment and its various attributes also place natural limitations and constraints on the types of human activities that can be pursued within any given physical space. For example, shrimp farming on an industrial scale requires the presence of certain physical geographic features that exclude the possibility of producing shrimp in most locations. Locating appropriate sites for large-scale cultivation of shrimp was one of the objectives of the early industry, and the ecological characteristics of southern Honduras provided the ideal space the industry was seeking. Essential to aquaculture development "was a large saltwater source, local seed stock, reliable electrical power, the ability to purchase earth-moving and refrigeration equipment and the absence of large-scale agricultural activity" (Luxner, 1992). Of the region's ecological characteristics, described in detail in this chapter, five were particularly attractive to those interested in aquaculture development.

First, the prevalence of extensive salt flats, where mangroves are usually scarce and tend to be dwarfish due to high levels of salinity, made it fairly easy to clear the land for the construction of shrimp ponds. Second, during the rainy season, the salt flats tend to be inundated due to their close proximity to the estuaries, acting as natural retention areas that could easily be converted into shrimp ponds. Third, not only do the wetlands serve as a source of habitat and food for a number of species of birds, they are also directly linked to the life cycles of various marine species, including the same shrimp postlarvae needed to stock the ponds. Fourth, these areas were relatively isolated, unpopulated, and unsuitable for other types of agricultural activities because of the high levels of soil salinity. Finally, the salt flats in San Bernardo provided sufficient physical space for future industry expansion if the initial research and development proved successful in Punta Raton, a small coastal fishing village that lies on the south-western side of the Department of Choluteca,[14] where the first pond was to be constructed, and which eventually became one of the sites of contention between locals and the commercial industry. Without these specific ecological characteristics it would have been unfeasible for the industry to commence or develop on the scale that it did in the region of the GOF.

FIGURE 3.1: SALT FLATS
Source: Wilburn, 2005

As a result, there is an argument for developing a more nuanced appreciation of physical spatiality within the context of the social relations associated with productive activities. By so doing, my research aimed to remain conscientious of the ecology within political ecology. Davis has argued that the natural "space, or the 'setting' in which people live and act, establishes parameters on action even as it interacts with social forces, structures, and conditions that construct that action" and "influences the

---

[14] Personal interview with Dr. Daniel Meyer and site visit, July 2005.

formation, objectives, and strategies of citizens as individuals and collectively as social movements", institutions, or private businesses (Davis, 1999: 601). Consequently, how the environment influences the options available for people to pursue various productive activities is central to research within political ecology.

The example used illustrates how the natural environment had an inherent capacity to influence the types of productive activities people pursued in relation to this physical space, which placed natural limits and constraints that affected their options, particularly in terms of scale. Therefore, analysing environmental conflict requires the development of a better understanding of the influence that physical geographic space has on the positions, interests, and subsequent actions of actors. Dunford has argued that nature's affects on humans is always mediated through society and the multitude of institutions and social organisations that comprise it. Human activity undoubtedly has an influence on the environment; however "nature also changes in accord with natural laws" (Dunford and Perrons, 1983: 51). Consequently, my research takes seriously the social construction of the natural through the political sphere while keeping in mind the natural construction of the social and political sphere, and without falling into the trap of environmental determinism.

The Republic of Honduras

The Republic of Honduras is located in the Western Hemisphere on the Central American isthmus, which lies between the North American and South American continents. Honduran territory covers 112,492 km² making it the region's second largest country. The northern coast is 650 km long and runs from the border with Guatemala to the border with Nicaragua. The southern portion of the country borders the GOF, which is connected to the Pacific Ocean and is 185 km long and extends from the border with El Salvador to the Nicaraguan border (Honduran Secretariat of Industry and Commerce, 2002: 14).

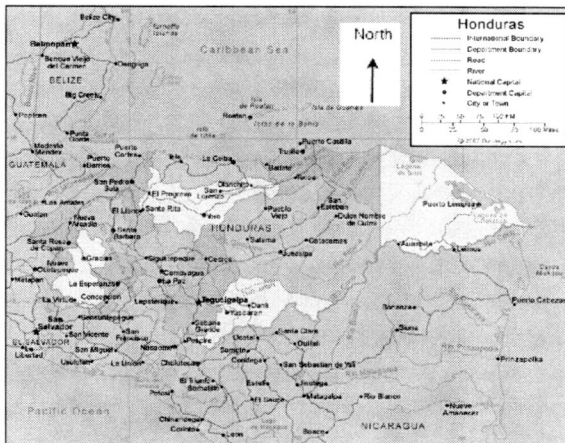

MAP 3.2: HONDURAS
Source: http://geology.com/world/honduras-map.gif

Honduras is divided into 18 Departments and 298 municipalities. In 2007, the population had reached 7.5 million people, a million of whom live in the capital Tegucigalpa (Central Intelligence Agency, 2007). There are seven distinct ethnic groups in Honduras: Los Tolupanes, Chortis, Lencas, Tawahkas, Pechs, los Misquitos, and Garifunas (Afro-Caribbeans). These various ethnic groups make up about one seventh of the Honduran population, or about one million people. None of these ethnic groups have any presence in southern Honduras, the population of which is primarily *mestizo*, mixed Amerindian and European, like much of the population (Honduran Secretariat of Industry and Commerce, 2002: 2). My thesis, therefore, does not focus on any specific indigenous group of people.

Ecologically, the country is diverse: characterised by savannas throughout; lush rainforests nearer the Caribbean coast; dry tropical hardwood forests in the south; pine, thorn-scrub, mesophytic, and broad leaf forests (Johannessen, 1963: 5). There are also cloud forests on the peaks of a dozen or so mountains throughout the country. The marine environment is defined by mangrove forests, sea-grass beds and coral reefs on the Caribbean coast, and extensive mangrove wetlands in the south.

The southern region of Honduras near the GOF comprises 5.2 per cent of national territory, and is defined by three geomorphic features: mountains, plains, and coastal area. The coastal plains of the Pacific and Caribbean are areas of high economic value and this is where most agricultural activities take place due to the regions' 'natural agency'. In the south, these plains are considered to be any area of land no higher than 150m, but are 41m in height on average covering an area of approximately 2,000 km^2 around the GOF. These coastal plains eventually merge with the foothills leading into the mountains that vary between 900m and 1,500m on average (Honduran Secretariat of Industry and Commerce, 2002: 18). The inter-tidal zone, extending from the coastal plains into the GOF, is approximately 1,000 km^2. This area is composed of estuaries, islands, and lagoons in the southern portion of the country that are important ecological features influencing the wetland ecosystem that characterises the GOF (PROMANGLE, 1997: 9). Wetlands are systems that have at least one of three characteristics: (1) at least periodically, the land supports predominantly hydrophytes, (2) the substrate is predominantly undrained hydric soil, and (3) the substrate is saturated with water or covered by shallow water at some time during the growing season of each year (Mitsch and Gosselink, 1986: 18). All three of these characteristics are present in the GOF.

Southern Honduras

It is estimated that there are over a million people living near the GOF, split between the three bordering countries, approximately 600,000 in Honduras; 240,000 in Nicaragua; and 160,000 in El Salvador. The southern coastal zone of Honduras comprises two Departments, Choluteca and Valle, which have different characteristics. See Table 3.1 on the following page.

TABLE 3.1: CHARACTERISTICS OF THE DEPARTMENTS OF CHOLUTECA AND VALLE, SOUTHERN HONDURAS

Southern Honduras	Department of Choluteca	Department of Valle
Population	413,148	167,375
Population Density	76-125 per km²	76-125 per km²
Area	4,360 km²	1,665 km²
Per cent of National Territory	3.87%	1.48%
No. of Municipalities	16	9
Municipalities bordering the Gulf Fonseca	Marcovia (482.3 km²) and Choluteca (1,069.10 km²)	Alianza (215 km²), Goascoran (200.5 km²), Nacaome (529 km²), Amapala (80.70 km²), and San Lorenzo (234.6 km²)
Largest Population Centres	Choluteca 98,000 inhabitants	San Lorenzo 25,000 inhabitants
Principal Economic Actvities	shrimp farming, sugar cane, cantaloupe, watermelons, corn, sorghum, cotton, cattle ranching, and	shrimp farming, cantaloupe, water melon, corn, sorghum, and sugar cane

Source: Honduran Secretariat of Industry and Commerce (2002: 14)

Southern Honduras is primarily rural, and has one of the highest indices of poverty in the country. Most of the rural population is dependent on the Gulf's natural resources for subsistence, especially those living in coastal communities. In 1990, the rural population in Honduras was estimated to be 51.4 per cent of the total Honduran population, 75.4 per cent of which were considered below the poverty line (Thorpe, 2002: xiii and 137). Honduras is the third poorest country in the Western Hemisphere, with a per capita GNP of US$1,030 (*Washington Post*, 2004: Deportees Bittersweet Homecoming).[15] Over half of all households live in poverty and one third in extreme poverty; this reflects the skewed distribution of wealth and access to land (Warren, 2000: 1). Rowlands estimated that, in 1992, the poorest 20 per cent of the population had 5.5 per cent of the income, whereas the top 5 per cent control 54.2 per cent (1995: 39).

Unemployment is high around the GOF, probably exceeding 40 per cent, and well above the national average of 30 per cent. Both Departments of the southern region have the second highest levels of unemployment in Honduras. In 1990, the largest proportion of the economically active population in 1990 (68 per cent) generally worked in agriculture and cattle ranching (Thorpe, 2002: 209). For most people in the south, less than 30 per cent of total income is generated by on-farm employment in both Valle and Choluteca

---

[15] When I use $ it always indicates US Dollars, unless otherwise noted. A table of exchange rates (Lempiras to US Dollars) from 1980 to the present is included just after the table of contents.

whereas off-farm income is extremely important, especially in Valle (59 per cent) (Thorpe, 2002: 141).

Access to land has been an important issue affecting how local people employ the Gulf's resources. In 1984, it was estimated that 75 per cent of Honduran farmers lacked full, legal title to the land they were working (Montaner, 1995: 3). Subsequently, local people derive their income from a variety of sources in southern Honduras making rural communities diverse and heterogeneous in terms of the types of economic activities that they pursue. Furthermore, 25.6 per cent of respondents (93/363) to the household questionnaire indicated that they received remittances from family members who do not live in the region to support those that do.

FIGURE 3.2 DEPICTION OF THE LOCATION FROM WHICH REMITTANCES ORIGINATE

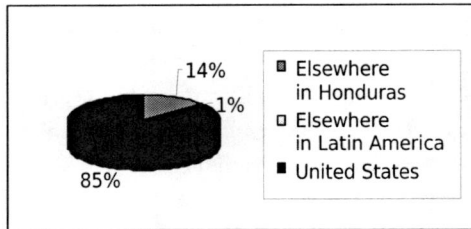

Deportees
IN THOUSANDS

13,789

25

20

15

10

5

0
'97   '99   '01   '03   '05   '07*

*through mid-June

Remittances
IN BILLIONS

3.0

$2.8 billion

2.5

2.0

1.5

1.0

0.5

0
'97   '99   '01   '03   '05   '07*

*estimated

FIGURE 3.3

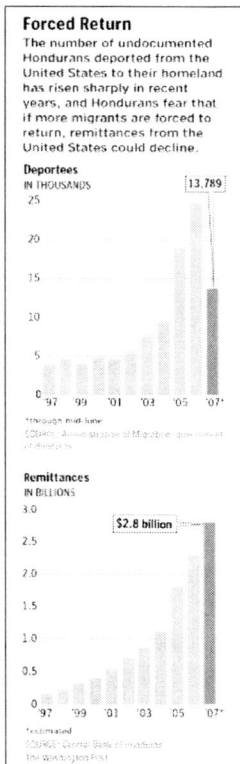

Source: *Washington Post*, 27 June 2007

Of those who do receive remittances, 58.2 per cent receive an average of $144.73 each month. Most of those that receive support from a family member, or a friend in the US live in Marcovia (46.8 per cent or 37/79) or Nacaome (22.8 per cent or 18/79). Marcovia is probably the highest as it was the most devastated during Hurricane Mitch and, therefore, more emigrants originated from the region post-Mitch. On 27 June 2007 The *Washington Post* reported, "40 percent of the populace earns less than $3 a day and just over half the workforce has a sixth-grade education. Money sent directly to Honduran families from relatives working in the United States, both legally and illegally, provides nearly one-third of the national income -- $1.8 billion in 2005, $2.3 billion last year" (*Deportees Bittersweet Homecoming*). The diversity of economic activities and remittances makes the interests of local actors associated with the physical geographic space of southern Honduras varied. This is most apparent in how local people discuss issues relevant to their livelihoods as presented throughout the thesis (Wilburn, Socioeconomic Data: July 2005).

Poverty in the region is also exacerbated by a high birth rate leading to increased pressure for subsistence resource use between 1973 and 2006, exemplified by the increase in artisan fishermen from less than 1,500 in the early-1970s to over 15,000 by 1998. Recent data reveals that women in rural areas have between six and seven children during their reproductive period, whereas women in urban areas have only four. My socio-economic data revealed that in southern coastal communities, on average, there are 5.22 family members per household (Wilburn, Socioeconomic Data: 2005). The national average is five children per woman (De Ferranti et al., 2000: 29).

The high fertility rate in rural areas has led to a sharp increase of the non-urban population above ten years of age between 1990 and 1997 (Thorpe, 2002: 208). Other indicators reveal the situation in Honduras: adult literacy 27 per cent (21 per cent for women), infant mortality at 35 per 1,000 live births, and maternal mortality rate of 221 per 100,000 births in 2000 (Warren, 2000: 1). Rates for the southern portion of the country are worse than the national averages, but there are no comprehensive studies that can be relied upon to depict the situation in southern Honduras accurately. Furthermore, less than 75.5 per cent of the respondents (272/360) to the household questionnaire have not completed primary school or higher (Wilburn, Socioeconomic Data: July 2005). It is apparent that unemployment, low income, high birth rate, and poor social infrastructure make the majority of people living near the GOF highly vulnerable and dependent on local resources for subsistence.

The majority of the coastal population in the GOF is dependent on local resources for subsistence use. The mangrove wetlands are situated in the inter-tidal zone to which this population has no legal right, *de jure*, since this zone is claimed by the state. In other words it is treated as a common pool resource; an open access system governed more by informal institutional arrangements rather than formal ones. Resource-dependent activities throughout coastal communities in southern Honduras include small-scale agriculture, salt production, artisan fishing, mangrove harvest for structural wood and fuel, and aquaculture. However, since the 1960s, there has been a decline in self-sufficiency among coastal communities (30-50 per cent unemployment) and an increase in emigration (50 per cent of the people born in the region leave), food insecurity, economic and political instability, and conflicts over land use (Vergne, *et al.*, 1993: 120). The government, international organisations, and NGOs have played an active role in trying to promote economic development and food security by creating jobs and further developing the region through the promotion of a number of agricultural schemes some of which have generated conflict in the region, one of which is analysed in this thesis.

Gulf of Fonseca

The GOF is a shallow depression located on the Pacific side of the Central American isthmus. This area is shared by Nicaragua, Honduras, and El Salvador and covers an area of approximately 3,200 km². The coastline of the Gulf extends for 261 km, of which 185 km are in Honduras, 47 km in Nicaragua, and 29 km in El Salvador. This marine environment is an integral part of the 'Pacific Central American Coastal Large Marine Ecosystem' (LME)[16] which extends along the Pacific Coast of Central America, from *Cabo*

---

[16] Large Marine Ecosystems (LMEs) are regions of ocean and coastal space that encompass river basins and estuaries and extend out to the seaward boundary of continental shelves and the seaward margins of coastal current systems. The LME as an organisational unit facilitates management and governance strategies that recognise the ecosystem's numerous biological and physical elements and the complex

Corrientes in Mexico to the vicinity of the equator. The extensive mangrove wetlands of the GOF are one of the most important shrimp nurseries associated with this LME. The climatic conditions in the GOF are influenced by atmospheric phenomenon typical of tropical and subtropical regions, which create two distinct seasons, the rainy and the dry, referred to locally as winter and summer.

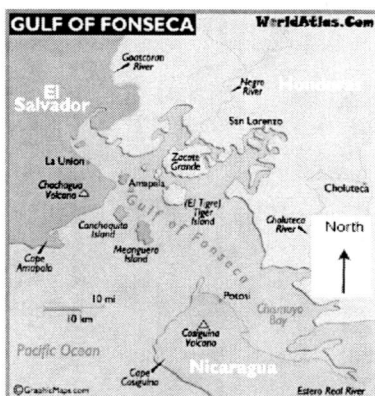

MAP 3.3: THE
GULF OF FONSECA
Source:
http://www.worldatlas.com/aatlas/infopage/fonseca.htm

The Gulf receives nearly 80 per cent of its total rainfall (1400–1600 mm of rain per annum) during the rainy season from May to November (Honduran Secretariat of Industry and Commerce, 2002: 23). This is particularly important for the formation of seasonal lagoons throughout the south, which are also believed to influence micro-climatic conditions during those periods.

The dry season occurs between December and May and contributes to an annual evaporation rate of 2800 mm. As a result of less water flowing into the GOF, the currents tend to flow inward from the Pacific Ocean, levels of salinity in the estuaries increase, and seasonal drought occurs (CODDEFFAGOLF, 2001: 10). Levels of salinity affect the population of shrimp postlarvae in the estuaries, which in turn influences the supply of postlarvae for the industry to capture to stock the ponds. This is another example of the 'natural agency' of the environment. Temperatures in the GOF average between 25°C and 30°C; March and April are the warmest months and November and December the coolest. Relative humidity varies between 65 per cent and 86 per cent depending on location. In contrast, the interior of the country is semi-tropical and cooler with an

dynamics that exist amongst and between them.
http://www.oceansatlas.org/servlet/CDSServlet?status=ND0xMjcyNyZjdG5faW5mb192aWV3X3NpemU9Y3RuX2luZm9fdmlld19mdWxs&19mdWxsJjY9ZW4mMzM9KiYzNz1rb3M~

average temperature of 26°C (Honduran Secretariat of Industry and Commerce, 2002: 14).

FIGURE 3.4A AND B: MANGROVE WETLANDS

(a)

(b)

The vegetation in the wetland ecosystem is dominated by species of mangroves, leading local people to refer to the areas as *manglar*[17]. Mangroves are evergreen trees found in inter-tropical latitudes in the inter-tidal zone between land and sea. Mangrove forests are generally located along sheltered coasts, estuaries, and in deltas, and are influenced by tidal regime, differing conditions of salinity, and rainfall regimes; they are also found around islands off-shore (Rollet, 1981: iii). The forests are comprised of halophytic trees, shrubs and other plants that grow in the brackish to saline tidal waters found along coastlines (Mitsch and Gosselink, 1986: 231). The defining feature is usually

---

[17] Throughout the course of this research *mangal* or *manglar* is referred to as 'mangrove ecosystem' or 'mangrove forests' and will be used synonymously with wetlands.

their dense and tangled prop roots that are periodically inundated by the high tide (see the figure below for different types of mangrove forests that can be found in the GOF).

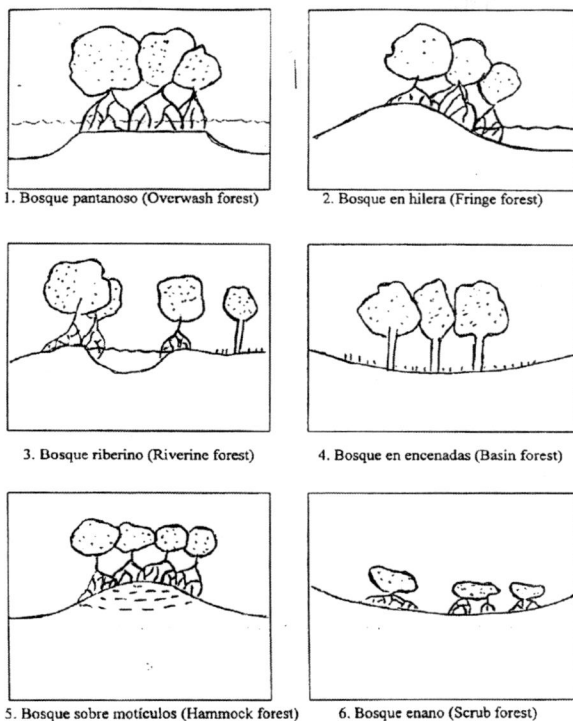

1. Bosque pantanoso (Overwash forest)   2. Bosque en hilera (Fringe forest)

3. Bosque riberino (Riverine forest)   4. Bosque en encenadas (Basin forest)

5. Bosque sobre motículos (Hammock forest)   6. Bosque enano (Scrub forest)

FIGURE 3.5: VARIATION IN MANGROVE FOREST TYPES
Source: Hutchings and Saenger, 1987: 123. Found in Sobreexplotacion de la Vegetacion Manglar y sus Efectos al Ecosystema de law Bahia de Chismuyo, Honduras, 1997.

Mangrove ecosystems are open systems, linked upstream to the land and downstream to the sea. As a result, mangrove species are able to survive in conditions of variable salinity and nutrient availability. They have developed specialised features that allow them to thrive in conditions in which salinity levels and nutrient availability vary (Ellison and Farnsworth, 2001: 425). In general, mangroves have a high level of tolerance for anaerobic soil conditions and high levels of salinity. As these factors change throughout the southern portion of Honduras, so do the structure, composition, and distribution of the mangrove forests located in the region (Calix Vindel, 1997: 25).

Mangrove ecosystems regulate the movement of nutrients between the upstream catchments and the marine system, making them the key to a complex detritus-based food web linking a vast amount of flora and fauna in the inter-tidal zone. This exchange area is one of the most important marine-based nursery grounds and habitat for a variety of land and sea-based species, a number of which local people are dependent upon to subsist (UNESCO, 1979b: 5).

Latin America is home to 10 of the approximately 70 or so species of mangrove known to exist (Mitsch and Gosselink, 1986: 256). Six species are found in the GOF (see table below). Covering 6.7 million of the total 17 million ha of mangrove forests found worldwide, Latin America has more than a third of the total on earth (UNESCO, 1979a: 1). Nearly 70 per cent of mangrove forests occur on the Atlantic coasts of Brazil, Mexico, Cuba, and Colombia (Calix Vindel, 1997: 17). In Central America, there are approximately 9,000 km^2 of mangrove forests. According to a study by Sherman and Tang, the GOF contains some 22 per cent of the entire mangrove area of the Pacific coasts of all the states of Central America, with the majority located within the territorial boundaries of Honduras (Sherman and Tang, 1997: 277). In 2001, it was estimated that the entire area around the GOF consisted of 78,000 ha of mangrove forests with approximately 49,500 ha located in the southern portion of Honduras. The largest densities of mangrove, 150 trees per ha, can be found in the Bay of Chismuyo and the Bay of San Lorenzo (CODDEFFAGOLF, 2001: 14).

Scientific Name	Common Name	Family Name
Rhizophora mangle	Mangle Rojo/Red Mangrove	Colorado Rhizophoraceae
Rhizophora harrisonii	Mangle Rojo/Red Mangrove	Colorado Rhizophoraceae
Aviccennia germinans	Curumo negro/Black Mangrove	Avicenniaceae
Aviccennia bicolor	Curumo blanco/White Mangrove	Avicenniaceae
Laguncularia racemosa	Curumo blanco/ Angeli	Jeli Combretaceae
Conocarpus erectus	Botoncillo/Buttonwood	Combretaceae

Source: Personal interview with Laura Sosa, Technical Specialist, COHDEFOR PROMANGLE, April 2005) *Second source:* Diagnostic and Preliminary Zonification of Mangrove Forests of the Gulf of Fonseca, Honduras; PROMANGLE, 2000

MAP *3.4:* M*ANGROVE WETLANDS OF THE* G*ULF OF* F*ONSECA.* The dark green areas depict the mangrove wetlands. The bottom right shows shrimp ponds in the San Bernardo region. *Source:* NOAA

Of the six species identified in the GOF, the last one, Conocarpus erectus, is not strictly a species of mangrove but is often found in the transition zone between the mangrove wetlands and the drier uplands (Mitsch and Gosselink, 1986: 240). The mangrove ecosystems of Latin America, as elsewhere, are characterised by zonation that reflects elevation above sea level, with sometimes a succession of species as elevation increases following sediment accretion; different species of mangrove are dominant in different zones.

In the GOF, red mangrove is the most common species, occupying mostly the areas permanently inundated by the tides towards the exterior border of the mangrove ecosystem nearest the sea. Black mangrove is the second most pervasive species and is generally found around the rivers where sediments are deposited along the shoreline. White mangrove is the third most dominant, followed by *botoncillo*; both are generally found further inland and are inundated by the tide less frequently. The dominance of different species over others correlates with the frequency of inundations, water quality, and levels of salinity (Sanchez, 1999: 13).

TABLE 3.3: SPECIES OF MANGROVE AND ANNUAL INUNDATIONS

Inundations (no. per annum)	Dominant Species of Mangrove in the GOF
530-700 and more	*Rhizophora mangle*
400-530	*Avicennia germinans*
150-250	*Laguncularia racemosa*
4-100	*Laguncularia-Conocarpus erecta*

Source: Sanchez, 1999: 13

The inundations and the discharges from the rivers are important in providing a constant source of nutrients and organic material that support not only the survival of the mangroves but also the large number of marine species dependent upon them for survival (Barnes and Hughes, 1999: 79). The cycle of tides is 2.3m on average per day in the GOF, causing the inundations to play an important role by creating a bi-functional system. During the low tides the soils are inhabited by crabs, conch, and other species that aerate the soil. In southern coastal communities 16.5 per cent (60/364 responses) of the respondents to the household questionnaire stated that they gather crustaceans and molluscs and use them for sustenance. 71.7 per cent of the 16.5 per cent, or a little less than 12 per cent (43/364 responses) of the respondents collect various species within the mangrove forests or in the estuaries. The majority of those who collect walk to various locations a few times per week and in less than half an hour (Wilburn, 2005). During the high tide the mangrove forests serve as a feeding ground and habitat for fish, shrimp, and other species. The root structure of mangroves provides a natural source of refuge for these species from larger predators aquatic and non-aquatic (such as birds) (CODDEFFAGOLF, 2001: 14).

Mangrove ecosystems are connected to the life cycle of numerous tropical fish species, making them an important link to the rest of the marine ecosystem (Nations and Leonard, 1986: 83). Mangroves also serve as an important breeding ground for fish, molluscs, and crustaceans, and are home to various species of fungi, bacteria, and protozoa that play roles in the overall functioning of the system (Barnes and Hughes, 1999: 79). The latter organisms play an important role, in conjunction with other crustaceans and molluscs, in the nutrient cycle through the decomposition of organic material, normally detritus – fallen leaves, branches, flowers, and fruits – produced by the mangrove forests, converting the cellulose into protein which offers an abundant source of nutrients that serves as the base of the rest of the marine food web (Ellison and Farnsworth, 2001: 428). Additionally, mangrove forests trap nutrients and sediments that stabilise the coastline, and act as both a natural water filter and barrier between the sea and land, which can reduce the impact of tropical storms or hurricanes on the coastal lowlands and the communities that lie near the sea.

Contested Representations of Mangrove Forests in the Gulf of Fonseca

Although I indicated that in 2001 there was an estimated 78,000 ha of mangrove forests around the entire GOF, with approximately 49,500 ha in southern Honduras, the precise area occupied by mangrove forests in the region is highly contested and has been a central issue in the conflicts surrounding them. First and foremost, contestation can be attributed to conflicts between social actors who want to represent deforestation in specific ways (see Appendix III 'Examples of Various Representations of Mangrove Cover'). For CODDEFFAGOLF, the more deforestation illustrated and attributed to shrimp farming, the more the organisation can substantiate one of its claims that the shrimp industry is the cause of it. Analysis of the open-ended interviews conducted throughout the Central American region revealed that mangrove loss due to shrimp farming had become the dominant explanation for the conflicts between CODDEFFAGOLF and ANDAH. However, over 65 per cent of locals use mangrove forests for a variety of purposes that must also be taken into consideration when studies are conducted:

- Fuel-wood
- Salt production
- Shrimp farming
- Construction materials
- Cleared for pastureland
- Fishing and gathering crustaceans and molluscs

The debates pertaining to mangrove loss are often situated around contested definitions of what 'mangrove forests' ought to be considered as *a priori* to a study. The particular method used when analysing satellite images or aerial photographs can also be a factor in various depictions of mangrove cover and, therefore, variations in representations as depicted in the tables below. In other words, the methods used to study mangrove deforestation and, therefore, the results are contested and a source of conflict.

TABLE 3.4: DIRECT USE OF THE AREAS OF THE MANGROVE FORESTS IN THE GULF OF FONSECA

Type of Use	Area in 1987 (Hectares)	%	Area in 1992 (Hectares)	%	Area in 1995 (Hectares)	%
Mangrove Forests	46,710	65.49	43,678	61	41,320	58.00
Salt Flats	14,240	19.96	3,000	4.2	3,000	4.00
Shrimp Farms	8,291	11.62	22,113	31	24,471	34.72
Salt Producers	1,292	1.81	1,842	2.58	1,824	2.50
Artisan Fisheries	624	0.87	375	0.52	375	0.50
Plantations	111	0.15	260	0.36	260	0.30
Sand	58	0.08	58	0.08	58	0.08
Total	71,326	100	71,236	100	71,326	100

Source: Adapted by CONGESA, in the database of AFE-COHDEFOR.

TABLE **3.5:** T**HE RATE OF CHANGE IN COVERAGE AND CORRESPONDING AREA BETWEEN**
THREE DATES

Coverage	Year 1989	Year 1995		Year 1998		
	Area (ha)	Area (ha)	Rate of change since 1989	Area (ha)	Rate of change since 1989	Rate of change since 1995
M. Rhizophora (Red Mangrove)	25,440	25,583	143	25,730	290	147
Other Mangrove	21,578	16,892	-4,686	16,282	-5296	-610
Salt Flats	20,278	16,354	-3,924	13,777	-6501	-2577
Sand	11,042	14,463	3,421	9,230	-1812	-5233
Shrimp Farming	2,611	12,222	9,611	15,677	13, 066	3455
Salt Production	174	718	544	746	572	28
Natural Lagoons	------	158	158	150	150	-8
Other Classes	489,464	484,197	-5,267	488,995	-469	4798
Total	570,587	570,587	------	570,587	------	------

Source: Sánchez, 1999, p. 21

TABLE **3.6:** L**OSS IN COVERAGE OF MANGROVE FOREST IN THE GULF OF FONSECA**
1965–1999

Department	Year 1965	Year 1999	Change in Coverage
Choluteca	57,300	21,800	35,500
Valle	34,500	25,400	9,100
Total	91,800	47,200	44,600

Source: CONGESA, 2001: 136 Adapted by CONGESA from Forestry Inventory of 1965 and Annual Statistics of COHDEFOR, 1999.

TABLE **3.7:** D**ISTRIBUTION OF MANGROVE LOSS IN HECTARES DUE TO SHRIMP FARMING**
AND SALT PRODUCTION

Coverage	1989-1995	1995-1998	1989-1998
Shrimp Farming and Salt Production	2,848	588	3,733
Other Coverage	3,385	380	3,293
Total	6,233	968	7,026

Source: Sánchez 1999, p. 24.

In some studies, mangrove forests may denote the entire wetlands; including the salt flats where dwarf mangroves grow. However, dwarf mangroves are sometimes not considered to be a part of the mangrove forests since they grow on the salt flats, are small, and non-existent in some locations. In some studies these are considered to be distinct from mangrove forests, but a part of the wetlands; however, salt flats are sometimes considered part of the mangrove forests since dwarf mangrove can be found in these areas. The point was to illustrate how competing definitions can affect the outcomes of studies in regard to mangrove cover and that the nuances matter.

Studies that rely on satellite imagery or aerial photography are unable to take into consideration local use of mangrove wood for fuel and construction because often they cannot see where the wood has been extracted. For several reasons, therefore, these types of studies are often unable to assess dispersed and sporadic local extraction. Local people do not generally clear cut mangrove forests for subsistence use. They use small amounts on a routine basis collected from varied locations. To depict how much mangrove wood locals actually use requires conducting a study that bridges both satellite and aerial imagery with agreed-upon parameters that one can ground truth through local research.

FIGURE 3.6: COHDEFOR/PROMANGLE REFORESTATION PROJECT IN THE **GOF**
Source: Laura Sosa, COHDEFOR/PROMANGLE, 2004

Consequently, understanding mangrove loss requires studies that take into consideration more than aerial satellite or photographic images to understand what is happening on the ground. Furthermore, mangrove conservation or restoration requires local-level research that analyses how much wood locals cut by measuring weight, length, volume (dry and wet) as well as where they extract it and what species they extract. Research of this nature would assist with developing a better understanding of local impacts on mangrove forests.

My research was not focused on conducting a scientific geospatial study of mangrove forests in the GOF or conducting a thorough study of local use of mangrove wood to compare possible sources of impacts. My point was to demonstrate that there are numerous factors that contribute to debates surrounding mangrove deforestation and how much mangrove has actually been lost and in relation to what types of human activities in the region. Subsequently, when reading documents produced in relation to land cover change, it is important to determine who contracted the study, why they contracted it, and in whose interests. Once the actors who have contracted the study have been identified, it is worthwhile to assess the Terms of Reference (TOR) developed, the parameters and variables used, and what the results of the study would be used to do.

The complexity behind research associated with land use change makes it apparent that the process of determining the area of mangrove forests in existence is a political process and one often associated with specific economic interests. Actors will generally use representations of mangrove deforestation that supports their interests in relation to the physical space in which they want to pursue various activities. These factors are important since I present data that differing actors have used to support specific positions historically. It is obvious that the amount lost and to whom that loss is attributed is highly contested and visible in my case study.

Summary

I began this chapter by clarifying the concept of physical geographic space, as used throughout the course of this thesis, and discussed the 'natural agency' of the environment. This was followed by a discussion of relevant epistemological and ontological issues in terms of human conceptions of the natural environment. The chapter then provided a geographic overview of the Republic of Honduras, socio-economic information related to the southern region of the country, a description of the physical area surrounding the Gulf of Fonseca, and sought to explain important features of the extensive mangrove wetlands in the region due to their role in the social construction of environmental conflict. The last section explained contestation in regard to mangrove deforestation in southern Honduras. Subsequent chapters will explain, as necessary, how certain characteristics of the wetlands influence human activities in relation to this physical geographic space in order to point to concrete examples of the 'natural agency' of the environment. Furthermore, the chapters that follow elucidate how conflict and competition associated with access to and control over this space has emerged and evolved between 1973 and 2006.

Part II
The Emergence of Neoliberalism
and Aquaculture
1973 – 2006

Chapter 4
Neoliberalism 1973 - 2006

Introduction: Neoliberalism in Geography

The rise of market economics in the Latin American region commenced in the late-1970s and was increasingly embraced in 1982 due to the 'debt crises'.[18] Countries in the region shifted from protectionist economic models towards favouring more open markets guided by export-led growth, neoliberal ideas. Neoliberal economic theory favours rolling back state control of the economy, free markets, the privatisation of national industries, financial liberalisation, elimination of barriers to trade and export promotion policies. It is based upon the Western cultural ideas of competition, freedom, and responsibility of the individual and has even been articulated as a necessary component for the promotion of democracy (Johnston et al., 2000: 547).

The ideas that underlie neoliberalism can be traced back to the University of Chicago, the birthplace of neoliberal economics in the 1960s. Liberal philosopher Friedrich von Hayek, and his students such as Milton Friedman, laid the ideological framework, which was eventually promoted through an extensive network of institutions, research centres, and foundations designed to research, market, and implement the neoliberal doctrine as early as the 1970s. However, the ideas underpinning neoliberalism originated much earlier amongst the Mont Pelerin Society in 1947, led by Friedrich von Hayek (Harvey, 2005: 19-20). Currently, some of the most important American institutions that have participated in the dissemination of neoliberal ideas are the American Heritage Foundation, parts of the Hoover Institution, the American Enterprise Institute, and other US organisations well financed by conservative corporations (Peet, 2002: 63-64).

This chapter shows why we need to understand the socio-historical contexts in which neoliberal reform processes have occurred. Harvey has argued that the uneven geographical development of these processes and their lopsided application from one state to another reveals the tentativeness of neoliberal ideology's application in differing historical and political contexts where diverse institutional arrangements have shaped

---

[18] The beginning of the debt crisis in Latin America is generally associated with the collapse of the Mexican economy in 1982 after Mexico defaulted on its debts.

its trajectory (Harvey, 2005: 13). Consequently, when used as a totalizing concept in scholarly analysis, neoliberalism is particularly problematic. The aim is to unravel neoliberalism's trajectory in the Honduran context by focusing on the effects of the ideas, policies, and practices that have underpinned it while elucidating their locally contingent forms (Perreault and Martin, 2005: 192). A further aim is to illuminate the dynamics of the capitalist world economy by assessing the effects of neoliberal policy in the Honduran context while illustrating the differing ways that social actors, environmental change, state and international policies mediate its path (Corbridge, 1986: 247). In particular, the case of aquaculture development and expansion throughout the 1990s in Honduras illustrates the particularities associated with regional economic integration in Central America, the rise of global capitalism, and, subsequently, growing grassroots resistance movements seeking to reconceptualise neoliberal futures.

In Honduras, these policies were not fully implemented until the end of the Cold War in 1989, but the transition from ISI had begun to emerge as early as the 1970s under President Oswaldo Lopez Arellano. Robinson argues that this early transition was the result of reforms under Arellano's administration, aligned with private interests, the military, and government elites, that "unblocked" capitalist development in the 1970s, establishing the basis to transform the economy under globalisation, and leading to neoliberal reform in the 1980s and 1990s (Robinson, 2003: 120) The aim here is to demonstrate that aquaculture development was heavily promoted as neoliberal policies were adopted in Honduras. Subsequently, I highlight the specific policies and legal changes that encouraged aquaculture, however, I first turn my attention towards the adoption of neoliberal policies throughout the Latin American region.

Latin American Neoliberalism and the International Context

Economic reform along neoliberal lines occurred throughout Latin America in the 1970s and early 1980s. Ushered into the region with the overthrow of Chile's President, Salvador Allende, by Augusto Pinochet in 1973, neoliberal reform slowly eclipsed the past policies of tariff protection and subsidies for national industries that most states in the region favoured (Harvey, 2005: 7-8). The US government supported the initial experiment with neoliberal state formation in Chile, along with Chilean economists known as the 'Chicago Boys', and their mentor Milton Friedman who taught them at the University of Chicago (Green, 1995: 25-26). Reform included establishing new institutional structures to convert countries from Import Substitution Industrialization

(ISI) strategies, until then the prevailing development paradigm, which, from a neoliberal perspective, were seen as protectionist and closed to the global economy. However, it is critical to recognize that neoliberalism was not necessarily the "inevitable successor to the political-economic policies of ISI in Latin America" (Perreault and Martin, 2005: 194). Export promotion policies, in addition to partial trade liberalisation, were one aspect of economic structural adjustment policies (SAPs) and played an important role in setting the stage for the emergence and proliferation of non-traditional agricultural export (NTAX) activities as a way to stimulate market based economic growth. NTAX are those products that a country has not traditionally produced and, in Honduras, include shrimp, melons, pineapples, cashews, and flowers, all of which began to be developed for export in the early 1980s.

As Bulmer-Thomas has demonstrated, before the emergence of neoliberal economics in the 1960s, most countries in Latin America had to make a decision to either pursue an inward-model of development, to avoid susceptibility to external factors, or to pursue export-led growth by intensifying and diversifying exports (Bulmer-Thomas, 1994: 276). Throughout the 1950s and 1960s, most countries in the region chose the internal-looking model of ISI, which was promoted by the Economic Commission for Latin America and the Caribbean (ECLAC), an operation under the auspices of the United Nations. ECLAC's structuralist[19] approach was primarily designed to reduce the external vulnerability of Latin America, while working towards internal market stability (Van der Borgh, 1995: 278)

Development practices in the region shifted away from ISI after the system of fixed exchange rates underpinning monetary discipline in the 'developed world' ended in 1971 with the collapse of the Bretton Woods system (Bulmer-Thomas, 1994: 325). This event led to a commodity price boom that encouraged developing countries, such as Honduras, to move away from ISI strategies and begin pursuing export-led growth as a means of increasing their foreign exchange earnings. The higher commodity prices that resulted from the demise of the Bretton Woods system eventually collapsed following the oil crises of the 1970s, and plunged the region into an economic tailspin that ushered in an era of neoliberal reform.

The oil crisis of 1973 led to a recession in the developed world that decreased exports from Latin America to developed countries such as the US. In turn, this event reduced

---

[19] Here I refer to structuralism within development theory vs. structuralism in social theory, which was previously discussed.

the amount of foreign exchange revenue generated by countries in the region, leading to a decline in economic growth rates, which tumbled from 5.2 per cent (1960-70) to 3.2 per cent (1970-1980). This trend was noticed globally as the expansion of world trade was halved after 1973 (Thorpe, 2002: 5). The second oil crisis of 1979, which sent prices of crude oil soaring, pushed many Latin American countries into further debt as their economies, already struggling from reduced export revenue, slid as the cost of production increased alongside oil prices, forcing them to turn towards international lenders to salvage their economies (Thorpe, 2002: 6).

To some extent, international lenders provided funds deposited by Middle Eastern oil producers and nation-states profiting from the higher oil prices. However, as developed countries slid into recession, interest rates climbed, making it more difficult for Latin American countries to repay their external debts, which increased from $27 billion to $231 billion between 1970 and 1980 and annual debt-service payments reached $18 billion (interest plus amortisation) (Skidmore and Smith, 2001: 58). Events like this sealed the fate of the region's ISI strategy, which dominated the region after World War II, and led to the conditions that made neoliberal reform perforce.

State interventionism and the ISI strategy throughout Latin America was blamed for the conditions that had led to unsustainable trade practices, fiscal imbalances, and inflation that was apparent in the 1970s and 1980s (Thorpe, 2002: xiv). The advent of neoliberal policies gained unprecedented support amongst international financial institutions (IFI), academics, and those governments in the developed countries that favoured the implementation of these policies as economic conditions worsened in the region (Bulmer-Thomas, 1994: 366-367). A number of institutions, including the multilateral development banks and the Organisation for Economic Cooperation and Development (OECD) supported neoliberal reform. As Thorpe has argued, this was due to "the success of supply-side economics (Monetarism, Thatcherism, and Reaganism) in curing inflationary pressures in the developed world" (2002: 345). However, it was not until the late-1980s and early-1990s that neoliberal economic policy recommendations, known as the 'Washington consensus', dominated a previously social democratic and Keynesian development discourse throughout Latin America (Peet, 2002: 65).

The 'Washington Consensus' favours promoting the reform of a country's macroeconomic policy, trade regime, and policies encouraging private sector development (Van der Borgh, 1995: 277). In addition to macroeconomic policy reform, driven by a desire to move away from ISI, reform of the trade regime was an important policy goal of the 'Washington Consensus'. This policy goal included pursuing a reduction in barriers to trade, regional economic integration within Central America, and the creation of the Free Trade Agreement of Americas (FTAA). Another important goal was agrarian reform. ISI had neglected agriculture as a part of its programme, which contributed to its demise due to the region's heavy dependence on agriculture for export to earn foreign exchange needed to service debts.

The push for neoliberal reform in the region was consolidated with the announcement of the Enterprise for the Americas Initiative by the first Bush administration in June 1990, which aimed to create an integrated market in the western hemisphere. Although Enterprise for the Americas was a US initiative, it was promoted in conjunction with international lenders. From this point forward overseas development aid, from the US and others, was outwardly focused on promoting economic growth through neoliberal reform as clearly articulated by USAID Director John Sanbrailo in a 1992 radio interview. He stated that "the new [US] strategy of the 1990s is based on, not so much aid, but more on trade and promoting trade and exports, promoting private investment" (Naylor, 1992).[20] These efforts coincided with the implementation of Structural Adjustment Programmes (SAPs), neoliberal agrarian reform, land titling efforts and the establishment of new legal frameworks to support these objectives.

The strategy was based on the establishment of Free Trade Agreements (FTAs) in line with neoliberal policies and called for governments to privatise national industries, implement austerity programmes, and promote exports to the US (Naylor, 1992). The Enterprise for the Americas Initiative is now most commonly pursued under the guise of the 'Summit of the Americas', which depoliticised where the Free Trade Agreement of the Americas (FTAA) was pursued. The FTAA has manifested itself in a number of forms including the North American Free Trade Agreement (NAFTA) 1993 and, more recently, the Central American Free Trade Agreement (CAFTA) 2005, which is discussed in Part IV.

---

[20] John Sanbrailo, Director, Agency for International Development, National Public Radio Interview Transcript: All Things Considered, *Vanishing Homelands: Honduran Shrimp*, B. Naylor, Host: 18 April 1992

Honduran Experiences of Neoliberalism 1973 – 2006: Adoption and Priorities

As elsewhere in Latin America, the transformation of the Honduran economy to neoliberal practices began to take shape in the 1980s, pushed forward by the economic crises of the time, and was fully consolidated in the 1990s. Neoliberal economic reform in Honduras was encouraged by the US, mostly through USAID, and coincided with the country's transition from a US supported military dictatorship to a democratic republic. The democratic transition and neoliberal reform were thought to go hand-in-hand since the neoliberal state favours "strong individual private property rights, the rule of law, and the institutions of freely functioning markets and free trade," which were thought to be more likely guaranteed under democratic rule (Harvey, 2005: 64). Here, I argue that the Honduran government's acquiescence to US policy was partially due to its interest in securing its territorial sovereignty. However, the larger driving force was its need to focus on market expansion to increase the amount of foreign exchange revenue generated as the state grew increasingly worried about the impacts of the looming 'debt crisis'.

By 1982 the oil crisis of 1979 had caused world commodity prices to shrink by 35 per cent, their lowest level in 30 years. As in other Latin American countries this event had a negative impact on the Honduran economy, and eventually forced the government, at the behest of international lenders, to make significant structural adjustments to deal with the external shocks and increased foreign debt that resulted, (Honeywell, 1983 in Walton, 1989: 306). In Honduras between 1977 and 1983 there was a fourfold increase in the country's foreign debt, the economy was shrinking by 1982, and there was a 20 per cent decline in per capita production between 1980 and 1984 (Perez-Brignoli, 1989: 153). Consequently, between 1977 and 1984 there was an acute economic crisis in the country as export revenues fell more than 13.7 per cent annually, where they had previously been growing at a rate of 9.8 per cent at the end of the 1970s (Thorpe, 2002: 66). As the Honduran economy contracted, real wages fell by 12 per cent between 1981 and 1984, open unemployment rose by 64 per cent, and the foreign debt increased to $2.25 billion: $314 per capita GDP in 1984 was just one dollar higher than it had been in 1970 (Dunkerley, 1988: 567).

As Honduran foreign debt grew creditors increasingly demanded that economic policies change in the country, eventually leading to the implementation of economic adjustment policies as advocated by the International Monetary Fund (IMF), the US government and various other international organizations allied to local elites (Calderon *et al.*, 1992: 31). The stated objective of these SAPs was to restore rapid economic growth while supporting both internal and external financial stability (World Bank, 1990 in Thorpe, 2002: 1). The export-led growth model encouraged investment in NTAX and had two particularly important objectives. The model encompassed two main approaches "(i) seeking to maximise economies of scale through land concentration and (ii) reducing the demand for rural labour through parallel improvements in labour productivity" (Thorpe, 2002: 61). The first is exemplified by the swathes of land conceded to the private sector for aquaculture development and, the second by encouraging the early-industry's transition from wild-caught to laboratory-produced postlarvae designed to reduce rural labour demand and increase productivity through technological improvements.

The US was an active advocate of neoliberal reforms in Honduras during the 1980s, providing significant support for changes through USAID. The interest stemmed directly from the US' political interests in the region and ushered in a period of militarisation, and political and economic transformation. The wider US strategy in Central America was to end the revolutions in Nicaragua (a proxy war against communist Soviet Union), El Salvador (a communist 'guerilla' uprising), and Guatemala (a brutal civil war). Simultaneously, the US sought political and economic stability by supporting democratic transformation whilst identifying ways to integrate the Central American economies. The US was particularly interested in defeating the leftist Sandinistas in Nicaragua, securing the GOF to prevent the shipment of arms from Nicaragua to El Salvador, and funding the strengthening of the Honduran military presence along the borders with Nicaragua and El Salvador.

Honduras was an essential staging ground for US military operations in Nicaragua, El Salvador and Guatemala, since all three borders Honduras. As such, securing support from the Honduran government was a priority and the US had the capital to ensure this support, giving a total of $1.6 billion between 1980 and 1990, $711 million of which were through Economic Support Funds (ESF). These funds provided the US with enormous influence over state allocations, macroeconomic policies and agricultural policy, and encouraged investment opportunities in new sectors such as aquaculture (Robinson, 2003: 125).

The full implementation of neoliberal economic reform in Honduras was all but assured when President Rafael Leonardo Callejas Romero won the 1989 election and came into office on 27 January 1990, representing the National Party of Honduras (NPH). The NPH's victory was a pivotal moment. It ushered in a decade of unparalleled neoliberal reform and solidified the alliance between conservative political elites, the private sector and the military, and the US (Robinson, 2003: 121). The reform efforts were supported by both Liberal and NPH administrations throughout the 1990s, coinciding with globalising processes and the rise of world-wide capitalism after the Cold War. Furthermore, this event coincided with a period of peace and stability throughout the Central American region after decades of revolutions and insurgencies came to a close.

In conjunction with the US, and with the support of the World Bank, the IMF, the WTO and others including private-sector elites and foreign economists that had ties to USAID programmes and later led to charges of corruption, President Callejas launched an economic reform package called *paquetazo* (Robinson, 2003: 129). The *paquetazo*, part of the Structural Adjustment Program, was pursued in alignment with several US initiatives designed to work toward the full implementation of neoliberal policies and included incentives to continue promoting NTAX.

Structural Adjustment in Honduras

Structural Adjustment in Honduras centred on reforming the country's agrarian and land tenure systems. The neoliberal perspective holds that without a proper title the land owner can not pledge his land as collateral and thus can not obtain capital to invest in it, thus the land is "failing to fulfil its productive potential" (Thorpe, 2002: 267) and thus not contributing to economic growth. A secure tenure arrangement was also seen as necessary for the producer to develop long-term production strategies and to increase the potential exchange value of the land by opening up the market (Thorpe, 2002: 267). Based on these perspectives there was broad agreement that in order to promote economic development it was necessary to address agrarian reform and land tenure issues as well as the inequalities associated with them.

At the national level, the Callejas administration (1990–1994) supported the passage of two laws that reformed the country's agrarian and land tenure systems and formed the central components of the Honduran SAP. The first was the Law for the Modernisation and Development of the Agricultural Sector (La Ley para la Modernizacion y Desarrollo del Sector Agricola LMDSA) in April 1992, which defined agrarian reform into the next decade. The second was Decree 18-90, Law for Structural Ordinance of March 1990 that created stabilisation programmes that "stressed (i) the regulation of domestic credit expansion, (ii) a reduction in the level of government debt financing and (iii) international reserve management" (Thorpe, 2002: 13). The privatisation of land was also emphasised, extractive activities geared toward the export market were encouraged, and foreign investment, leading to the continued conversion of wetlands for aquaculture, as the country began implementing structural adjustment policies.

In order to promote extractive activities, older agricultural institutions were either abolished and eclipsed by new institutions or realigned to promote neoliberal agrarian reform objectives, illustrating how SAPs were used strategically to transform existing institutions, and sometimes by creating new institutions. In particular, the passage of the LMDSA exemplified the government's interests in pursuing agrarian reform in the neoliberal tradition. Section 3 of the LMDSA reaffirmed the Honduran government's commitment towards developing agro-industrial enterprises by creating state run institutions (see Chapter 6) to support the sectors development (Article 32) and eliminated both NTAX taxes and agro-export permits thus facilitating the growth of the industry by lowering costs of operation (Article 33) (Thorpe, 2002: 95). Thus, "neoliberalism does not involve a necessary decrease in the state's functions or size, but rather its reconfiguration and reinstitutionalization (Peck, 2004 in Perreault and Martin, 2005: 193). For example, the passage of the LMDSA favoured the large-scale private sector, particularly those investing in NTAX, rather than small producers. These changes coincided with the implementation of policies congruent with the Caribbean Basin Initiative Act (CBI) passed by the US Congress in 1983.[21] The effects of these policy changes are explicated in subsequent chapters.

In addition to internal elements that contributed to the promotion of aquaculture during this period, there were also a number of external factors that influenced the development of the shrimp industry: the actions of the EU in relation to banana exports from Honduras, and the response of the government, illustrate the power of external actors in relation to the political economy of the country. For example, unknown to the early Callejas administration, the imperative to promote aquaculture expansion increased in February 1993 when the Honduran government was hit hard with new

---

[21] Caribbean Basin Economic Recovery Act (CBERA) was enacted by the US Congress on 5 August 1983 (Public Law 98-67, title II). The objective of the Act was to expand private sector opportunities and investment in non-traditional sectors of the Caribbean Basin beneficiaries to assist with the diversification of their economies and expansion of their exports.

policies on the export of bananas into the EU, its second largest export after coffee. The imposition of quotas on banana exports from Latin America forced the government to support alternative means of generating export revenue.

The purpose of the EU resolution was to decrease banana exports into the Eurozone to prevent market saturation, in addition to the French desire to protect the banana market in its former colonies. Consequently, the EU imposed a 20 per cent tax on the first two million tons of bananas, and 170 per cent on any quantity over that amount (*La Tribuna*, 19 February, 1993). The political decision made by the EU was a direct threat to the original 'Banana Republic'.[22] The potential impacts led the Minister of Natural Resources, Mario Nufio Gamero, to announce that the government would have to rely increasingly on NTAX, such as shrimp and melons, to maintain or increase foreign exchange revenue. The recent passage of the LMDSA and the Structural Ordinance Law was believed to be the government's best hope (*La Tribuna*, 19 February 1993). Due to the potential impact of the decision, the EU and OLDEPESCA agreed to work with the Honduran government to reduce the effect of the new policy and expand their support for the aquaculture and fisheries sectors to mitigate the impacts.

The external pressures on the Honduran government to promote NTAX were particularly relevant to the rise of contestation against aquaculture development between 1988 and 1998 (see Chapter 8). There is no doubt about the fundamental role that powerful actors such as USAID, the World Bank, and multinational corporations played in reshaping the coastal landscape of the GOF (Collinson, 1997: 4). Each propagated the free market model, a central feature of neoliberal economic policies, to encourage the government's promotion of NTAX and, subsequently, aquaculture.

The promotion of NTAX, and the eventual expansion of shrimp aquaculture, were only a small part of the capitalist restructuring of the entire region and have had consequences for the marine and coastal environment of southern Honduras. These impacts were the result of powerful forces working to establish the appropriate political and economic structures to facilitate closer economic ties with the US and within the region (see Chapters 5 and 6). Capitalist restructuring has played a significant role in terms of how the physical geographic space of southern Honduras has been occupied, categorised, labelled and transformed for NTAX over a number of years. In particular, a shift in

---

[22] The term was coined by O. Henry, an American humourist and short story writer, in reference to Honduras. "Republic" was often a euphemism for a dictatorship, while "banana" implied reliance on basic agriculture.

agricultural practices in the south transformed large tracts of land that had been used for cotton and cattle production after 1950 to the production of NTAX crops such as melons, cashews, and, in the coastal wetlands, shrimp in the 1980s and 1990s.

Neoliberalism in Southern Honduras: Agriculture, Land Tenure, and Aquaculture

As early as April 1990, the government issued statements confirming its belief that the LMDSA and the Structural Ordinance Law would improve the 'disaster in the south' - the collapse in traditional commodity prices, environmental degradation, food insecurity, and shortages of basic grains. By the time these were published in *La Tribuna* (1990), several government officials had admitted that the 'disaster in the south' could be directly attributed to the past agrarian reform processes of the 1960s and the 1970s. President Callejas, Minister of Natural Resources, Mario Nufio Gamero, the Director of the National Agricultural Institute (INA) Juan Ramon Martinez, the Mayor of Choluteca, Omar Guillen Bellino, and the Governor of the Department of Choluteca, Alfredo Galo Rodriguez, admitted in conjunction with Roberto Gallardo, the President of the Federación Nacional de Agricultores y Ganaderos de Honduras (Honduran National Federation of Agriculturalists and Ranchers, FENAGH) that past agrarian reform had led to ecological problems, but agreed that their best hope to improve the situation in the south was through the Structural Ordinance Law, which was passed in 1990 just before the LMDSA in 1992 (*La Tribuna*, 2 April 1990).

Although structural adjustment policies were considered harsh due to cuts in social expenditures (e.g. education and health), some government officials argued that they helped reverse the negative trends that plagued Honduras in the early part of the 1980s. The change was most visible in Gross Domestic Product (GDP). During the latter part of the 1980s, GDP increased by 21.6 per cent, 4.3 per cent per capita and coincided with improved performance in the export sector which rose by 22.5 per cent after 1985. However, per capita GDP was still 23.7 per cent below the 1980 figure, which mirrored the economic situation throughout the Latin American region (Lustig, 2000: 3). Furthermore, the statistics used to justify continued growth did not identify the beneficiaries of agro-export growth during this period.

As described above, the Structural Ordinance Law and LMDSA led to increased investment in NTAX, which in southern Honduras manifested as industrial shrimp farming. Although development of the industry commenced in 1973 and underwent some expansion in the 1980s as part of early neoliberal reforms, it did not really begin to boom until the 1990s after the introduction of these two new laws. By the 1990s, the government and outside supporters (mainly IFIs) portrayed the industry's expansion as an important factor in encouraging economic development and food security. Eventually these discourses and the expansion of the industry were contested by local people, most notably through the Committee for the Defense and Development of the Flora and Fauna of the Gulf of Fonseca (CODDEFFAGOLF), who claimed that industry growth led to destruction of the local mangrove habitat and blocked access to traditional fishing grounds that they relied on for their livelihoods.

NTAX and Aquaculture in the South

The government began focusing attention on the southern region in the 1970s as a response to the decline in its traditional agricultural sector.[23] The need to focus on increasing exports of agricultural related products and natural resources was directly related to a weak manufacturing base and a lack of comparative advantage within traditional export sectors. The advent of market economics only served to intensify the push to export new products (Bulmer-Thomas, 1994: 371). First, there was lowered profitability in traditional agro-export sectors such as cotton and beef cattle.[24] Second, the production of basic grains was affected by prices and drought, leading to food insecurity among the local population. The government sought to address these issues by promoting NTAX.

A national newspaper, in an article titled: "El Ministerio de Recursos Naturales Interesado en el Cultivo de No-Tradicionales", revealed the initial discourses surrounding the government's intention to promote NTAX (*El Cronista*, 30 January 1975). In this 1975 newspaper article, the Secretariat of Natural Resources (SRN) explicitly stated that the government's goal was to promote economic development by further expanding into national and international markets. The SRN also argued that NTAX would assist with

---

[23] The transition from cattle and cotton production in the 1970s can be attributed to agrarian reform processes and a decline in traditional commodity prices in the 1970s and 1980s for the former products that were heavily promoted in the 1950s and 1960s. These agrarian transformations played an important role in terms of migration into the lower parts of the coastal wetlands as new agricultural activities were pursued. In turn, this also led to changes in the ways in which local people interacted with the resource.

[24] Traditional agricultural exports in Honduras have primarily been coffee, bananas, cotton, beef, and sugar. The latter three dominated agricultural production in southern Honduras throughout the 1960s and 1970s and, at one point, contributed to nearly two-thirds of all export earnings (Nations and Leonard, 1986: 67). However, the beef and cotton industries were heavily affected by the commodity price collapse in the 1970s and their contribution to export revenues declined. Sugar cane production remained relatively stable in comparison.

addressing shortages of basic grains that were affecting the south and, thus promote food security (*El Cronista*, 30 January 1975). Discourse on food security and economic development was consistently used by the Honduran government as the rationale for promoting aquaculture development between 1973 and 1988. These discourses were underpinned by six actions taken by the government of Honduras and IFIs in conjunction with the early investors:

1. knowledge acquisition for research and development including training and technical assistance;
2. provision of finances;
3. formation of cooperatives;
4. infrastructure development;
5. institution building to promote non-traditional agricultural exports; and
6. creation of international and domestic incentives for the promotion of aquaculture

These 'strategies' of power employed by the Honduran and IFIs and are demonstrative of the enactment of neoliberal policy reforms. Each action is analysed in Chapter 6 as a strategy of power that defined the physical-spatial reality tactically in line with the state's interests.

The Nascent Shrimp Industry in Honduras

The cultivation of laboratory-produced shrimp first occurred in Japan in the 1930s, but it was to be another 40 years before aquaculture would be pursued in Latin America after a long process of research and development in Asia and the US. Aquaculture development in Latin America, particularly Honduras, Panama, and Ecuador, commenced between the 1960s and 1980s when investors began partnering with local entrepreneurs to build farms, hatcheries, feed mills and processing plants (Rosenberry, 2004: 5). The industry matured rapidly in Panama and Ecuador, and was fully operating in these countries by the 1970s. The Honduran industry successfully emerged by the mid 1980s.

After a failed attempt by The United Fruit Company/Armour Company to bring shrimp farming to the Caribbean coast of Honduras in 1967, new investors explored aquaculture development in southern Honduras for the first time in 1972 and commenced activities in 1973 (Rosenberry, 2004: 9-10). Shrimp Culture Incorporated, financed by Pittsburgh Plate and Glass, began exploring the cultivation of shrimp in Florida. When their initial attempts in the laboratory proved unsuccessful, they teamed up with other investors to explore shrimp farming in southern Honduras. A joint venture called Sea Farms of

Honduras was created in 1972 (Rosenberry, 2004: 7). Sea Farms of Honduras was the result of negotiations between Shrimp Culture Inc., incorporated in the US in 1966, and Banco de Occidente, owned by Jorge Bueso Arias, one of the principal Honduran investors in the early shrimp industry. Establishing relations between them was one of the first formal partnerships between private investors from the US interested in aquaculture and the Honduran elite, both backed by the US government.

Despite these early explorations it was not until the mid 1980s that shrimp began to emerge as one of the most profitable products in Honduras as state, corporate, and donor agency efforts were made to transform 'unproductive' wetlands that traditionally had 'no commercial value' into extensive shrimp farms in order to pursue foreign exchange earnings (Hall, 2003: 252). Expansion of the industry peaked in the 1990s when these goals were facilitated by the creation of a neoliberal legal framework, embodied in the LMDSA and the Structural Ordinance Law in the early 1990s.

USAID and other international organisations helped create the necessary national institutions to implement export promotion policy objectives during the 1980s. The institutions created to promote NTAX in the south were continuously involved in promoting aquaculture production. Members of the ruling National Party of Honduras (NPH) and the military were often the primary beneficiaries of aquaculture development, as they took advantage of USAID funds and invested their own resources into the sector's development. Increased political attention on the south, and the growth of aquaculture, led to a number of accusations over an NPH party leader's, Ricardo Maduro, use of export earnings to fund the NPH and bring President Callejas into power (*La Tribuna*, 19 April 1989).

At the time, Maduro held about a 5 per cent share in the largest shrimp company, and had interests in several other lucrative businesses in Honduras. Maduro and the NPH actively supported the IMF and World Bank efforts to devalue the Lempira in line with SAPs pursued under neoliberal reform. As with other NTAX crops, the economic incentives associated with SAPs were geared toward development of private sector activities such as aquaculture. The NPH's interest in devaluing the currency was directly associated to its benefits to exporters of shrimp and other commodities as it would increase profit (*La Tribuna*, 19 April 1989). The effects of the changes related to the implementation of the SAPs are discussed more extensively in Chapter 10.

As indicated in Chapter 3, the salt flats in the GOF were identified by the industry as suitable for aquaculture development; once that was done investors had to acquire concessions to the land. The first concessions of just over 200 ha of wetlands were issued in 1983 to Sea Farms of Honduras, which became known as Grupo Granjas Marinas San Bernardo (GGMSB) in 1984[25] when five separate operations consolidated. The following table indicates the concessions issued or in the process of adjudication between 1984 and 1988. Since the government had the power to issue the concessions, the laws relevant to development in the marine and coastal zones were particularly important and are addressed in Chapter 5.

TABLE 4.1:   SHRIMP FARMS: CONCESSIONS ISSUED OR IN PROCESS OF ADJUDICATION BETWEEN 1984 AND 1988

Size (ha)	Number of Beneficiaries	Area (ha)	Percentage Conceded	Percentage of Beneficiaries
1-70	25	722.66	2.6	43.87
100-900	24	7,955.75	28.2	42.10
1000-6000	8	19,534.56	69.2	14.03
Total	57	28,212.97	100.00	100.00

Source: SECPLAN, 1989.

Impacts on Local People and the Rise of Contestation:

The promotion of aquaculture development affected local people in the region of the GOF by changing how the coastal wetlands were utilised by various social actors. The rise of commercial shrimp farming in the 1980s and 1990s changed the labour market dynamics of the fisheries sector leading to conflicts between factions of the artisan fishing community, those who captured wild postlarvaes to supply the industry against those who fished for other species in the estuaries, seasonal lagoons, and the open sea. The fishermen providing the wild postlarvaes were often in agreement with industry expansion, as it would potentially increase their profits, since the industry's demand for the seed stock would increase. However, the fishermen who lived by fishing for species other than shrimp post-larvae were negatively affected by the expansion of the industry (see Chapters 6 – 10). Thus, the early conflict with the industry (between 1973 and 1988) was fundamentally about access to, and control over, the mangrove wetlands due to its relationship to the fisheries resources.

---

[25] By 1996 the industry consolidated even further and became known as Sea Farms International (SFI) in 1996,

In 1988 the affected fishermen joined forces to create CODDEFFAGOLF and resist the expansion of the industry. Their discourses focused on the industry's impacts on fisheries resources and diminished access to traditional fishing grounds for local people. As expansion ensued, impacts on the wetlands increasingly became another concern as the natural salt flats were conceded by the government to the industry and, in some cases, mangrove forested areas. The organisation began to link mangrove degradation to the shrimp industry and focused less attention on the impacts associated with local use of the resources that the AHE brought forth early in the period (see Chapter 7).

Ultimately, it is argued that the conflicts that emerged around the coastal wetlands of the GOF were directly attributed to the Honduran government's efforts to promote NTAX as it sought to pursue export-led growth, a fundamental attribute of neoliberal reform. The resistance to aquaculture can thus be seen as a "rejection of the inevitability of the mastering of space by either states or markets" through the enactment of neoliberal ideology (Agnew and Corbridge, 1995: 227). Consequently, the conflict represents an articulation of new representations of space as civil society actors began to make their own claims about how the mangrove wetlands ought to be used, managed, and conserved.

Summary

In this chapter, I have provided a brief overview of the macro-economic and political factors that led to the development of aquaculture in southern Honduras. I addressed neoliberalism as a concept, defined it, and explicated its roots as it provided the model for Honduran economic reform in the 1980s and 1990s. I explained the justifications for neoliberal reform and the effects that the promotion of NTAX, specifically aquaculture, had on the fisheries sector, thus leading to conflicts within it. I introduced the six actions that were taken to promote shrimp farming and demonstrated their relationship to both of the discourses of economic development and food security. The socio-economic impacts within the fisheries sector were discussed briefly to clarify the roots of the conflict and the reason contestation emerged once these actions were taken (each is illustrated further in subsequent chapters).

I explained the actors behind the promotion of aquaculture and their early discourses on the wetlands; including a narrative that viewed the land as 'unproductive' and having 'no commercial value'. I argued that a number of actors discovered that the physical space around the GOF was ideal for shrimp farming due to five characteristics of the coastal wetlands, which were particularly favourable for the nascent shrimp industry. Once it seemed possible to develop the industry on these lands, businesses had to acquire the necessary concessions to them. In Chapter 5, I discuss state discourse and the legal construction of the mangrove wetlands, while discussing the legal powers conferred upon the state to issue concessions to the early industry. The goal is to elucidate how the legal framework that existed prior to the passage of the LMDSA and Structural Ordinance Law was used as an instrument of power to achieve the actions outlined in this chapter, and subsequently led to more permanent changes in the legal framework (e.g. reformation of the 1959 Fisheries Law). In the following chapters, I illustrate how these actions resulted in the formation of new social and economic networks between the state, private sector, and civil society and led to the emergence of environmental conflict in the GOF.

Chapter 5

State Discourse and the Legal Construction of the Mangrove Wetlands

Introduction

In this chapter I argue that the government's interpretation of the legal language in various laws was aligned with its objectives to pursue economic development and food security, two discourses that remained dominant between 1973 and 1988. The Honduran government's actions, in conjunction with private industry and IFIs, will be analysed in relation to the state's legal discourses.

Legal discourse is important to review because the language exposes the key characteristics of state agencies and their powers. Legal decrees are the mechanisms by which the state confers decision-making jurisdiction to its agencies. The power conferred on an agency is two-fold: first, authority is given to enquire into the matter in question, for example to assess how mangrove wetlands may be used. Second, an agency is given discretion to decide when and when not to use the power it has been given in relation to a specific matter, such as whether the mangrove wetlands *ought* to be exploited or conserved.

The passage of laws are key events in terms of their potential effect on the discursive positions and actions of any given set of actors with interests in a physical geographic space affected by laws. I focus particular attention on the laws pertaining to the geographic area of this case study since laws establish (at least on paper) the rules associated with state interventions into a specific physical geographic space. Legal language is important to assess not only as a discursive intervention, but also as an attempt to establish the basis of state 'legitimacy' to intervene in a particular physical geographic space where the law permits. I demonstrate that a key source of the conflict is directly associated with the inconsistencies between the legal framework and the actions of the state in the case study area, often due to overlapping legal jurisdictions between state agencies in relation to the same physical space. The following chapters illustrate and validate my point utilising the empirical information acquired.

I show how the law was a product of the imaginings of the state in relation to the land. How the state chose to interpret the laws demonstrates that the actions to promote economic development took precedent over environmental conservation in the GOF during the period between 1973 and 2006. Assessing the government's legal discourse in relation to its actions assists in the task of revealing the performative function of state representations of public and physical space, shedding light on the interconnections between legal, physical, and social spaces (Blomley, 1994). The need to consider more carefully legal constructions of nature or space within legal discourse and practice is an important objective in current geographical research (Johnston et al., 2000: 438) and is analysed in relation to aquaculture development in southern Honduras. Of particular relevance is the contestation that emerges from within civil society as the state pursued a number of actions to realise their images of how the land ought to be used.

The industry's access to selected sites for aquaculture development relied on the government issuance of concessions, providing the industry with quasi-private property rights over state-owned coastal lands (Stanley, 1996: 25). The areas conceded to more powerful actors – government, military officials, foreign and private Honduran investors – replaced local people's informal claims to the wetlands. Previously these areas had been treated as a common pool resource for fishing, the collection of crustaceans and molluscs, and cutting of mangrove wood for fuel and construction. Access to the coastal wetlands was determined not by a free land market, but by an administrative transfer of claims, resulting in over 25,000 ha in renewable long-term land leases being granted by the government of Honduras to private investors, at a cost of $5 per hectare, between 1983 and 1996 (Stanley, 1996: 25).

In this chapter, I argue that the legal language that provided the SRN and other agencies with the power and authority to provide concessions to the wetlands was particularly important. Although the National Agrarian Institute (Instituto Nacional Agrario, INA) was thought to have the power to concede lands, its legal basis to issue concessions to the wetlands in the early-1970s was doubted. Consequently, the government interpreted laws selectively to achieve its ends. My argument is that the legal discourse in the 1959 Fisheries Law, the 1972 Forestry Law, and the 1974 Honduran Corporation for Forest Development Law was interpreted selectively in line with the interests of the state, private investors, and IFIs in order to provide the legal justification for intervention into the wetlands of the GOF for the purpose of aquaculture development by 1984.

I also address the 1980 Marine Resources Law (Ley Sobre El Aprovechamiento de los Recursos Naturales del Mar) and the 1980 Declaring, Planning, and Developing Tourist Zones Law (Reglamento de la Ley Para La Declaratoria Planeamiento y Desarrollo de las Zonas de Turismo) to illustrate how the government legally conceived of the coastal wetlands. Later, aspects of these laws were contested and subsequently changed as civil society actors compelled the government to redefine how the coastal wetlands *ought* to be used, managed, or conserved. Due to the indeterminate, equivocal, and enigmatic character of law, contestation arose as social actors within civil society sought to redefine it so that it was locally more pertinent. Consequently, the conflicts surrounding access to and control over the mangrove wetlands expose law as intimately bound in the local politics of resource struggles.

The legal discourse in each law analysed revealed relevant state agencies' key characteristics and interests in relation to the mangrove wetlands of the GOF. Laws defined the problem and how the government planned to address it. Laws also conferred certain rights and responsibilities on government agencies to see that the law was followed. Subsequently, legal language was an important source of information to characterise the government's position and interests in relation to the mangrove wetlands. However, the government's actions were not necessarily aligned with the legal language. For example, in cases of corruption or where laws were overlooked for the benefit of more powerful actors, actions were sometimes taken that were contrary to the law. During an interview with Jorge Varela he stated that "It is difficult to obtain a concession; when the people have money, good lawyers, it is much easier, this is the easiest way, it's the easiest path that people seek" (Personal interview, 22 July, 2005)[26].

Although the law established legal procedures for the state, government agents did not necessarily comply with them. Varela implies that acquisition of concessions required taking steps that were not actually legally defined. Consequently, state actions pertaining to the issuance of concessions to the wetlands were sometimes contrary to the established *modus operandi* of the law. It was, therefore necessary to compare the government's actions in relation to the language in the law to determine if they were congruent, assisting with the elucidation of the government's true interests in relation to the coastal wetlands, and on whose behalf. The implication is that legal discourses and

---

[26] The original quotation was taken from transcript 31, a personal interview with Jorge Varela on 22 July 2005: "es muy difícil para obtener una concesión, cuando la gente tiene mucho dinero, buenos abogados, es mas fácil, ese es el camino mas fácil, es el camino mas fácil que se busca".

actions are inherently embedded in power relations. Finally, it is important to note that law is not inert. It is continually interpreted and reinterpreted within the context of negotiated order, as power is negotiated and contested through political or judicial processes.

When I analysed the laws, I focused on the government's legal conceptions related to the wetland resources, specifically how they *ought* to be allocated, managed, conserved, exploited, developed, and protected, and so on. Legal discourse is therefore the source of the normative claims of the state in relation to the wetlands. Normative claims were of particular interest because they state what *ought* to be done in relation to a physical space, creating certain images, social categories and distinctions between people, their environment and events that can often lead to conflicts between actors with varying interests in relation to the same physical space (Blomley, 1994: 12). In this regard, the meaning and significance of legal discourse had to be analysed in relation to a specific socio-cultural setting, physical space, or issue in order to elucidate its relevance.

The interpretation of legal discourses regarding the allocation of, access to, and control over the wetlands in the GOF constituted the basis of the government's power in relation to this space. In other words, legal language is viewed as a discourse of state power. Each law analysed provided government agencies with jurisdiction over the mangrove wetlands. In some cases, these jurisdictions overlap and became a source of conflict.

The capacity of an agency to accept or deny access to the physical space for any purpose demonstrated the state's power, the implications of which were both political and social in terms of the effects of power produced. Consequently, the state's choice to issue legal concessions for the promotion of aquaculture development constituted decisions that were political in both nature and effect (Blomley, 1994: 13). The contestation of and resistance to aquaculture development that emerged at the end of the period covered in Part II is used to illustrate their political nature and effects.

The rise of contestation against the Honduran government and the industry also illustrates that law cannot be viewed separately from material interests or social struggles over meaning in relation to the law (Blomley, 1994: 40-42). Laws pertaining directly or indirectly to the wetlands were understood varyingly at different social and spatial scales. Often, an actor's knowledge in relation to the law affected his or her interpretation of it and, therefore, the meaning which framed how s/he debated elements of a law associated with the coastal environment, revealing the politicised nature of physical space in legal discourse. Furthermore, I argue that legal interpretations and decisions have both material and discursive effects and that the apparatuses of the state perform a mediating role in both, revealing the state's role in the institutional mediation of discourse.

Although law is an instrument of state authority, it is also a source of power for the dispossessed when local social actors collectively organise to influence it. The actions of the government, private industry and IFIs in relation to the wetlands were directly related to the emergence of counter-discourses from within civil society in the mid 1980s and, subsequently, reactions against the state and the shrimp industry. In the latter part of this period, the Honduran environmental movement took a more active role to contest various impacts on the mangrove wetlands resulting from actions pursued by the state and the industry.

The ideas underlying contestation against the state and industry revealed the consolidation of environmentalism in Honduras and coincided with the strengthening of the environmental movement worldwide as social actors within civil society sought to influence outcomes related to neoliberal policies. One aim of environmentalists was to change laws and policies thought to affect local people and natural resources while providing suggestions to change policy that pertained to the governance and management of marine and coastal resources. Consequently, conceptualisations of law and space are historical and political processes negotiated within the public arena. In regard to the latter, legal representations were utilised strategically and contested through a variety of tactics, as is demonstrated throughout my thesis. Finally, the conflicts associated with the law were not divorced from the larger debates pertaining to environmental governance in the context of neoliberal reformation, of which the global aquaculture and environment debate is but a single facet. Overall, the analysis provided valuable insight into processes of both globalisation and resistance in the context of neoliberal reform in Honduras.

The 1959 Fisheries Law

There was no legal framework giving the state rights and responsibilities to pursue aquaculture development in the 1970s. The 1959 Fisheries Law was the only legislation regulating fisheries, and provided the SRN with the power and the authority to oversee any activities associated with the conservation, commercialisation, and industrialisation of the fisheries sector. The activities included everything from extraction, possession, conservation, and the general exploitation of the fisheries or elements affecting the fisheries. The SRN was thus one of the agencies that had the authority to grant the permission and concessions necessary for aquaculture development in the southern region of the country (1959 Fisheries Law, Chapter III, Concessions and Permissions, Article 11).

There were two articles, 52 and 70, of the Fisheries Law relevant to aquaculture conflicts. Article 52 prohibited the deforestation of mangroves or other trees that border the sea, the margins or mouth of the rivers: including areas such as canals, creeks, lagoons, inlets, coves, and any other location that served to shelter fish, areas where aquaculture development was planned to take place (1959 Fisheries Law, Chapter V, Prohibitions, Article 52).

Article 52 was particularly interesting when viewed as a state discourse on the protection of forest resources. Although the state was articulating its interest in protecting mangrove forests by including Article 52, it was undermined by Article 70, which provided an important exception to Article 52. Article 70 stated that there was an exception when permission had been conceded for the exploitation and industrialisation of vegetation, trees or plants, and therefore these normative claims took precedence over those in Article 52. In other words, economic concerns were given greater priority over environmental protection, which was consistent with the state's discourse and interest to promote economic development in the south.

However, Article 70 was vague because it did not specify whether the state was addressing direct exploitation and industrialisation of vegetation (such as logging or bio-prospecting), or indirect activities that might negatively affect trees and plants due to exploitation and industrialisation (such as the construction of ponds for aquaculture development) (1959 Fisheries Law, Chapter VIII, Penal Dispositions). Regardless, the political choices made by the SRN in relation to the 1959 Fisheries Law revealed that the legal discourse of Article 70 took precedence over that in Article 52 during this period. Permitting concessions to the industry aligned the government and industry's interests and later led to future aquaculture expansion.

The legal authority in the 1972 Forestry Law and the 1974 Honduran Corporation for Forest Development Law overlapped with the authority provided in the 1959 Fisheries Law. The legal language in both illustrated the state's interests in economic development and environmental conservation as it did in the 1959 Fisheries Law. How these laws were interpreted and enacted was also aligned with the state's interest to promote economic development and provided the justification for intervention in the south. The six actions taken by the state in pursuit of aquaculture development substantiate this claim in Chapter 6. The laws are addressed chronologically and reveal the overlapping legal jurisdictions between different agencies responsible for performing various roles at the local level in the GOF. Overlapping jurisdictions became a key source of inter-agency conflict and conflict between agencies and social actors in the region later in the period, revealing the complexity of micro-politics in a conflict scenario.

The 1972 Forestry Law

In 1972, the SRN's authority over coastal areas was expanded when it was given primary responsibility for the new Forestry Law. The law directed SRN to establish and guide the appropriate conservation, restoration, and propagation of forest resources including their rational use, industrialisation and commercialisation (Forestry Law, Chapter I, Article 1, *La Gaceta*, 4 March, 1972: 7). Mangrove forests were included under the purview of this newly created law, so SRN had authority over both fisheries and forests, including mangroves.

The agency was delegated the responsibility of defining, classifying, declaring, and planning the administration of zones and forested areas (Chapter III, Forestry Law), including the creation of Protected Forest Zones (Article 11a). These zones were defined as public and private areas of significant importance for the conservation of the country, its waters and soils, and were to be managed for limited use in accord with arranged forestry plans formulated or approved by the State Forest Administration (AFE) (Forestry Law, *La Gaceta*, 4 March, 1972).

The mangroves of the GOF in the Departments of Choluteca and Valle were recognised as Protected Forest Zones both in the Forestry Law (Chapter XIII, Article 138 (1a)) and by Legislative Decree (No. 117, 17 May, 1961), (Chapter XIII, Article 138 (1a), Forestry Law, *La Gaceta*, 4 March, 1972: 18). Chapter VII of the law outlined the powers of the State Forest Administration in relation to the protection of these areas. Chapter VIII specifically pertained to the conservation of soils, waters, and the protection of fluvial margins. In this section, Article 64 prohibited the cutting, damaging, burning, or destruction of the trees, shrubs, or in general the forest or woods within 250m of any source of water and in a band of 150m, or any course of permanent water, lagoon or lake, inside the drainage area of the current. Any human activities that might negatively affect these forested areas required that the State Forest Administration (Administración Forestal del Estado, AFE) also approve concessions (Forestry Law, Chapter VII, *La Gaceta*, 4 March, 1972: 10). The power of this agency to issue concessions was provided two years later when the Honduran Congress passed the 1974 Honduran Corporation for Forest Development Law. Later, civil society, the state and private actors violated nearly every aspect of these laws in relation to the mangrove wetlands.

The 1974 Honduran Corporation for Forest Development Law

In 1974, the Honduran Congress passed another law that created the Honduran Corporation for Forest Development (Corporación Hondureña de Desarrollo Forestal COHDEFOR) as a part of AFE (Decree-Law Number 103, 10 January, 1974). The law created new powers for the SRN, giving COHDEFOR the authority to issue concessions and approve any activities associated with designated forested areas, including activities that might affect forested areas such as aquaculture development (Article 14).

COHDEFOR was designed as a semi-autonomous institution with both judicial and patrimonial characteristics (Chapter I, Article I). The new institution's objective was to optimise the use of forest resources in the country's interest; to secure their protection, improvement, and conservation, and to generate funds for the financing of state programs with the ultimate interest or goal of accelerating the social and economic development processes of the nation (COHDEFOR, Chapter I, Articles 1 and 2, *La Gaceta*, 10 January, 1974: 1).

The new agency was responsible for forest policy, and was directed to facilitate programmes and projects aligned with the National Development Plan, which outlined the state's strategy for economic growth and development (Article 3). To accomplish the stated goals, the law gave COHDEFOR the authority to approve, constitute and operate forestry-related commercial or industrial businesses and linked its programmes and activities to the National Corporation for Investment and the National Bank for Development (Banco Nacional de Fomento, BNF, Chapter II, Article 8(f)). It was also responsible for granting the finances and guarantees to businesses for the extraction of forest products (Article 8(g) *La Gaceta*, 10 January, 1974: 2).

Although conservation and economic development were both included in the legal language, the issuance of concessions revealed that economic development took priority. This was consistent with a statement made by Joaquin Aguero, senior advisor to the Minister of Environment in 2004, who was involved in the creation of COHDEFOR in the 1970s. During a personal interview he stated that, although COHDEFOR included conservation in its language, creating protected areas and wildlife refuges was not a priority, but only permitted by the newly-created law. The agency, from his perspective, was more oriented towards economic development, mostly focused on extractive activities, such as logging for state profit or other activities that would benefit the military dictatorship during the period.[27]

These newly-created laws led to overlapping legal jurisdictions within the SRN. The jurisdiction of SRN's General Directorate for Renewable Natural Resources (Dirección General de Recursos Naturales Renovables RENARE), Department of Hunting and Fishing (Departmento de Caza y Pesca, DCP), the agency responsible for the 1959 Fisheries Law, overlapped with that of AFE-COHDEFOR. The conflicts surrounding access to and control over the wetlands were partially rooted here. The interests of these agencies within the SRN clashed as each attempted to exert their authority in different manners. The power of AFE-COHDEFOR to issue concessions in relation to the wetlands did not require it to take into account potential impacts on fisheries related resources. Legally, COHDEFOR should have sought to comply with the 1959 Fisheries Law. The 1959 Fisheries Law, however, did require RENARE to take these impacts into account. The issue of overlapping legal jurisdictions in relation to the wetlands was further exacerbated with the passage of the 1980 Marine Resources Law and the creation of the Ministry of Tourism, which provided new legal jurisdictions over the coastal wetlands.

The 1980 Marine Resources Law

In 1980, the National Congress enacted the Exploitation of the Natural Resources of the Sea Law (Ley Sobre El Aprovechamiento de los Recursos Naturales del Mar). Although it was partially in preparation for the British relinquishment of the Bay Islands in the Caribbean, which had been a part of British Honduras until 1980, it had implications for the marine and coastal environment of the GOF. This law was one of the first that clearly established the state's sovereign right to exploit, explore, conserve and administer living and non-living marine and coastal resources within the state's exclusive economic

---

[27] Personal interview with Joaquin Aguero at the White Water to Blue Water Conference in Miami, Florida, March 2004.

zone (EEZ) (Article 1(a)). Article 1(c) extended the state's authority, jurisdiction and control of regulations, authorisations, and activities related to marine science research. Article 1(ch)[28] provided the state with the authority to preserve the marine environment and to prevent, reduce and control contamination from any source. Article 3 gave the state the right to determine what was permissible in relation to the capture of living marine resources (*La Gaceta* Number 23, 127, 13 June, 1980 Decree Number 921). This article was important because early aquaculture development relied on the capture of wild shrimp postlarvae to stock the ponds and fell within the jurisdiction of this law.

The law clearly articulated that the priority was to optimise the use of marine resources with priority given to satisfying the nutritional needs of Hondurans and the National Congress thus wrote a food-security discourse into the law. The state was also given the authority to provide concessions, licenses, and permission related to fishing in Honduran territorial waters. Agencies that already had this authority, such as the Departmento de Caza y Pesca (DCP) [Department of Hunting and Fishing] and COHDEFOR, were not directly referenced, although RENARE/DCP became the lead authority in relation to the law. Article 4(b&c) allowed the state the authority to determine which species could be caught, and made provisions for regulating the fishing season and the zones in which fishing occurred. This article was axial as it provided the state with further authority over the artisan fishing community's activities and their access to marine and coastal resources (*La Gaceta*, Numero 23, 127, 13 June, 1980)[29]. In the same year, another law created the Secretary of Culture and Tourism and was passed, further confounding the issues in relation to jurisdiction over the wetlands.

The 1980 Law Declaring, Planning and Developing Tourist Zones

In 1980, the post of Secretary of Culture and Tourism (Secretaría de Cultura y Turismo SECTUR) was created through the Law for Declaring, Planning, and Developing Tourist Zones (Ley para la Declaratoria, Planeamiento y Desarrollo de las Zonas de Turismo, 22 July, 1980). SECTUR's authority fell under the Secretariat of State, Office of Culture and Tourism. In Article 1, the law established the southern region of the country as Tourist Zone 6 and it was defined as the total area of the GOF bordering the Nicaragua and El

---

[28] In Spanish ch is the fourth letter of the alphabet and therefore was used by the Honduran government to denote this section of the law.

[29] The law was the predecessor to the establishment of DIGEPESCA in 1991. The passage of this law was precipitated by the approval of Agreement No. 14 written on 31 July 1979 in London, England to adhere to the Convention on the Dumping of Waste into the Sea which had been established in London, England on 13 November 1972.

Salvador, and covered the Departments of Choluteca and Valle, including the islands of the GOF.

In Article 4, SECTUR was given authority to approve requests to construct hotels, spas, recreational centers, sports grounds, and any other tourist destinations lying within the tourist zones, including the coastal zones in southern Honduras. Article 5 further extended its power to authorise any activities that required deforestation and excavation in the tourist zones declared; and any industrial, commercial, mining, fishing or forestry related activities established in the zones. Article 6 extended the tourist zone 2km into the sea from the highest tide, placing any of the aforementioned activities in this area of the coastal zone under the authority of this agency. SECTUR thus became responsible for the issuance of concessions in the coastal zone surrounding the GOF. Its authority coincided with the previous power given to the DCP and AFE-COHDEFOR to issue concessions to the wetlands, once again illustrating the overlapping legal jurisdictions between state actors.

The power of one agency relative to another is critical to understanding conflicts of interest between agencies responsible for conservation of forest resources and the issuing of concessions for aquaculture development. Contradictions in legal language led to impediments in acquiring concessions needed to pursue various development activities in the coastal zone, especially in relation to aquaculture. Firstly, in some cases, those wishing to obtain concessions were not familiar enough with the laws to know how to acquire the needed concessions or that they had to be approved by multiple agencies. In these cases, aquaculture development commenced without the necessary concessions and was more closely associated with small-scale producers who converted their lands for aquaculture production. In other words, those less knowledgeable of the law were disadvantaged in terms of their capacity or agency to comply, whereas those with the knowledge or connections to obtain concessions were in a more powerful and advantageous position, as indicated by Varela.

Secondly, the power conferred on various agencies to issue concessions placed government bureaucrats in competing positions. The issuance of concessions resulted in bureaucrats being able to make decisions that favoured those with whom they had relations, as a means to demonstrate their power. Often the Honduran business elite, military officials, and other high-level government officials were those able to obtain the concessions. Most of the land suitable for aquaculture development was conceded early on, leading to land speculation in the coastal region.

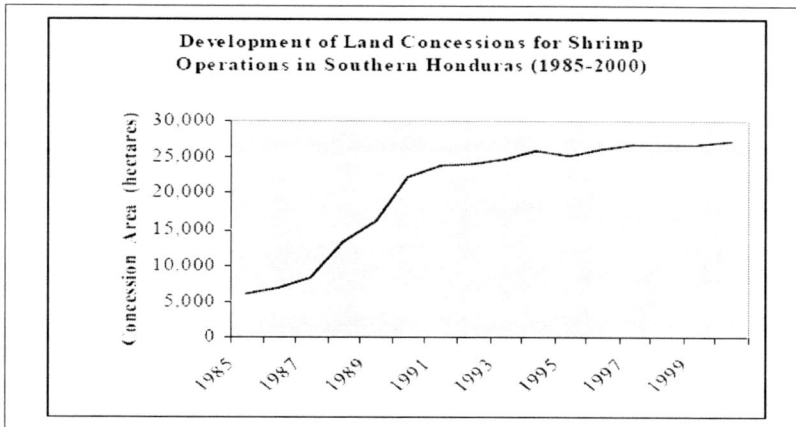

FIGURE 5.1: DEVELOPMENT OF LAND CONCESSIONS FOR SHRIMP OPERATIONS IN SOUTHERN HONDURAS (1985–2000)
Source: Stanley and Alduvin. 2002 (Taken from ppt presentation)

A number of those acquiring concessions were speculating on the future value of the land in areas that could potentially be developed for aquaculture, but they did not necessarily intend to pursue aquaculture.

Finally, overlapping legal jurisdictions and the lack of clarity associated with acquiring concessions generated confusion within the public realm and contributed towards the conditions that led to conflicts between different social actors and the state. Those who did not have access to the political authorities that issued concessions were marginalised. Ultimately, these factors associated with overlapping legal jurisdictions and the application of the law contrary to the *modus operandi* became important to the conflicts around access to and control over the wetlands as I demonstrate in Chapters 7, 8, and 9. Consequently, by 1988 these actions led some local fishermen to organise to contest the emerging industry due to the threats to their livelihoods. In turn, contestation led to further changes in the legal framework by 1990. The industry's impact on the mangrove forests was a concern but not the initial focus of the legal changes.

By 1990, the government announced efforts to revise the 1959 Fisheries Law to consolidate the powers of a number of agencies, ultimately leading to the creation of a new agency and the elimination of older ones like RENARE. The initial focus on developing the fisheries sector further led to a series of consultations with various groups to develop the legislation. The result was the creation of the General Directorate of Fisheries and Aquaculture (DIGEPESCA) (*La Gaceta*, No. 26,493, 18 July 1991 Decree Number 74-91) that coincided with neoliberal reform efforts to consolidate government agencies and eliminate those that were no longer needed under the implementation of the SAP for Honduras (See Chapter 4). Article 1 of the new law placed DIGEPESCA under the authority of the SRN.

The agency was given the authority to oversee the use and protection of continental and maritime fisheries resources, as well as overseeing all functions related to aquaculture, research and fisheries policy that promoted economic development. The law also transferred the functions of the Forestry Department of RENARE to COHDEFOR, eliminating RENARE. AFE-COHDEFOR became the agency with lead authority for the conservation of mangrove forests.

Summary

Between 1973 and 1988, new government-supported institutions emerged for the purpose of promoting economic development and environmental conservation. Although the laws relating to the marine and coastal resources of southern Honduras included language directed towards conservation, economic development took precedence. This outcome was consistent with state discourse and newly initiated neoliberal reform processes during the period, further demonstrated by the commencement of large-scale private industrial shrimp aquaculture in 1984.

Each of the laws addressed established the legal basis of the state's power to intervene in the region to promote aquaculture. I argue that the interpretation of the legal discourses on forestry, fisheries, and marine resources changed at various points throughout the socio-historical trajectory of the conflict. Different presidential administrations and the National Congress attempted to respond to demands articulated from within civil society as neoliberal interpretations and their effects were continually contested. The state's discursive strategies as well as those of various social actors changed as access to and control over the mangrove wetlands was negotiated and contested within the public arena. The demands that pertained to the use, management and conservation of the wetlands began to be articulated more frequently in the early 1980s as aquaculture expanded. In Chapter 6 these demands are presented chronologically in order to show how the debates evolved.

Chapter 6

Actions of the State, USAID, and International Financial Institutions for Aquaculture Development

Introduction

In this chapter I address the actions taken by the state in conjunction with private investors, USAID, and IFIs for the promotion of aquaculture development. I will demonstrate that these actors pursued six actions (see chapter 4), viewed as techniques of power, in relation to aquaculture development that were aligned with the discourses of economic development and food security. The state's and international actors' actions to promote the development of aquaculture led to the formation of new social and economic networks (e.g. the creation of fishermen's cooperatives) within the traditional fisheries sector, sometimes leading to competing interests and conflict. Competition was likely to occur between the aquaculture industry and factions of the local fishing communities, which was where the main opposition to future industry expansion emerged.

I argue that the emergence of the conflict was not initially associated with mangrove degradation but was directly associated with the changing dynamics of the fisheries sector as aquaculture development influenced the socio-economic conditions and labour market in southern Honduras. Here, the actions that led to conflict are revealed. For example, government-sponsored training programmes for aquaculture development benefited some factions of the artisan fishing community as they were taught to catch the wild postlarvae and then sell them to the industry to stock the ponds. Some of these groups of fishermen were legally organised into cooperatives and federally recognised, making them eligible to receive aid and funds from international institutions. These actions represented the deployment of state power and the reordering of social relations in the south in pursuit of export-led growth. Others in the wider artisan fishing community were excluded from these efforts and continued to fish for other species or for adult shrimp in the open sea. As industry expansion ensued, the latter group perceived that the capture of postlarvae was affecting the population of the fisheries as a whole due to extensive by-catch. They also became concerned that pond construction was reducing access to their traditional fishing grounds in the seasonal lagoons that formed on the salt flats. Each of these points became central issues in the conflict and is discussed throughout. The analysis reveals that although neoliberalism is linked with anti-state rhetoric, the actions of IFIs and the Honduran government illustrate how neoliberal policies actually "mobilized and created an activist role for the state in promoting the privatization of goods and services and in opening up 'market opportunities'" (Perreault and Martin, 2005: 193). In sum, the point is to demonstrate that the root of the conflict initially had less to do with mangrove degradation and more to do with changes in the fisheries labour market.

The actions of the state and IFIs were later contested by the emerging Honduran environmental movement. Although the Honduran Ecological Association (AHE) played the most active role in regard to issues surrounding the degradation of the wetlands in the early-1980s, it was initially focused on threats to the natural system from local use of mangrove wood for fuel, construction, and salt production and the associated impacts. The AHE did not focus on the impacts of aquaculture development until 1985 when Jorge Varela, the future leader of CODDEFFAGOLF, began working for the organisation. In 1988, he established CODDEFFAGOLF in conjunction with twelve fishermen who were opposed to the industry. The links between aquaculture expansion and mangrove degradation were eventually forged and used successfully to further politicise this group's interests in relation to the fisheries sector and the wetlands from 1988.

Knowledge Acquisition for Research and Development, Training, and Technical Assistance

Most early training and technical assistance directed towards aquaculture development focused on local fishing communities and was a part of the state's strategy to further exploit fisheries resources as legally mandated in the 1959 Fisheries Law and further codified in the 1980 Marine Resources Law. These efforts coincided with regional efforts under the direction of the Latin American Organisation for Fisheries Development (OLDEPESCA) to develop the export market associated with fisheries resources throughout the entire Latin American region. In its efforts to further exploit fisheries resources, the Honduran government began providing technical assistance and training to local fishermen with two goals in mind.

The first goal was to create more localised social and economic networks to increase capacity in the traditional fisheries sector and expand the market. The second was to create opportunities for local fishermen in the emerging aquaculture sector. The state began to work towards these goals through the formation of cooperatives and the construction of hatcheries, and by providing training and technical assistance so that fishermen would have the necessary skills to participate in the emerging aquaculture market, and expand the scope of their activities in the traditional fisheries market (*El Cronista*, 9 June 1975).

Both the hatchery and the provision of technical assistance were congruent with the government's intentions to develop the infrastructure for aquaculture production and acquire the knowledge and skills to make it successful. Publicly, the government reiterated their goal to increase employment opportunities for the people of the south, particularly in Amapala (a traditional fishing village), whilst ensuring their access to a stable food source in a region prone to malnutrition and food insecurity (*El Cronista*, 9 June 1975).

The project was coordinated in conjunction with the National Port Council (Empresa Nacional Portuario, ENP), which was legally mandated to promote economic development. The ENP's goal was to support the cooperative's expansion into the markets in Choluteca and Tegucigalpa and create employment in the region. Supporting the ENP was another government institution, the National Supplier of Basic Products (Suplidora Nacional de Productos Basicos BANASUPRO), which was responsible for facilitating the opening of agricultural markets and identifying ways to increase production in various sectors (*El Cronista*, 9 June, 1975). BANASUPRO was eventually eliminated in alignment with neoliberal reform becoming BANADESA, which was privatised with the passage of the 1992 Agricultural Modernisation Law (LMDSA). Until 1988, SRN, DCP, ENP, and BANASUPRO all worked to achieve the state's goals in relation to economic development and addressing food security issues in the south as revealed in newspapers throughout the 1970s (*El Cronista*, 13 July 1979). The government was particularly interested in fisheries resources due to the high levels of protein available in aquatic animals, and stated that supporting aquaculture could lead to heightened profits.

The government subsequently announced that it would construct 36 tanks/laboratories by the end of 1979 in order to develop further the necessary infrastructure for the aquaculture industry (*El Cronista*, 13 July 1979). Development also required investment in the formation of fishermen's cooperatives since skilled labour was critical to future industry expansion. When commercial shrimp farming was introduced in 1983 and 1984, new private labs were constructed in Cedeño and elsewhere. Consequently, the emergence of government-sponsored private hatcheries created tensions as the artisan fishermen running the cooperative hatcheries were relied upon less and less as the private commercial industry expanded and used its own laboratory-produced larvae. Furthermore, reliance on government hatcheries was reduced when they discovered that wild postlarvae could be harvested from the sea to stock the ponds. As a result, artisan fishermen in Amapala later joined forces with CODDEFFAGOLF to contest the industry and establish a base for actions against the industry from Tiger Island (Isla del Tigre). Future dependence on privatised hatcheries was in line with neoliberal policy, illustrating the effects of increased competition linked with privatisation efforts and capitalist transformations in the region.

Formation of Fishermen's Cooperatives

State intervention in the local fisheries sector was the direct cause of divisions within the labour market and created the conditions for conflict. Examples are used to illustrate that some cooperatives were formed as a result of direct government support, whereas others were independently organised to represent artisan fishermen, and later, small-scale shrimp producers in various communities.

Cooperatives were designed to represent the collective interests of different factions of the artisan fishing community through establishing set prices, purchase of captured fish, and, subsequently, its sale to markets in the region. This resulted in the formation of new social and economic networks in the emerging aquaculture and traditional fisheries sectors. Most of the cooperatives formed in the south were naturally located in communities on or near the coast. Cooperatives sprung up in Amapala, La Brea, Playa Grande, Coyolito, Punta Raton, Pueblo Nuevo, Cedeño, Guapinol, and others (refer to Table 2.1 Coastal communities selected for the socio-economic analysis). Some of the cooperatives established during this period later became the main suppliers of shrimp postlarvae to the industry.

The 1959 Fisheries Law provided the authority for fisheries cooperatives to be legally organised (Chapter III, Article 15) as long as their formation complied with the 1954

Association of Cooperatives Law.[30] As a result, the government took two important steps to support their organisation. First, the government financed the ENP to create the first Regional Fishermen's Cooperative of the Gulf of Fonseca (Cooperativa Regional de Pescadores del Golfo de Fonseca DIFOCOOP). Second, it began planning a training course for fishermen associated with this cooperative based in Amapala which was to be held on 5 January 1980. The purpose of the course was to provide training and technical assistance in a number of areas relevant to both the aquaculture and fisheries sectors (*El Cronista*, 11 January 1980).

DIFOCOOP was designed to represent a number of smaller cooperatives that had formed throughout the 1970s. Technical specialists associated with the newly-founded DIFOCOOP worked in conjunction with employees from the National Institute for Professional Formation (Instituto Nacional de Formacion Profesional, INFOP)[31] in order to address issues in relation to both sectors (*El Cronista*, 11 January 1980). Early investors supported these efforts through the establishment of aquaculture extension programmes which were designed to train individuals to work with the hatcheries to ensure seed stock for the industry's ponds and later to train them in larvae gathering.

Another fishermen's cooperative was founded independently by 16 fishermen in San Lorenzo in August 1981. The Association of Shellfishermen of the South (ASOMAR) was created in order to promote the well-being of its members and ensure their access to existing and emerging markets. They also wanted to take advantage of financial opportunities, technical assistance, and training being provided by national and international organisations supporting the fisheries and aquaculture sectors (*La Tribuna*, 6 August 1981). The cooperative was federally recognised with the passing of Resolution No. 122 on 28 September, 1983.

In sum, financial and technical support for the development of fishermen's cooperatives as well as legal recognition was how the state exerted its power to achieve its objectives to promote aquaculture. Training and technical assistance were used as techniques of power, strategies of the state to ascribe roles within the aquaculture commodity chain, whether the cooperative was formed by the government or, independently. In turn, cooperatives' acceptance of financial incentives, technical assistance and training

---

[30] 1954 Ley de Associaciones de Cooperativas No. 158-1954

[31] INFOP was created through the Ley de Instituto Nacional de Formacion Profesiona Decree Number 10 Passed on 28 December 1972 Printed in La Gaceta No. 20873 on 6 January 1973. INFOP was created to promote national productivity, economic and social development through the formation of professional organisations (article 2).

aligned them with the interests and actions of the state as it pursued economic transformation in the south and correlated with neoliberal reform. Consequently, cooperative formation in southern Honduras must be viewed as a technique of power leading to the political construction of new economic networks and social identities associated with the emerging aquaculture sector.

USAID and other IFIs played an important role in the development of these new networks through an institution-building process. In particular, Economic Support Funds (ESF) provided by the US was partially used by USAID to finance the establishment of private-sector associations and elite think-tanks in order to accomplish five objectives:

1. cultivate new business leaders;
2. reorganise the private sector;
3. spread US influence within the state and civil society;
4. diversify economic activities by supporting NTAX; and
5. work with the government to develop new neoliberal social and economic policies (Robinson, 2003: 125).

The next section focuses on how the early industry learned new methods for capturing wild shrimp postlarvae, which changed the course of the development of the industry and created new employment opportunities for artisan fishermen associated with various cooperatives, but also led to conflict.

Gathering of Wild-Caught Shrimp Postlarvae

The conflict that arose was initially associated with the capture of wild postlarvae. Once the practice of capturing the wild postlarvae was established, it created competition within the artisan fisheries market and allowed the industry to expand. Consequently, the industry's development did not take off until it discovered that stocking, feeding, and pumping the brackish water from the estuaries into the ponds created higher profit, encouraging early investors to work more closely with artisan fishing communities to acquire the postlarvae to stock them (Heerin, 2004: 19). After 1981, more hatcheries were constructed and fishermen were trained to supply the necessary shrimp postlarvae to stock newly-built ponds. There was a noticeable shift from state-run to private-sector hatcheries. State-funded cooperatives that did not have access to the private sector postlarvae market were often cut out of the commodity chain. In some cases, they were enrolled in the commodity chain as suppliers, while the industry constructed new laboratories and hatcheries in the region.

Sea Farms of Honduras, which was producing shrimp in Punta Raton, relied on the postlarvae from its hatchery in Key West, Florida, and its newly-constructed hatchery in Cedeño in the Department of Choluteca. The Cedeño facility was constructed to reduce the transaction costs associated with importing shrimp from Crystal River and Key West as the industry sought to be more profitable and efficient (Romero, 2004). It also allowed the emerging commercial industry to be less dependent on other cooperatives for postlarvae produced in the hatcheries initially funded and established by the government. It is worthwhile to point out that a number of powerful government officials had invested in the private industry. As it sought to integrate vertically, it was less important for them to rely on cooperative-run hatcheries. Because they had invested in the hatcheries and labs it was beneficial to promote private capital gain. However, the private industry did not become profitable until Ralph Parkman, one of the investors associated with Sea Farms of Honduras, took a trip to Guayaquil, Ecuador to learn from their successes with aquaculture development.

Their objective was to assess the reasons for the industry's successes. Previously, Parkman and others thought that the entire seed stock for the ponds had to be controlled, and consequently the early industry in Honduras had relied on hatcheries. During the trip to Ecuador, they learned that wild-caught postlarvae were feasible to use in the ponds (Meyer, 2005). They also learned how postlarvae collectors in Ecuador were trained to use dip nets from canoes in the local estuaries rather than push nets out in the open ocean (Heerin, 2004: 22). Subsequently, the early-Honduran commercial industry learned that the postlarvae could be collected in the rivers and estuaries to stock the ponds, decreasing mortality rates and making large-scale production more feasible.

Technical experts associated with the Ecuadorian industry were brought to Honduras to train local fishermen in their methods, which gave Sea Farms of Honduras an early advantage. Meanwhile, other shrimp farmers still relied either on the hatcheries, pumping in water from the estuaries and using postlarvae naturally present in the ponds, or using push nets out in the open ocean to catch wild postlarvae. Local artisan fishermen were then trained as larvae gatherers. They were taught where to look for the postlarvae, how to judge quality, identify, count, acclimatise, and stock them (Heerin,

2004: 22). This knowledge changed the socio-historical trajectory of aquaculture development in southern Honduras, leading to the first significant industry profits in less than a year, and made it feasible to expand operations.

More importantly, stocking the ponds with wild-caught postlarvae significantly changed the labour market dynamics in relation to the emerging aquaculture industry. It created competition between different groups in the artisan fishing community interested in supplying the industry with seed stock. The artisan fishermen trained to gather the postlarvae had an advantage in the emerging labour market. Proximity to the sites where production was taking place made it more likely that artisan fishermen in these areas would be those trained. Other factions of the artisan fishing community continued to fish in the estuaries, seasonal lagoons, or out in the open sea. As expansion ensued later in the period, the issues of by-catch and enclosure of the seasonal lagoons became more prominent and, eventually, politicised by twelve fishermen that formed CODDEFFAGOLF in conjunction with Jorge Varela in 1988.

Stanley identified four options that became available for the industry to gather the necessary postlarvae for shrimp farm expansion in her research, revealing the divisions within the artisan fishing community. First, the industry could hire full-time employees to undertake seed gathering. Second, the industry could buy "ready-made" seed-spawned in hatcheries. Third, contractors could be retained to provide postlarvae on-call. Last, businesses could buy postlarvae on the spot-market from contractors. She found that physical geography affected the choice of technique and equipment utilised by the larvae gatherers once again demonstrating the 'natural agency' of the wetland ecosystem (Stanley, 1996: 95).[32] The third and fourth options created were less stable than the first two, creating uncertainty and increased competition among groups of larvae gatherers vying for access to the market.

The relationship between the early industry's development and the artisan fishing community is particularly important given that CODDEFFAGOLF was founded by fishermen who contested industry expansion from 1988 and thus revealed divisions

---

[32] Stanley found that "geography affects the choice of technique and equipment in gathering." She found that "gatherers face a very unstable natural environment causing variation in catch productivity and incomes. Ecological differences across southern Honduras contribute to different types of equipment being used for natural larvae gathering. Many contractors in the muddy San Bernardo zone, who have acquired estuary access rights, use boats docked on company property to travel down estuaries to the gathering sites. The Laure gatherers (at a drier site with larvae swimming up to the source of the estuaries can walk to the sites taking only their buckets and nets" (1996: 96). All are examples of the 'natural agency' of the environment.

within the artisan fishing community. The discussion now turns to industry expansion in the San Bernardo region, which occurred after the acquisition of new techniques to acquire postlarvae.

Aquaculture Expansion into the San Bernardo Region, Choluteca

The United Nations Food and Agricultural Organisation (FAO) completed a study around 1965 to assess which areas in the south were feasible for aquaculture development. It was agreed that the region of San Bernardo in the Department of Choluteca was the best site for future expansion due to its extensive salt flats.[33]

Leonel Guillen, the Director of AFE/COHDEFOR/PROMANGLE in 2003, stated that the 1965 FAO study declared 20,000 ha of salt flats, out of 100,000 ha of the wetlands, were suitable for aquaculture development around the entire GOF (Nicaragua, Honduras and El Salvador). He also stated that no one knew exactly how many hectares of mangrove there were in relation to the number of hectares declared as wetlands during this period, but that these initial estimates provided the baseline information for the early industry's development.[34]

The FAO supported USAID efforts in relation to food security and played a technical role in evaluating options for various productive activities such as aquaculture. The importance of this study is three-fold. First, it was one of the initial assessments on the total area of mangrove wetlands in southern Honduras and was used by a number of social actors making claims in regard to mangrove degradation in the future. Second, political and economic decisions were based on the information in the report and led to the state's issuance of concessions for aquaculture development on these lands. Third, the study demonstrated the role that international institutions perform in processes of economic transformation; they influence the choices as to how a particular physical space *ought* to be exploited, conserved, or managed.

The FAO study, coupled with the knowledge acquired in Ecuador led Sea Farms of Honduras to look towards expansion on the salt flats of San Bernardo. By 1984 or 1985, the industry had acquired around $4 million in initial investment, concessions, and permits to commence operations on a larger scale and expand into the San Bernardo

---

[33] Silviagro, 1996 cited the 1965 FAO Study. His study was titled: "Análisis del Sub-Sector Forestal de Honduras". His work was on behalf of Cooperación Hondureña Alemana, Program Social Forestal. AFE-COHDEFOR. Honduras. The numbers cited by Silviagro were also stated by Leonel Guillen, Director or AFE-COHDEFOR in 2003, during a personal interview in southern Honduras.
[34] Personal interview with Leonel Guillen, San Lorenzo, Honduras, January 2003.

region (Meyer, 2005). Honduran investors and the World Bank (one of the largest traditional lenders to governments) were invited to join these efforts. The World Bank's private-sector lending arm, IFC, loaned to and invested directly in Sea Farms of Honduras to expand into the San Bernardo region. Eventually, Sea Farms of Honduras became known as Grupo Granjas Marinas San Bernardo (GGMSB) in 1984 as the expansion of the commercial industry led to further consolidation. The goal was to increase the market power of the investors (Heerin, 2004).

The initial development and expansion of the industry coincided with efforts of USAID and the Honduran government to promote NTAX. Collectively, they created the necessary institutions to open markets in line with neoliberal reform and influence the conditions that would permit the shrimp industry in Honduras to become competitive and successful. The satellite image below depicts shrimp pond construction in the San Bernardo Region in 1985, primarily GGMSB.

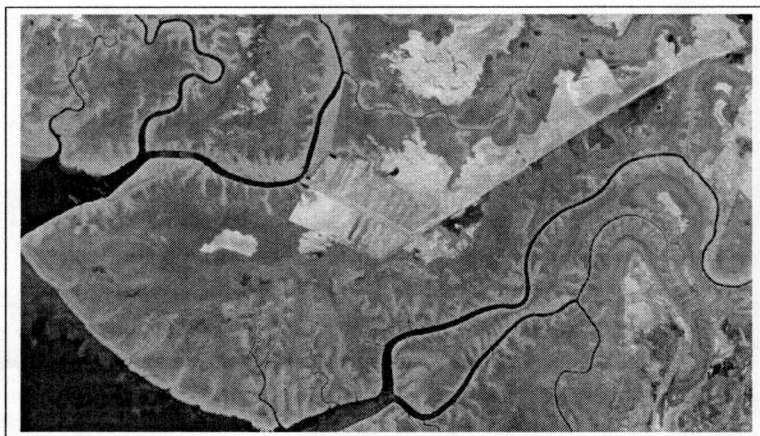

FIGURE 6.1: SATELLITE IMAGE OF GGMSB's OPERATIONS IN 1985
Source: NOAA, 1985. First Phase of Construction of Granjas Marinas San Bernardo on the salt flats in the Department of Choluteca (grey area) of Punta Guatales (southern part of the Gulf of Fonseca), behind the mangrove forest (green areas along the estuaries).

International Engagement and the Emergence of New Institutions: US Involvement in the Early Development of the Industry

International actors such as USAID, World Bank, and the IDB supported the state's position in relation to aquaculture development on the wetlands. Crucial to the promotion of NTAX were the provision of technical assistance and the establishment of new institutions and the transformation of older institutions. By 1983, it became clear that the expansion of shrimp farming was feasible, and a number of international actors were ready to commit the resources and expertise to assist industry development. USAID played the most important role.

US support increased in 1983 when USAID decided to expand further its involvement in aquaculture through its Collaborative Research Support Program (CRSP). CRSP's mission was to link the US land grant universities' capacity to the international food and agricultural research mandate of the US government[35], which was aligned with neoliberal agrarian reform strategies. The CRSP in Honduras funded technical specialists, such as Dr Claude Boyd and Dr Bart Green (University of Georgia), to travel to the country to provide expertise on pond dynamics in order to assist the industry's development as the government lacked the capacity to transfer the necessary knowledge and skills (Green, 2004). The technical assistance was directed to Sea Farms of Honduras, financed by US investors, which later became GGMSB/SFI (Boyd, 2004). USAID thus supported shrimp farming's development by supporting American technical experts in order to promote aquaculture in southern Honduras. Most of the actions were directed towards supporting US foreign investments in aquaculture, such as those of Sea Farms of Honduras. By 1990, USAID had loaned over $11 million for shrimp pond construction in southern Honduras (Naylor, 1992). USAID also began investing large sums to establish new institutions to support, not only aquaculture, but the promotion of NTAX in general as a part of its neoliberal reform strategy.

---

[35] (http://crsps.org/)

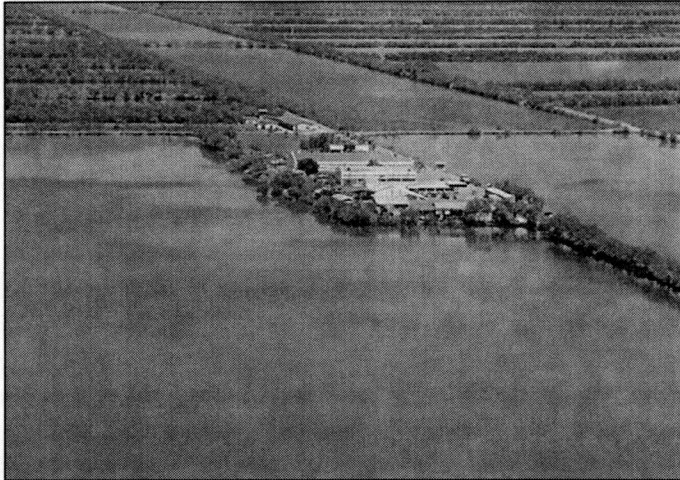

FIGURE 6.2: ***GGMSB's MAIN OPERATIONS IN SAN BERNARDO***
Source: GGMSB, SFI 2005

Institution Building for Aquaculture Expansion

Institution building processes initiated by USAID were directly related to promoting the neoliberal strategy and the focus on export-led growth a central element. Economic development and food security remained the rationale and the various actions taken by USAID and IFIs illustrate how aid was directed towards linking fishermen's livelihoods to the emerging aquaculture industry. The actions benefited some over others. The artisan fishermen who received benefits had a more advantageous position in the aquaculture market due to support from external actors, whereas others were often inadvertently marginalised. Consequently, the divisions created as a result of the introduction of this new economic technology led to conflicts amongst a number of social actors within the fisheries sector.

Over two dozen new institutions were established in the 1980s, a number of which focused exclusively on development of the agro-export sector and the creation of new social and economic networks associated with NTAX. In less than five years, USAID had successfully worked with the Honduran government to create the:

• Fundación Hondureña de Investigación Agrícola (Honduran Foundation for Agricultural Research, FHIA);

- Honduran Federation of Agricultural and Agro-Industrial Producers and Exporters (FEPROEXAAH/FPX);
- Foundation for Investment and Export Development (FIDE);
- National Council to Promote Exports and Investment (CONAFEXI);
- National Association of Honduran Exporters (ANEXHON);
- National Association of Industrialists (ANDI);
- Associated Managers and Entrepreneurs of Honduras (GEMAH);
- National Development Foundation of Honduras (FUNDAHEH); and
- Honduran American Chamber of Commerce (HAMCHAM).

Each of these institutions had an effect on the Honduran state and society and was a part of a larger nation-building exercise that sought to shift political and economic influence towards groups supportive of the USAID strategy (Robinson, 2003: 125). Each was important to industry development but the first three played the most prominent role.

By 1984, the first large shrimp ponds began to be constructed in San Bernardo (Heerin, 2004) and in that year USAID began working with the Honduran government to establish several institutions for the development and expansion of the industry. The first institution, FHIA, was founded on 15 May. FHIA was established as "a private, non-profit foundation dedicated exclusively to agricultural research - with an emphasis on NTAX crops" (Thorpe, 2002: 89). FHIA's principal function was to transfer research results and the technologies necessary for the cultivation of export-oriented crops and diversify production throughout the country.[36] Aquaculture was thus a priority sector for FHIA.

A second institution was the Federación de Agroexportadores de Honduras (Federation of Agro-Exporters of Honduras, FPX), also known as the Honduran Federation of Agricultural and Agro-Industrial Producers and Exporters (FEPROEXAAH/FPX). FPX's mission was to develop occupational groups representing producers in existing sectors as well as in emerging sectors. Its primary goal was to convert Honduras into a competitive agro-export model using neoliberal economic principles. To accomplish this mission, FPX's role centred on generating local businesses, establishing commercial contacts outside of the country, and assisting with the development of agro-businesses by securing the finances and then marketing the sectors.[37]

[36] (http://orton.catie.ac.cr/bfhiainf.htm)
[37] (http://www.mayanet.hn/fpx/fpx.htm)

FPX was created and initially funded by USAID, with an operational budget of around $600,000. Peter Hearne, Environment Officer with USAID, has argued that FPX was designed because there were a number of opportunities for the cultivation of non-traditional crops. Southern Honduras, in particular, had several areas suitable for cultivation, but the knowledge and capacity did not exist to move the country into new productive activities such as aquaculture. FPX was designed to fill this gap (personal interview with Peter Hearne, USAID, 2005). According to Mrs. Ruiz, who served as commercial attaché at the Honduran Embassy in Washington DC, FPX became the primary institution assisting farmers from everything between post-harvest handling to marketing (Luxner, 1993: 5a).

Finally, another institution was created to provide financing to support the activities and missions of FHIA and FPX. The Fundación para la Investigación y el Desarrollo Empresarial (Foundation for the Research and Development of Business, FIDE) was specifically created to provide loans to support the development of the industry in Honduras. For example, once the fishermen's cooperative ASOMAR was recognised in 1983, FIDE provided a loan totalling nearly $150,000 for the cooperative to construct an ice manufacturing facility in San Lorenzo. This plant was developed specifically for the benefit of the members of ASOMAR, but also to supply ice to the wider market (*La Tribuna*, 6 August, 1981). Subsequently, ASOMAR aligned with industry development and was involved in several aspects of the commodity chain, including postlarvae collection.

International and Domestic Incentives for the Promotion of Aquaculture

Aside from the creation of new institutions being central to the process of neoliberal reform and capitalist transformation in Honduras, the provision and creation of international and domestic incentives for aquaculture development was just as important to ensure that the nascent industry was competitive in the global market. USAID and other institutions made use of several techniques of power to alter the macroeconomic and political structures to facilitate NTAX crops reaching the US market.

In addition to the newly-created institutions providing technical expertise and resources as an incentive to promote shrimp production, the Honduran government, at the behest of, and in conjunction with, other organisations, made political decisions to provide the "fiscal incentives to export producers in its drive to open up the Honduran economy"

(Stanley, in Collinson, 1997: 140). One agreement in particular, the US Caribbean Basin Economic Recovery Act (CBERA), or the Caribbean Basin Initiative Act (CBI), was passed in 1983 as the USAID/CRSP commenced. CBERA provided duty-free entry of NTAX and funds for agricultural development and diversification projects through FHIA and FPX. Legal reform within Honduras, such as the passage of the 1983 Export Promotion Law, was commenced in order to smooth the country's transition towards a capitalist-oriented agricultural market. A number of other laws and incentives were also provided, and Mendes has summarised them nicely:

> The Honduran government also has provided incentives such as tariff exemptions for input materials, tax rebates, special foreign exchange purchases, and customs benefits. One primary measure is the Temporary Import Tax Exemption (RIT) created under Decree 37 of 1984. This allows the suspension of customs, consular duties, sales taxes, and administrative fees on imported goods used to produce exports sold outside of Central America. It also exempts exporters from paying taxes on profits and the common 1 per cent FOB tax on exports for a period of 10 years to firms developing a nontraditional export product and providing a specified number of jobs. Shrimp qualified as a nontraditional export since in the mid-1980s mariculture was new to the country and provided less than 5 per cent of total export receipts. Despite the sector's current position as a large source of foreign exchange, the RIT exemptions remain in place. Generally, medium and large firms have completed the necessary paperwork to participate in the RIT program (Mendes, 1999 in Stanley and Alduvin, 2002).

Meanwhile, USAID was also pursuing strategies to invest in infrastructure development for the benefit of the NTAX and traditional export sectors. The actions taken were again linked with the discourses of economic development and food security. A 1984 *La Tribuna* article addressed a shortage of basic grains in southern Honduras, which was thought to be the cause of food insecurity and malnutrition. The Honduran government and USAID signed an agreement that led to the provision of $6.5 million for an emergency programme to address these issues. Over 750 projects took place, targeting 40 communities and 25,000 people. The majority of funds were directed towards infrastructure development and food security related activities, which were linked to the promotion of NTAX and economic development (*La Tribuna*, 9 March 1984).

These two objectives – infrastructure development and food security – also coincided with larger development goals for the country as articulated in the National Development Plan (NDP) that covered its objectives between 1983 and 1986 and were later linked to President Azcona's Food Security Doctrine, which was between 1986 and 1990. USAID played an important role in the development of the NDP and the Food

Security Doctrine during this period for two reasons: to identify those areas where the two countries could collaborate bilaterally in pursuit of their mutual interests. Of the $1.2 billion allotted to Honduras between 1980 and 1990, $173 million was directed towards 'food aid' under USAID's Public Law 480, and in 1982 alone $31.2 million was provided in military aid, more than the entire period between 1946 and 1981 (Robinson, 2003: 123 and 126). As a result, the US wanted to ensure that the funds were spent wisely and in relation to US interests. The US worked closely with the Inter-American Development Bank (IDB) and other institutions to achieve mutual development goals in Honduras.

As one of its top priorities, the NDP, between 1983 and 1986, was assigned the development of natural resources and the country's raw materials, specifically in the agricultural, fisheries, and mining sectors. In order to support these goals, USAID announced its intentions to invest in infrastructure development and food security related issues (*La Tribuna*, 25 June, 1984). The estimated cost of the fisheries component of the NDP was $1,033,240 with 55.1 per cent originating from IDB, 26.1 per cent from the United Nations Development Program (UNDP) and its counterpart the FAO, and 18.8 per cent originating from the Honduran government. The goal of the programme was to work towards the systematic development of fisheries resources which would permit Honduras to export to industrialised nations and stimulate social and economic development within the country, a discourse shared by USAID (*La Tribuna*, 25 June, 1984). The emphasis was placed on aquaculture development but also included identifying ways to exploit other marine resources (*La Tribuna*, 16 January, 1987). Significant financial resources were allotted to these efforts through BANADESA when the US and Honduras signed an agreement to address food security issues in the south, which was aligned with President Azcona's Food Security Doctrine (*La Tribuna*, 4 February, 1987).

The importance of these initiatives lies in the links between different institutional strategies and discourses that demonstrated how international actors in the region supported the agenda to achieve US interests that were justified on the basis of promoting economic development and food security. The result led to the rise of market economics in Honduras as neoliberal reform gained pace in the late 1980s. The discourses underlying the strategies and the various techniques of power utilised to accomplish the stated objectives were continually employed to justify intervention in the coastal wetlands of the GOF. Finally, the discourses surrounding the objectives articulated by numerous international actors were also linked to the larger processes of neoliberal agrarian reform that were being encouraged in Honduras and throughout the entire Latin American region, eventually leading to the passage of the LMDSA in 1992 which was discussed in Chapter 4.

Growth of the Industry and Employment

The early industry's successes became apparent when RENARE announced on 6 March 1987 that over 125,000 lbs of shrimp had been produced by a single business in the south for national and international markets (*La Tribuna*, 6 March, 1987). GGMSB, based in Choluteca, was responsible for more than 200 ha, 800 employees and 15 ponds that were under production at the time and was one of the businesses targeted by RENARE according to the manager of the business, Carlos Alfredo Lara (*La Tribuna*, 6 March, 1987). Private investors were recognised for making the food security programme a success, according to Adan Antonio Benavides, Director of RENARE. The shrimp sales generated over $400,000 profit, which provided evidence of the potential success of the shrimp industry in the south (*La Tribuna*, 6 March, 1987).

A group of economists affiliated with the RENARE's Planning Department completed a report titled: *Status of the Cultivation of Shrimp in Southern Honduras* (*La Tribuna*, 30 October, 1987). For the economic study, the government completed a survey of 29 shrimp farms in the south to determine their contribution to President Jose Azcona Hoyo's Food Security Doctrine. This report demonstrated that the shrimp industry was providing around 1,230 direct jobs by 1987, which allowed the government to demonstrate that it had been successfully promoting economic development in the south by focusing on new sources of employment in relation to the collection of larvae, production, harvest and commercialisation. The report emphasised that over 800,000 lbs of shrimp were produced in 1986, earning over $90 million; 672,000 lbs were for the international market (primarily the US), while 128,000 lbs were for the domestic market. Benavides reiterated that the government had identified at least 15,000 ha that could be put under production (*La Tribuna*, 30 October, 1987).

At this point, there were four large-scale producers with the potential to expand: GGMSB, CIFAR, Cultivos del Mar, and Salinas de Honduras. These four companies provided direct and indirect employment to around 3,000 people, in comparison to the smaller producers that collectively employed around 300 people (*La Tribuna*, 30 October, 1987).

RENARE reassured the small-scale producers that it still had the intention to focus on providing technical assistance. The goal was to protect them from larger market forces driven by powerful US and Honduran military, political and commercial interests that received the major benefits associated with industry development. In essence, the smaller-scale producers were marginalised from access to capital markets, exacerbating further the reasons for conflict.

FIGURE 6.3: THE GGMSB PROCESSING FACILITIES IN SAN LORENZO
Source: GGMSB, SFI 2005

Finally, an important aspect of this economic study was the suggestion that the shrimp mortality rate could be reduced if the industry shifted from wild-caught postlarvae to the production of laboratory-produced stock. It was argued that laboratory-produced postlarvae yielded better results than wild-caught postlarvae, by minimising the risk of predation in the early stages of shrimp development, and would also allow the industry to control species produced, both of which could lead to increased profits (*La Tribuna*, 30 October 1987). Following this study and the decisions that resulted, sections of the artisan fishing community began to be impacted as the commercial industry began to shift to laboratory-produced postlarvae. Nevertheless, the industry continued to rely on wild-caught postlarvae.

The reliance on wild postlarvae had implications for the fishing communities that provided them to the industry, and were a partial factor in the division that took place in the fishing communities of the south. Those in support of the laboratory-produced postlarvae were generally fishermen who thought it would decrease pressure on other fisheries resources by reducing by-catch (CODDEFFAGOLF is later aligned with this position). Also, fishermen's cooperatives that controlled production in hatcheries were supportive of producing more postlarvae in laboratories, as it would give them an advantage within the market when selling to the shrimp-farming operations. At the time, operations relied on multiple sources of postlarvae, sometimes their own laboratory-or hatchery-produced postlarvae, although they also continued to use wild-caught postlarvae.

The artisan fishermen opposed to laboratory-produced postlarvae were those dependent on the capture of wild postlarvae as a livelihood strategy, or cooperatives that ran hatcheries, but were threatened by the larger companies' pursuit of vertical integration. Their interest was in maintaining their position within the commodity chain from which they would be removed if the suggested policy change were fully implemented. Although the industry was supportive of laboratory-produced postlarvae, it continued to rely on wild-caught postlarvae and employed a number of artisan fishermen to continue supplying the industry. CODDEFFAGOLF, once established, worked assiduously to reduce by-catch by advocating a ban on the use of drag-nets, which were used to capture wild postlarvae and were thought to be the cause of the problem.

It is important to make it clear that the issues surrounding the capture of wild postlarvae are the roots of the conflicts that initially emerged between CODDEFFAGOLF and the industry, after this organisation's establishment in 1988. Although at first organised to resist industry expansion, because it was an advocate for laboratory-produced larvae CODDEFFAGOLF aligned itself with the industry's interests. The paradox is easily explained.

First, although there was an increased use of laboratory-produced larvae, the industry continued to rely on wild-caught postlarvae for more than a decade. For CODDEFFAGOLF, laboratory-produced postlarvae would reduce by-catch affecting other species. For the shrimp industry, it would allow them to control their seed stock and complete efforts to vertically integrate. Although in this respect their interests were aligned, since the industry continued to rely on wild-caught postlarvae,

CODDEFFAGOLF's concerns remained. Finally, private commercial producers were able to afford the transition to the use of laboratory-produced larvae, while small and medium-scale producers remained more dependent on the wild-caught postlarvae to stock their ponds. Capital was thus a limiting factor for small and medium-scale producers, making it difficult for them to shift their practices.

As actors involved in local fisheries, fishermen's cooperatives controlled access to different markets depending on what species they caught and to whom they provided the product.[38] Impacts on other species due to the capture of wild postlarvae increasingly became a bigger threat to various groups within the artisan fishing community. Some of these fishermen supplied adult shrimp caught in the open waters and other species of fish to the local markets; thus the postlarvae gatherers were a threat to their livelihoods due to the by-catch. CODDEFFAGOLF originated and initially represented this group of artisan fishermen, and consequently it represented only a fraction of the artisan fishing community in southern Honduras. The voices of certain segments of the fishing community were therefore diminished as this organisation became more dominant and increased its power to advocate policy changes, especially in regard to the capture of wild postlarvae and aquaculture expansion. By the 1990s, the state attempted to reduce the by-catch associated with the capture of wild postlarvae by outlawing the use of drag-nets.

Summary

In this Chapter, I argued that as neoliberal economic reform commenced in Honduras, a component of which was aquaculture development in the south, the state used the twin discourses of economic development and food security to justify the promotion of NTAX. I also argued that the state and international actors used various techniques of power – which were represented by six actions that led to the material manifestation of new institutions to promote NTAX and, subsequently, aquaculture – to reorder the landscape of southern Honduras to produce shrimp for the global market.

The institutions established throughout the 1980s became agents of neoliberal change as economic reform gained pace throughout the decade. To facilitate and promote new exports, the Honduran government and other powerful social actors provided incentives to the producers to encourage future development and expand the industry. Power was deployed to form new social and economic networks to support shrimp production. I

---

[38] For more information on the capture of wild postlarvae see the thesis of Pacheco, (1999), University of Zamorano

argued that the formation of cooperatives and the training of larvae gatherers was the most significant event, as it changed the socio-economic situation and led to divisions within the artisan fishing community. Consequently, I argued that the divisions that took place during this period led to the formation of a resistance group opposed to future aquaculture development as well as the negative impacts on the wetlands; this is illustrated in more detail when I discuss CODDEFFAGOLF's emergence in subsequent chapters.

Although additional concerns were generated around the impacts on the wetlands, such as mangrove degradation and enclosure of common pool resources, I will continue to argue that the conflict is consequent on the Honduran government's adoption of neoliberal policies. I will illustrate that the conflict actually emerged as a consequence of the divisions that took place within the artisan fishing community as a result of the promotion of NTAX and the introduction of aquaculture. The livelihood issues were later successfully linked to the negative impacts on the wetlands, mangrove deforestation, and enclosure of common pool resources to generate further political and international support to protect the coastal wetlands of southern Honduras.

I now argue that a number of civil society actors had alternative conceptions as to how the coastal wetlands *ought* to be used, conserved, and managed, and so on. These alternative conceptions were incongruent with the objectives shrouded under the veil of neoliberal reform and became a further source of tension. The propagation of neoliberal ideology occurred through the discourses and actions of more dominant social actors, such as USAID, IFIs, the Honduran government, and private sector as they competed to influence public perception. Their aim was to realise their neoliberal images, the future of the physical space around the GOF, an image that commoditised and capitalised the wetlands for shrimp farming and becomes increasingly contested.

In Chapter 7 I will show how the issue of access and the socio-economic and environmental changes that resulted from the introduction and expansion of aquaculture in Honduras led twelve fishermen and a former government employee, Jorge Varela, to create CODDEFFAGOLF by 1988. I shall argue that CODDEFFAGOLF was intended to create a new social, political, economic and physical reality in the south, and one that was not necessarily aligned with the contemporaneous neoliberal transformations. The driving force of this organisation was directly connected to the impacts that aquaculture was perceived to have on the livelihoods of a faction of the local artisan fishing community and on the coastal wetlands. Once formed, this new environmental social movement increasingly challenged the future expansion of the industry and its impacts on the wetlands and fisheries resources, which I discuss in more detail in Chapter 7. For now, I intend to establish what the discourses on the wetlands were throughout the 1980s prior to the formation of CODDEFFAGOLF in 1988 and how they related to the conflict that emerged later.

Chapter 7

The Honduran Environmental Movement, Local Actors and Alternative Discourses on the Mangrove Wetlands

Introduction

In this chapter, I discuss the emergence and strengthening of the Honduran environmental movement in the 1980s. I focus attention on the formation of the Honduran Ecological Association (AHE) to illustrate how new actors were emerging within civil society as the democratic transition in Honduras was taking place. I have selected this organisation because it was the first to address mangrove deforestation in the south. The links that this organisation later developed with Jorge Varela eventually led him to establish CODDEFFAGOLF, the primary opposition to the shrimp industry. The AHE first brought the issue of mangrove degradation into the public arena in 1984. The organisation was concerned about the impacts on the mangrove wetlands due to the use of mangrove wood for salt production, fuel wood, and construction. I explain the characteristics of these production processes and how mangrove forests are used, and highlight the AHE's early discourses in relation to their degradation in the south.

I then illustrate how the actions of civil society actors brought attention to these issues within the public arena, and led the state to assess the situation. The government admitted that it was responsible for the deforestation, identifying the contributing factors, and began to articulate a discourse of enforcement, although the actions that followed were inconsistent with the public rhetoric. Nevertheless, the state did collaborate actively with the AHE to address mangrove deforestation.

By 1985, Jorge Varela, had joined the AHE and began focusing on the impacts of the shrimp industry on the local fisheries sector. AHE politics encouraged him to leave so that he could focus his energy on artisan fishermen affected by the development of the shrimp industry. How Varela ended up working with these fishermen to establish CODDEFFAGOLF is explained. I describe the history that led to CODDEFFAGOLF's creation (A biographical sketch of Varela is included prior to Chapter 1). It was a slow process, and the ideas behind it had incubated since the 1970s, eventually leading to CODDEFFAGOLF's creation in 1988. I then turn my attention to the early discourses surrounding mangrove degradation and illustrate how they were increasingly linked to aquaculture development around the time that Varela joined the AHE.

I then turn to the AHE's campaign to protect the coastal ecosystem of the GOF, Varela's role, the links to the fishing communities affected by aquaculture development, and the socio-economic changes taking place within the fisheries sector. The intention is to assess how civil society actors started to contest both state and private representations of this physical geographic space and characterise fishing communities. I then explain their key interests and how their goals were counter to those of the other segments of the artisan fishing community, wherein lay the source of the conflict. I describe the characteristics of CODDEFFAGOLF, and its strategies. However, I reserve a more in depth discussion for Part III which covers between 1988 and 1998.

The Formation of the Honduran Ecological Association: Discourses, Interests, and Actions

As neoliberal reform was under way, a number of new social actors began to emerge within civil society. In the early 1980s, after the democratic transition in Honduras started, a number of NGOs began to play an active role. By 1988, it was estimated that there were over 200 Private Development Organisations (PDOs) operating in Honduras, primarily funded through large sums of foreign aid originating from a variety of sources, most of which was from USAID (Rowlands, 1995: 46). In this regard, USAID involved itself in nearly every aspect of Honduran society (Norsworthy in Robinson, 2003: 122).

The numerous environmental organisations that established themselves during this period were geographically dispersed but eventually began to consolidate, leading to stronger links between movements in Honduras and outside the country: in 1987 a number of the newly created NGOs in the region had come together in Managua, Nicaragua and created the Regional Network of NGOs for the Sustainable Development of Central America (REDES). REDES exemplified new forms of environmental governance that began to spring up as local movements collectively organized on a regional scale. The scaling up of the environmental movement in Honduras and internationally represented the emergence of a new environmentalism that paralleled the rise of neoliberalism in the region. Increasingly, the government institutions and private interests that were the forces behind neoliberal change became the target of these organisations: the shrimp aquaculture industry was no exception.

Middle-class professionals who became concerned with the unfettered degradation of natural resources formed a number of the organisations. The Honduran Ecological Association (Asociación Hondureña de la Ecología AHE)[39] was one such organisation, established by Jaime Gustillo with support from a US Peace Corps volunteer named Raymond Dodd[40] and USAID funds. It quickly became one of the most important organisations fighting for sound ecological policies in Honduras. AHE primarily targeted *campesino*[41] organisations, trades unions and other interest groups in order to convince them to include environmental concerns in the policies they advocated, since these organisations already had a strong track record in lobbying the government.

The AHE's focus was national, but the organisation was particularly interested in government actions affecting the southern region. Two issues of particular relevance to the south were deforestation and the prevalence of unsustainable agricultural practices.[42] In southern Honduras, clearance of hardwood forests for agriculture and deforestation of mangrove forests were focal points. Their attention on these two issues aligned with the wider-environmental movement emerging in Central America. The AHE was the first organisation to highlight the issues surrounding mangrove degradation in the GOF, and on 7 July 1984 publicly issued a formal complaint regarding the affects of human activities on the region's wetlands.

The same year, the industry was expanding into the San Bernardo region and making use of the extensive salt flats for the construction of shrimp ponds. However, the industry was not the initial target of the AHE since the region's production was in its infancy, and the construction of the ponds was taking place on the salt flats with minimal impact on the mangrove forests. An article published in *La Tribuna* summarised the concerns of the AHE, warning that the GOF's marine environment and the mangrove ecosystem were in danger of extinction due to the practices of salt producers, use of mangrove wood for fuel, construction, and tanneries (*La Tribuna*, 7 July 1984).

---

[39] "As a consequence f the expasion of the environmental consciousness, the AHE was founded in the 1980s. Following the example set in the foundation of the AHE, many other groups formed with the stated purpose of promoting ecologically sound policies. (http://lcweb2.loc.gov/cgi-bin/query/r?frd/cstdy:@field(DOCID+hn0068) US Library of Congress Country Studies).

[40] http://quest.nasa.gov/projects/spacewardbound/mojave2007/bios/dodd.html Peace Corps Volunteer named Raymond Dodd helped to set up the AHE.

[41] *Campesino* refers to rural person(s), or to 'peasant' a term seldom used in Honduras as it has a derogatory connotation.

[42] (http://countrystudies.us/honduras/60.htm, US Library of Congress)

FIGURE 7.1: SALT PRODUCTION
Source: CODDEFFAGOLF

In 1984, there were approximately 130 small, medium, and large businesses dedicated to the production of salt in southern Honduras (AFE-COHDEFOR/OIMT, 2001: 7).[43] Most were family operations[44] (personal interview with Consuela, April 2005). However, the percentage of local people involved in salt production is currently less than 1 per cent (3/358 responses) (Wilburn, 2005). Salt producers are active primarily during the dry season and generally operate on the salt flats that were identified for potential conversion to shrimp farming. Furthermore, as profits declined in relation to the production of salt, some producers shifted to the production of shrimp during the rainy season when the salt flats were inundated and unavailable for salt production. This transition was most apparent in the 1980s and 1990s, and was visible in the areas around La Brea, San Lorenzo and El Laure, three communities situated in south-central Honduras (Vasquez-Cristobal and Wainwright, 2002: 11).

---

[43] CONGESA (2001). Valoración Económica de los Manglares del Golfo de Fonseca, Honduras. Choluteca, Honduras, Administración Forestal del Estado Corporación Hondureña de Desarrollo Forestal (AFE-COHDEFOR) Organización Internacional de las Maderas Tropicales (OIMT) Manejo y Conservación de los Manglares del Golfo de Fonseca, Honduras (PROMANGLE) Consultores en Gestión Ambiental (CONGESA): Page 136.

[44] During a conversation with a local salt producer, Consuela, I was told that it costs nearly $11-$12 to produce 45 kg of salt that sells, often, for less than $11 on the national market (personal communication, April 2005). In contrast, imported sea salt can be purchased for $3 to $5 dollars per 45 kg bag in the capital (personal communication with Dr. Daniel Meyer, May 8, 2006). Most salt produced in the south is sold locally and not exported.

The AHE was concerned about the use of mangrove wood for fuel in the salt production process. In 1984, it was determined that approximately 81 per cent of salt producers were using mangrove wood as fuel, whereas 19 per cent were using solar energy[45] in the production process (AFE-COHDEFOR/OIMT, 2001: 7).[46] Salt producers use mangrove wood as fuel to evaporate the brine in order to extract the salt from seawater. Mangrove wood was readily available, as the ponds lay in close proximity to the mangrove forests. As the wood is fed into ovens and burned, the sodium chloride fraction is separated from the brine through the evaporation process, and the salt is then bagged, stored, and sold (*La Tribuna*, 7 July 1984).

The first comprehensive study of the extent of mangrove wood use in salt production was not completed until 1988. In a study completed by the Secretaría de Planificación, Coordinación, y Presupuesto (Secretariat of Planning, Coordination and Budget, SECPLAN), and published in USAID's 1989 Environmental Profile, it was estimated that there were at least 500 salt-making operations using mangrove wood for fuel in ovens used in the production process: in 1988 these businesses used 50,000m[47] of mangrove and other wood (Stonich, 1993: 88).

Mangrove Extraction for Structural Use, Fuelwood, and Tanneries

AHE was also concerned about the impacts of mangrove extraction for fuel, mostly used for household consumption and associated with small-scale markets (see Chapter 3 for per cent of usage).

---

[45] The percentages have changed but no one has completed a comprehensive study to determine these changes precisely.
[46] CONGESA, 2001
[47] To clarify, 50,000m is equal to 164,042 feet or 31 miles of mangrove wood if you were to place it in a row along a highway.

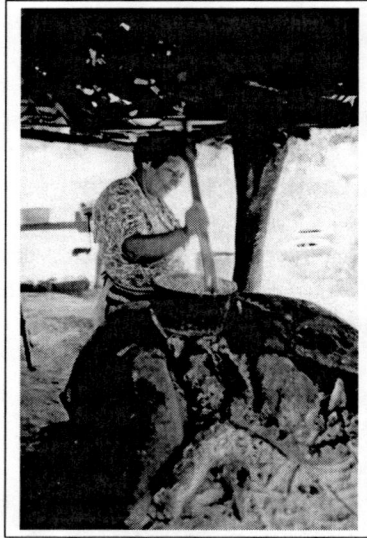

FIGURE 7.2: ***WOMAN COOKING WITH MANGROVE WOOD***
Source: CODDEFFAGOLF

Most mangrove cutting for local consumption occurs in Muruhuaca, Llano Largo, Pozo Largo Cubulero, El Aceituno, El Capulin, Los Guatelas, Playa Grande, La Brea, Puerto Soto (El Relleno), Pueblo Nuevo, and Guapinol (Vasquez-Cristobal and Wainwright, 2002: 12).

People have been coming to the area around Guapinol to collect mangrove wood for years. The wood is then sold to tanneries in two communities in the northern part of the Department of Choluteca (personal interview with VP of Guapinol, April 2005). Red mangrove, the most dominant of the six species found in the south, seems to be the most widely used (personal interview with Laura Sosa, Technical Coordinator, AFE-COHDEFOR, 30 July 2005). Subsequently, the AHE demanded that the government visit the region to assess the impacts of salt production, tanneries, and local people on mangrove forests. The government's response was fairly rapid, demonstrating that the concerns being articulated from within civil society were heard, and action followed.

FIGURE 7.3: RESTAURANT CONSTRUCTED WITH RED MANGROVE WOOD
Source: Photograph No. 30 in Zonificacion de los Bosques de Mangle del Golfo de Fonseca, Honduras, C.A. 2002

Government's Response to the AHE

The government's response to the AHE's campaign illustrated that collective action was a successful means to bring attention to important environmental issues. It led the state to admit that its actions had contributed to mangrove deforestation due to poor decisions of government employees responsible for the conservation of marine, coastal, and forestry resources. An Inter-Ministerial Commission comprised of representatives of the Honduran government traveled through the south to make observations in regard to impacts on the mangrove ecosystem. During this trip, the Commission identified a number of areas denuded and determined that mangrove wood was being illegally cut as the areas where it was being extracted were state-owned land and required a concession or a permit. Furthermore, the Commission stated that locals were taking advantage of the fact that there was a lack of governmental oversight in the region (*La Tribuna*, 7 July 1984). However, it was likely that most people in the south were either simply unaware of the law or too dependent on the resources for sustenance to care. The government admission that there was very little oversight in the region was indicative of the lack of state capacity and weak institutions in relation to natural resource management. The admission was probably the first time the government publicly recognised, via the newspaper, its role in the degradation of mangrove forest resources (*La Tribuna*, 7 July 1984).

The Commission admitted that the lack of governmental capacity to manage the resources was partially the result of overlapping legal jurisdictions, which made it unclear which agencies were directly responsible for protecting mangrove forests. In some cases, while one agency was working to conserve mangrove forests, another was issuing permits to allow locals to cut mangrove wood. In turn, the large number of people dependent on mangrove wood for a variety of uses made it almost impossible for local authorities to prevent illegal cutting. It was suggested that the government take steps to understand the connection between local livelihoods and the marine and littoral environment (*La Tribuna*, 7 July 1984).

Other factors complicated resource management as the government dealt with on-going disputes in relation to the territorial boundaries of the three countries bordering the GOF as well as transboundary resource disputes associated with use and conservation. The Inter-Ministerial Commission realised that foreigners, primarily Salvadoreñans and Nicaraguans, were increasingly extracting common-pool resources from the mangrove forests from Honduran territorial waters. Accordingly, the Commission encouraged the government to secure its territorial waters, to establish closed seasons, and to address the issues that the Honduran environmental movement claimed would lead to the extinction of species in the GOF (*La Tribuna*, 7 July 1984).

In accordance with neoliberal policy to strengthen the rule of law, SRN and COHDEFOR were urged to use their legal mandates to limit the amount of mangrove wood that could be cut. It was suggested that they either issue fewer permits or enforce laws against illegal cutting, since mangrove forests provided habitat to a diversity of species, including molluscs, shrimp, crab, and fish. The Commission recognised that locals relied on a variety of these species for sustenance. Recognising the links between local livelihoods and the health of the ecosystem, the government was urged to work with the private sector and international conservation organisations to protect living resources (*La Tribuna*, 7 July 1984). The Inter-Ministerial Commission's visit to the region resulted in collaboration between the AHE and the SRN/RENARE, demonstrating efforts between the state and newly formed organisations within civil society.

Emergence of Opposition to Aquaculture Development

As the AHE became more active in southern Honduras, it began to recognise that artisan and industrial aquaculture development could damage the mangrove wetlands. As a result, it began working more closely with Jorge Varela who had played a role in early aquaculture development but was critical of the government's policy. The concerns that he later raised developed throughout the 1970s and 1980s as he served in several capacities associated with fisheries and, eventually, aquaculture. He developed the founding principles of CODDEFFAGOLF during this period. Varela stated that the AHE hired him to coordinate and execute a project in the GOF with two objectives – environmental education and to complete the first inventory of the mangrove ecosystem of the GOF, funded by the World Wildlife Fund for Nature (WWF). His work with the AHE resulted in collaboration with SRN/RENARE to develop the first coastal resource management plan for the GOF in 1987[48], a plan that was never actually implemented (Varela, 2006a).

Emergent Discourses on the links between Aquaculture Devebpment and Mangrove Degradation

After Varela began working with the AHE in 1985, the organisation focused on the links between aquaculture development and mangrove degradation. The impacts of salt production and local mangrove woodcutters remained central, but aquaculture increasingly became a concern as pond construction became more frequent in areas vegetated by mangrove and hardwood forests. Honduran investors began visiting the region to explore potential sites for aquaculture development, locals began attempting small-scale operations, and salt producers who were facing a decline in profits began shifting their practices towards shrimp cultivation.

FIGURE 7.4: EXAMPLE OF MANGROVE CLEARANCE FOR SHRIMP POND CONSTRUCTION
Source: CODDEFFAGOLF

At this point, there was a notable shift in strategy as environmentalists increasingly focused attention on the destructive practices of the emerging aquaculture industry: a material manifestation of neoliberal change. Subsequently, AHE changed its focus and expanded the scope of its campaign in southern Honduras, and increasingly focused on

[48] Corporación Hondureña de Desarrollo Forestal. 1987. Inventario Forestal de los Manglares de la Zona Sur, Golfo de Fonseca, Honduras. COHDEFOR, Tegucigalpa, Honduras.

the clearing of mangrove forests for shrimp pond construction as Sea Farms of Honduras and other investors began to expand. The organisation was also concerned about small-scale operations, as they were often more closely associated with clearance of mangrove forests for pond construction. The salt flats were more often conceded to individuals and businesses with connections, financial resources, and good lawyers rather than the rural poor.

The result of AHE's efforts was two inter-institutional agreements with the government. One of the agreements established the basis for collaboration on the conservation of coastal resources. The second agreement focused on the protection and control of the use of mangroves in the GOF and on raising the awareness of salt producers, mangrove cutters, and the shrimp industry. Both documents were put forth to President Azcona for ratification illustrating the government's recognition of the demands from within civil society (*La Tribuna*, 12 July 1986). The agreements were an initial effort towards developing something that resembled a coastal management plan (*La Tribuna*, 12 July 1986). The coastal management plan being developed was based on the AHE and COHDEFOR findings financed by the WWF, published in 1987, and was authored by Jorge Varela in conjunction with COHDEFOR.

Despite the INA's participation in the discussions to protect and control the use of the mangrove wetlands, it announced new steps to promote NTAX less than two weeks after the meeting with the AHE and other Honduran government agencies. On 12 July 1986, *La Tribuna* published an article explaining the INA's intentions to diversify export production throughout the country. In particular, it discussed INA's efforts to open market channels in line with neoliberal reform by developing contacts with international corporations with the goal of identifying new markets for the export of NTAX produced in the south (*La Tribuna*, 24 July, 1986). AHE was concerned that increasing growth and market access would inevitably lead to further degradation of the wetlands as capitalist modernisation continued.

The period between 1973 and 1988 was defined by the emergence of neoliberal reform and coincided with the emergence of a number of newly created NGOs, several of which were focused on environmental issues. Although successes were demonstrated through the creation of public fora to address shared resource issues and the development of a framework for a coastal management plan, the government continued to accelerate its neoliberal reform efforts by the end of the decade. The increased attention on industry practices led it to organise to consolidate and respond to rising opposition. The consolidation of the industry, continued support for its expansion, and the emergence of new social actors contesting the industry were indicative of the nascent global aquaculture and environment debate within the context of an emerging environmentalism opposed to these neoliberal transformations.

The Formation of ANDAH and the Shrimp Industry's Response to the Rise in Opposition against Aquaculture

In response to the increased attention being placed on the industry's development, shrimp producers organised and formed the National Aquaculture Association of Honduras (Associación Nacional de Acuicultores de Honduras ANDAH) in 1986. Three factors can be attributed to the creation of ANDAH. First, as the industry expanded, issues between large, medium, and small-scale producers arose, and a forum was needed to organise the industry. Second, it was beneficial for the industry to work under a single institution in order to consolidate its power and advance its interests collectively. It also created a formal institutional mechanism for industrial organisation as ice-packing plants, processing facilities, hatcheries, shrimp farms and other relevant aspects of the industry were developed. Third, the bad publicity resulting from the AHE's campaign made it advantageous for the industry to collectively constitute, in order to respond publicly to the claims being made against the industry. Sea Farms of Honduras played a leading role in its creation as the company sought to be completely vertically integrated. It also placed them in an advantageous position to keep their finger on the pulse of the industry.

ANDAH characterised itself as an institution created for the purpose of providing for the security and the development of the aquaculture industry nation-wide. Its objective was to encourage aquaculture nationally and to contribute to the strengthening of the industry globally. ANDAH is structured to develop strategies and successfully achieve goals that benefit the economy and society of southern Honduras and the Honduran community generally (Personal Interview with Hector Corrales, 2004). The founding of ANDAH made it easier for the businesses to interface as a group with government officials and other international representatives. Taking into consideration the issues that were raised when the Inter-Ministerial Commission traveled to the region, Rafael Molina, the first President of ANDAH, demanded that the government take measures to prevent the cutting of mangrove forests. The industry directed attention to the impact of salt producers on the wetlands, and argued that the government should consider establishing protected areas in the region (*La Tribuna*, 9 August, 1986). This proposal was one of the first to propose the creation of protected areas and coincided with the study funded by WWF to develop a coastal management plan for the GOF.

At the time, the discourse surrounding mangrove degradation was still dominated by the impact that salt producers and local consumers of mangrove wood had on the ecosystem. The statements published by Molina, as an independent shrimp farmer and salt producer were an attempt to demonstrate ANDAH's concern for the wetlands, while trying to prevent the industry being blamed for mangrove deforestation. It had a direct interest in the protection of the wetlands due to the relationship between the lifecycle of shrimp and the mangrove ecosystem.

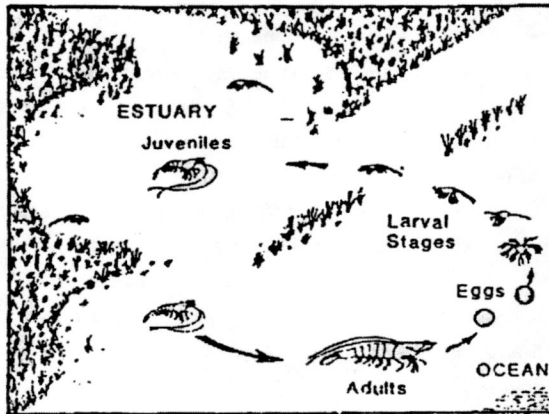

FIGURE 7.5: SHRIMP LIFECYCLE PENAEIDOS
Source: Reprinted in Sobreexplotacion de la Vegetacion Manglar y sus Efectos al Ecosystema de law Bahia de Chismuyo, Honduras, 1997

The industry relied on a healthy ecosystem to supply the wild-caught shrimp postlarvae needed to stock the ponds. Molina stated: "The government should prohibit the cutting of mangrove forests because they are the breeding grounds of the shrimp, without taking urgent measures we will have problems that we will lament in the future" (*La Tribuna*, 9 August, 1986).

ANDAH also announced that the industry should continue to expand since 15,000 ha of salt flats remained available for production. To continue expansion, ANDAH argued that the government needed to guarantee investments and provide access to the 'salt water lands' or salt flats for the continuing construction of ponds (La Tribuna, 9 August, 1986). The demand was consistent with land tenure issues advocated by the private sector and individuals supportive of neoliberal reform (see Chapter 4). The model favours individual over collective land rights, since they are considered to be the most likely to maximise behaviours geared towards capital accumulation and economic efficiency (Deere and Leon, 2000: 78). ANDAH highlighted that the industry was already generating export revenue (see Table 7.1) and providing employment. Finally, the organisation argued that the government should urge salt producers to use their lands for shrimp production for two reasons: salt was confronting declines in profit and the shift to shrimp farming would reduce demand on mangrove wood used in the salt production process and therefore address one of the concerns of the Honduran environmental movement (La Tribuna, 9 August, 1986).

TABLE 7.1: AMOUNT OF SHRIMP EXPORTED, BY WEIGHT AND VALUE: 1986-1988

1986	1987	1988
1,450,000 lbs	5,000,000 lbs	7,500,000 lbs
$5,000,000	$18,000,000	$26,000,000

Source: La Tribuna (1988). En Peligro Produccion Camaronera por Desastroso Estado de las Caretteras de Sur. 1 November.

AHE Launches Campaign to Protect the Coastal Ecosystem of the Gulf of Fonseca

By the end of August 1986, USAID provided additional funds for industry expansion. The AHE immediately made a public statement calling on the government to protect the coastal ecosystem of the GOF. The article announced the AHE's intent to commence a campaign to protect the coastal ecosystem in the southern zone of the country. Jorge Varela was involved in bringing it together (La Tribuna, 25 August 1986).

TABLE 7.2: LANDS CONCEDED FOR THE DEVELOPMENT AND PRODUCTION OF SHRIMP IN SOUTHERN HONDURAS, 1985-1996

Year	Land Conceded		Land Under Production (ha)		
	Amount	% Change	Semi-Intensive	Extensive	% Change
1985	5,800	—	750	0	—
1986	6,800	17	1,500	0	50
1987	8,200	19	1,800	100	20
1988	13,030	61	3,000	250	67
1989	16,115	24	4,300	250	43

1990	22,200	38	5,500	250	40
1991	23,777	7	6,000	250	20
1992	24,000	1	8000	250	11
1993	24,500	2	9000	250	13
1994	25,780	5	10,000	250	11
1995	26,000	1	11,000	250	10
1996	26,000	0	12,000	250	9

Source: Ministry of Natural Resources, Honduras 1997.

The AHE continued to focus on a wider set of issues affecting the coastal ecosystem of the GOF, indicating lack of focus in the movement's strategies. Attention was directed to the impacts of burning and cutting hardwood and mangrove wood, the use of gunpowder and dynamite for fishing, the impacts of certain types of fishing gear, and the destruction of the coastal vegetation for the construction of shrimp ponds (*La Tribuna*, 25 August 1986). The organisation stated that protecting the ecosystem was important to protect the coast, and coastal communities, from natural disasters. Finally, the AHE also called on the government to protect "sweet land" or agricultural land due to over-grazing, over-cultivation, and deforestation consistent with its national campaign (*La Tribuna*, 25 August 1986).

With the AHE, Jorge Varela had been working with artisan fishermen to transition to improved fishing gear and educating them about the impacts of gunpowder and dynamite on the fisheries. He became interested in several concerns that fishermen had begun to raise while working with COHDEFOR on the WWF-funded baseline assessment to develop a coastal management plan. At the time, some groups within the artisan fishing community became increasingly concerned about their role in the emerging aquaculture sector. Specifically, they were concerned about increased competition surrounding the capture of wild postlarvae, access to traditional fishing grounds, and securing their role in the fisheries market (*La Tribuna*, 21 August 1987). It was these issues that were to become central to the conflict in its early stages.

First, the fishermen wanted to stop the capture of wild postlarvae due to the by-catch, an issue with which Varela was familiar. Second, they wanted the government to ensure that they had access to their traditional fishing grounds in the estuaries and seasonal lagoons, areas where aquaculture production was developing. These concerns were later debated in policy circles and within the public arena, which subsequently led to changes in the laws. The role of civil society in contesting state legal representations of physical and socio-economic space was evident in the legal changes and reveals their vital role in the formation of law and policy.

In 1987, the largest fishing communities were located in Amapala, Coyolito, El Relleno, Punta Raton, Cedeño, and Guapinol. It was estimated that these towns were home to 1,500 fishermen[49], with nearly 600 operating out of Cedeño, 500 out of Guapinol, and 400 fishermen in Coyolito (*La Tribuna*, 21 August 1987). Guapinol was one of a number of communities with which Varela had been working during this period and later became one of the *loci* of conflict between the industry and artisan fishermen due to its proximity to the largest farms built in San Bernardo. The seasonal lagoons on which GGMSB built its ponds are easily accessible by boat from both Cedeño and Guapinol, with the *Estero San Pedregal* being the only barrier for the fishermen's access to this space. This area also became one of the most highly contested spaces throughout the conflict due to the proximity of these communities as well as others to the shrimp ponds being established on the salt flats of San Bernardo, illustrated in the chapters that follow.

Aquaculture Expansion Continues

By June 1988, over 6 million lbs of shrimp were produced on about 3,300 ha of land leading to foreign revenues of more than $19 million, which contributed positively to the balance of payments of the country. Most of the shrimp exported were produced by Sea Farms of Honduras/GGMSB. ANDAH estimated that the industry was providing direct and indirect employment for about 6,000 people in the south. ANDAH called for placing an additional 1,000 ha per annum under production. The local and national debates regarding the protection of mangrove forests became more prominent as aquaculture expansion continued, providing the impetus for the formation of CODDEFFAGOLF in the latter part of 1988.

In the late-1980s, there were several companies producing shrimp in the south – Sea Farms of Honduras, GGMSB, Culcamar, Cultivos Marinos S.A., Agrointernacional, Aquacultivos de Honduras, S.A. – and around 15 operators producing shrimp on a small scale using less than 70 ha for production. At the time, GGMSB was the largest farm,

---

[49] The number of fishermen in the region exceeded 15,000 by 2000. It is probable that the increase in the number of fishermen had to do with (1) population increase, and (2) increase in the number of migrant workers that came to the region to work as postlarvae gatherers for the industry. This is important because if the fishermen were working on behalf of the industry as postlarvae gatherers, they would have likely been aligned with the industry against CODDEFFAGOLF which claims to represent the fishermen of the south.

operating on more than 1,000 ha in the San Bernardo region of Choluteca and expected to expand into an additional 500 ha of salt flats in 1988–1989 (*La Tribuna*, 14 November 1988). ANDAH communicated the economic benefits of shrimp aquaculture and its contribution to food security, mirroring the state's discourse in the south. ANDAH also claimed that it was revitalising national and regional economies by supporting the emerging fishmeal, boat construction, and fishing equipment industries, as well as other commercial businesses in the south.

The AHE Demands that the Government Assess the Impacts of Aquaculture on the Mangrove Forests

Toward the end of the period the AHE shifted its arguments towards those of the state, industry and USAID. Some of its financing originated from USAID. One factor that may have influenced the organisation's change in position may have been their constituencies as well as their funding bodies. The AHE was founded by middle-class Hondurans living in the capital, and was not a localised environmental movement.

In contrast, CODDEFFAGOLF was founded by local artisan fishermen in conjunction with Varela and its concerns centred on the changes in the labour market, enclosure of traditional fishing grounds, and by-catch associated with the use of drag-nets for the capture of wild postlarvae. The AHE's primary interests were unsustainable agricultural practices and deforestation. The AHE did not consider industry practices to be unsustainable or contributing to deforestation. Furthermore, the AHE pursued a more collaborative strategy with the government than that of CODDEFFAGOLF, resulting with an assessment and a joint statement by RENARE and the AHE indicating that the industry was not a threat to the mangrove forests and that pond construction was taking place on the salt flats (*La Tribuna*, 14 March 1988).

RENARE's position was heavily influenced by the actions of the AHE and local politicians, such as Liberal Diputado, Antonio Ortez Turcio, from Choluteca. (*La Tribuna*, 14 March 1988). Consequently, public pressure led to a shift in the agency's discourse and position in relation to wetlands. It began exerting powers provided in the 1959 Fisheries Law pertaining to the protection of mangrove forests. As a result, RENARE started to work more closely with the Honduran environmental movement throughout the country and in the south.

RENARE admitted that they had issued permits and concessions for activities that may have impacted the coastal wetlands. The agency announced that it would begin sanctioning any businesses that degraded mangrove forests under the authority provided in Article 52 of the 1959 Fisheries Law (*La Tribuna*, 14 March 1988) and the 1972 Forestry Law, which considered them to be Protected Forest Zones (Forestry Law, *La Gaceta*, 4 March, 1972 (Chapter XIII, Article 138 (1a)) and by Legislative Decree (No. 117, 17 May, 1961). However, the state never clearly defined a Protected Forest Zone (PFZ), nor did it establish any management regimes. RENARE's Director General, Adan Antonio Benavides Herrera, stated protecting the mangroves is important as they provide protection from hurricanes and tropical storms, and are an important habitat linked to the life cycles of a number of species of shrimp and fish that are critical to regional food security (*La Tribuna*, 14 March 1988). This was the first statement to link regional food security with the protection of the mangrove ecosystem, turning RENARE's discourse on food security in favour of the emerging movement to protect the ecosystem.

In June 1988 Licenciado Mario Galeano Burges, the manager of the aquaculture company Acquacultivos de Honduras, S.A., stated that shrimp farmers were not actively or consciously participating in the destruction of the mangrove forests. Clearing mangrove forests to construct ponds was considered expensive and too great an investment and, ultimately, mangrove degradation presented a risk to the industry and its survival since water quality and intact habitat for the wild postlarvae were critical elements the industry relied on to succeed (*La Tribuna*, 17 June 1988).

The issue in relation to the large-scale industry was actually the increased enclosure of common pool resources, and conflicts generated around the collection of postlarvae. Mangrove degradation was associated more with small- and medium-scale operations established by individuals who did not have the power to acquire concessions to gain access to the salt flats, which were primarily conceded to more powerful actors such as foreign investors, previous and current military leaders and political elites. However, the blame for deforestation was placed on the commercial industry as expansion on the salt flats was politicised.

Within a few months a statement appeared in *La Tribuna* announcing that ANDAH would begin reforesting the canals, the perimeter of the shrimp ponds and the drainage areas, and reiterating the links between intact mangrove forests and the life cycle of the postlarvae. The industry was also interested in keeping the mangrove forests intact to protect their operations against storm surges, which was one of the reasons that it was interested in supporting the creation of protected areas, especially in areas surrounding the salt flats, such as in San Bernardo (*La Tribuna*, 1 November 1988).

The rapid growth of the industry was a further impetus for the creation of CODDEFFAGOLF when a small group of fishermen identified the need for the environmental movement to have representation in the south. These fishermen were opposed to private sector involvement in the management of marine and coastal resources.

The Rise of Contestation: the Formation of the Committee for the Defense and Development of the Flora and Fauna of the Gulf of Fonseca

CODEFFAGOLF's creation was supported by, Turcio, the Liberal Party member who represented the Department of Choluteca. He wanted to highlight the issues associated with aquaculture expansion, especially in the San Bernardo region. Several investors in the industry were associated with the opposition party, the National Party, such as Ricardo Maduro and supported candidates in opposition to Turcio. The politics surrounding aquaculture development became increasingly important in the period between 1988 and 1998 as politicians actually began to raise the issues in the political arena in order to win votes.

In November 1988, in his capacity as an extension agent with UNAH, Jorge Varela organised a General Assembly with nearly 2,000 fishermen to discuss fisheries-related issues and the expansion of aquaculture. Most of the fishermen were either involved in the capture of wild postlarvaes for the industry or fished for other species which some thought were being affected due to by-catch (Varela, 2006). The meeting of 2,000 fishermen was one of the largest convened since the commercial shrimp farming commenced and was an attempt to foster opposition against the powers behind aquaculture development and its subsequent impacts.

This assembly directly threatened the shrimp industry by organising fishermen to discuss the issues Varela had raised. Politicians and military leaders investing in the industry were inclined to stop the formation of this group. Varela's superiors at UNAH approached him shortly thereafter and explained that a number of powerful actors with interests in the region had approached the university and requested that he cease his association with the fishermen in the GOF, which was consistent with past state practices that opposed organising (e.g. unions). He would therefore be reassigned to Tegucigalpa away from the south. Varela resigned, and for nearly a year worked directly with the fishermen to consolidate CODDEFFAGOLF (Varela, 2006a).

FIGURE 7.6: GODDEFFAGOLF's JUNTA DIRECTIVA
Source: CODDEFFAGOLF

Varela made it clear to the group of fishermen that the inclusion of 'development' in CODDEFFAGOLF's name was vital, given the issue's importance in the south. The organisation created a space within the public arena for parts of the artisan fishing community to articulate their concerns to the general public and the government of Honduras. CODDEFFAGOLF's governance structure included a central Board of Directors that consisted of the founding members and smaller governing bodies at the community level, represented by an elected body with the President at the head of the organisation in each community. CODDEFFAGOLF attracted international support and eventually became the most powerful opposition to the shrimp industry (Varela, 2006a).

The early neglect of the potential impacts of aquaculture production on the landscape and local coastal communities thus played a role in the emergence of the conflicts surrounding access to and control over the wetlands, as new social actors contested the actions of the state and the industry on several grounds. The following chapters focus on the emergence of CODDEFFAGOLF as a powerful actor and explain how it was able to link up successfully with international actors who shared common discourses and interests in relation to the protection of the mangrove wetlands, leading to actions against the shrimp industry in southern Honduras and the formation of new transnational environmental political networks.

Summary

By perpetuating the twin discourses of economic development and food security, the state successfully justified and legitimated their intervention into the wetlands of southern Honduras in order to promote aquaculture development. In turn, the actions taken to promote NTAX, which were directed towards promoting economic development and food security, had unintended consequences for local social actors, such as groups within the fishing community, as the state and private sector's actions contributed to their marginalisation. This event subsequently led to the formation of CODDEFFAGOLF which began to contest the views perpetuated by the Honduran government, IFIs and the emerging aquaculture industry.

The government's use of economic development and food security to justify aquaculture development exemplifies how actors selected the discourses that they wished to set out, often in order to pursue specific actions related to their interests in southern Honduras. Representations of what 'needs' or ought to be done in the south were utilised to connote specific meanings around the promotion of shrimp aquaculture, such as, 'shrimp production will lead to economic development and create food security'.

Framing an outcome – economic development and food security – congruent to the interests of the local people in the region was a technique of power employed to influence local perceptions of what 'needed' to happen, regardless of whether or not anyone ever actually received any of the benefits associated with the actions that followed. However, when the actions that followed did not bring the hoped-for benefits for various groups of social actors, the conditions for contestation and conflict emerged. By creating specific representations or certain images of what has, is, or ought to happen in relation to the environment, the government and international institutions were able to successfully acquire support in relation to their interests, or justify and create legitimacy for their position in relation to aquaculture development on the coastal wetlands of the GOF. Although the actions were consistent with their interest in promoting aquaculture, the effects of the power exercised ultimately led to certain segments resisting future aquaculture development due to the socio-economic and environmental changes that followed.

The capitalist restructuring of the Honduran economy and the democratic transition that began in the early 1980s led to the emergence of social actors of all types pursuing their actions within the context of globalisation (Calderon *et al.*, 1992: 31). Later in the decade, new actors within civil society, such as the AHE and CODDEFFAGOLF, came to signify opposition to the future of neoliberalism in Honduras, as the socio-economic and environmental impacts of capitalist transformation became more visible. Increasingly, movements focused attention on labour issues, the right to participate in development processes, and environmental degradation (Peters, 2001: 122). In southern Honduras, these debates focused increasingly on aquaculture expansion, a material manifestation of neoliberal reform.

In Part III, I illustrate how organisations such as CODDEFFAGOLF came to symbolise 'development from below', grassroots development, or 'alternative development' that gained pace throughout Latin America in the later-1980s and throughout the 1990s and, in some cases, in direct opposition to neoliberal objectives. Most of the organisations resistant to neoliberal policies in the 1990s blamed economic policies and external development interventions for environmental degradation and socio-economic impacts on subsistence livelihoods.

The AHE and CODDEFFAGOLF, allied organisations, and government officials felt that the LMDSA and the Structural Ordinance Law passed in the early-1990s would lead to repetition of the mistakes associated with past agrarian reform processes, including further environmental degradation. Thus the perceived impacts associated with the conversion of the wetlands became one of CODDEFFAGOLF's central claims against the state and the industry throughout the 1990s and is illustrated in the following chapters.

Finally, I illustrate in Part III that CODDEFFAGOLF's goal, like that of other NSMs in Latin America, was to place politics at the centre of the debate and 'politicise' environmental conservation and poverty alleviation through alternative development strategies, with the goal of illuminating the difference of this approach in contrast to the approaches of the IDB and the World Bank (Van der Borgh, 1995: 284). The period between 1988 and 1998 analysed in Part III focuses on the debates central to the conflict that arose in relation to aquaculture. I accomplish this by directing attention to the actors behind the creation of neoliberal futures while thoroughly addressing the rise of contestation against them. Chapter 8 begins by addressing CODDEFFAGOLF's three main claims in the early part of this period, the tactics they used to advance their claims, and the government's response.

Part III

Neoliberal Futures and the Rise of Contestation 1988–1998

Chapter 8
The Rise of Contestation to Shrimp Aquaculture

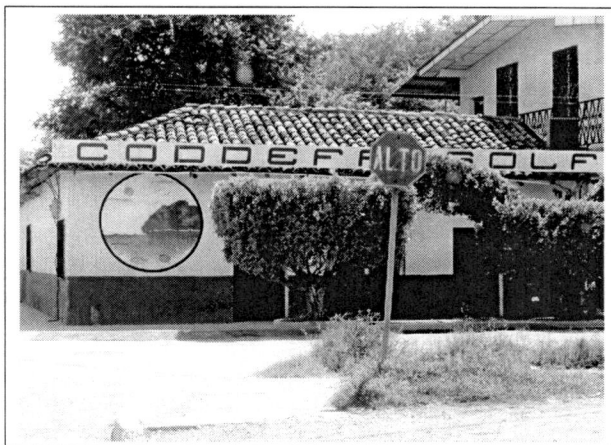

FIGURE 8.1: CODDEFFAGOLF HEADQUARTERS IN SAN LORENZO, SOUTHERN HONDURAS

Source: Wilburn, 2005

Introduction

Here, I discuss the growth of contestation to shrimp farming, which intensified between 1988 and 1998. I begin by outlining CODDEFFAGOLF's platform, provide an overview of its characteristics, and explain the actions taken by the organisation to resist future expansion. CODDEFFAGOLF's actions throughout the 1990s symbolised opposition to neoliberal economic policies and resistance to the socio-economic and environmental impacts attributed to aquaculture promotion due to divisions created in the artisan fisheries sector as a result of the introduction of aquaculture (Varela, 2006b). In this chapter I illustrate how CODDEFFAGOLF was able to link the debates regarding mangrove degradation successfully with the wider environmental movement.

Two tactics were pursued to bring political attention to CODDEFFAGOLF's claims against industry expansion. First, CODDEFFAGOLF focused more national and international public attention to the various impacts on the mangrove ecosystem of the GOF by taking direct action against shrimp-farming operations. Second, it sought financial resources that could support its campaign and activities in the south by linking more closely with national and international actors that could advance their three claims more successfully. As a result of local action, community leaders aligned with CODDEFFAGOLF and the environmental movement successfully increased their power and agency to make demands upon the state to address their claims. Thus, external links were important to CODDEFFAGOLF as it sought to solidify its collective power to focus attention to the socio-economic and environmental issues. These early ties represented the initial emergence of transnational environmental political networks that further politicised the issues and are discussed more extensively in Chapter 9.

As the election cycle commenced in 1988, the AHE, CODDEFFAGOLF, and several other environmental organisations concerned about the possible impacts of the macroeconomic and political changes taking place, contested more aggressively the neoliberal futures unfolding as a result of the actions of the state, IFIs and private industry. Subsequently, the Honduran environmental movement consolidated the position of the environment on the political agenda. In the south, the issues associated with aquaculture expansion brought these issues to the fore as their concerns were effectively politicised. National and global environmental organisations highlighted the impacts of various export-related activities, the latter influencing the former to take action in conjunction with CODDEFFAGOLF and the AHE.

Both organisations were able to place the environment on the political agenda, while USAID and ANDAH continued to emphasise both food security and economic development as the rationalisation for expansion and growth of the sector. Although the effect of capitalist transformation on the labour market within the fisheries sector was more important in the debate, it was the politicisation of mangrove deforestation that gained political attention for the Presidential candidates in 1989 and, eventually, each placed the environment on their campaign agendas.

The Rise of Contestation - CODDEFFAGOLF

By 1990, CODDEFFAGOLF claimed to represent 5,000 people dependent on fishing in more than 13 poor coastal communities. CODDEFFAGOLF immediately obtained support from WWF due to Varela's collaborative work with COHDEFOR to prepare the first coastal management plan for the GOF. Their links to WWF and other organisations such as

Friends of the Earth (FOE) and Greenpeace provided an avenue to acquire additional support to advance their claims against the government and the industry. In order to bring political attention to the issues they successfully highlighted the discourse on mangrove degradation strategically.

Although the AHE and CODDEFFAGOLF continued to work together in the late 1980s and early 1990s, their target audiences were different. To some degree, the AHE was representative of the wider-Honduran environmental movement and worked on a national level to change policy in relation to a number of environment and development issues. However, the AHE continued to interact with CODDEFFAGOLF and smaller environmental organisations throughout the country that emerged throughout the 1980s.

The AHE's campaign in the GOF continued to focus on the impacts of salt production and local use of mangrove wood for fuel and construction. Periodically, it worked with the government to assess the issues associated with the growing shrimp industry. The AHE's efforts continued until accusations of corruption and embezzlement[50] saw it begin to dissolve in the early-1990s, positioning CODDEFFAGOLF to take the lead on environmental issues in the south (Varela, 2006a).

The actions of both organisations in the late 1980s and early 1990s led the government to take specific actions to address their concerns by increasingly pressuring local and national politicians in an attempt to hold them accountable. The intention was to politicise the issues to bring attention to mangrove degradation and affects on local livelihood's due to USAID's and the Honduran government's aggressive promotion of aquaculture expansion. Ultimately, their efforts resulted in the creation of the first wildlife areas and refuges in the GOF around the region of San Bernardo, where most early aquaculture production commenced.

---

[50] The founder of the AHE, Jaime Gustillo, left the organisation around 1987 and was succeeded by Rigoberto Romero and then Roberto Vallejo, both accused of corruption, and the latter of embezzlement. They were succeeded by the final Director, Elena Fullerton, who effectively shut down the organisation in the early 1990s after she fled to La Ceiba with the AHE's assets and finances before leaving the country altogether. As a result, the AHE quickly disappears from any literature after 1993.

Politics of Aquaculture Expansion and the Politicisation of Mangrove Deforestation

The continuing mangrove deforestation led to attention being focused on the management of forest resources. The AHE and others put these issues at the centre of the environmental debate, and got them on the political agenda of the four dominant political parties. Consequently, the environment became a priority by 1988. The following meetings and debates were indicative of increased attention on environment and development issues as neoliberal transformation was taking place.

Local Liberal Party members backed the AHE and a number of local actors contesting the degradation of mangrove forests. One of the most vocal politicians during this period was Liberal Diputado Antonio Ortez Turcios. He opposed the expansion of shrimp farming and had raised awareness of several issues affecting the south – including mangrove deforestation – at the National Congress. Turcios linked the impacts on the wetlands to the shrimp industry and local use of the resources. He argued that the government needed to take measures to reverse the trends of the last twenty years, as local people had indiscriminately destroyed the wetlands. Furthermore, he claimed, the problem was being exacerbated by the expansion of shrimp aquaculture (*La Tribuna*, 9 March 1988).

Despite the intentions, aquaculture did not promote food security in southern Honduras. Turcios argued that over the previous decade the region had suffered from a decline in the production of basic food items such as corn and beans – staples of the Honduran diet – and that this was contributing to hunger in the south. During this time, shrimp aquaculture had expanded, but most shrimp produced were exported, providing very few economic and food security-related benefits for local people, Turcios suggested (*La Tribuna*, 15 April 1988). We can conclude that the relationship between economic development and food security discourses that was used to justify aquaculture development was counter to the reality experienced, which demonstrated the inconsistency between state discourses and local outcomes.

To address environment and development issues in the south, in April 1988, Turcios and others sponsored legislation to create the Commission for the Development of the Southern Region (CODESUR) (*La Tribuna*, 15 April 1988). In conjunction with the AHE and CODDEFFAGOLF, the government brought together numerous organisations to discuss the issues; they participated under the banner of the Federation of College and

University Professionals of Honduras (FECOPRUH). The group articulated the problems affecting the southern region, and proposed solutions to the 'crises' of unemployment and food scarcity in the south. The formation of FECOPRUH was indicative of the emergence of a multitude of civil society actors organising in order to elevate their interests in the nascent democracy.

The first meeting of FECOPRUH was held in Choluteca on 10 and 11 March 1988, the second the following year. Over 26 organisations participated in the meetings, including representatives of the rural poor, the labour movement, magistrates, students, fishermen and municipalities.

Government Response

The effectiveness of the group in gaining political attention was apparent when the four main political parties participated in the discussions. The political parties'[51] delegates had agreed with FECOPRUH that the 'crisis' in the south needed to be addressed. Calling for the passing of a number of resolutions in the National Congress, they began to develop an appropriate legal framework to address environmental, poverty, and economic issues, as well as to provide state agencies with the power and legitimacy to act more effectively in the region (*La Tribuna*, 15 April 1988). The laws and policies that emerged later were aligned with the neoliberal reform strategy already under way.

USAID, the government, IFIs, and the newly created institutions promoting NTAX, continually emphasised the need to create jobs in the region to address the unemployment level, which stood at over 40 per cent. Collectively, they continued to argue that the industry was needed to generate foreign exchange revenue to assist with the repayment of Honduras' over $3 billion of foreign debt. Their claims remained consistent with their discourse to emphasise economic development and food security (*Newsmagazine of Friends of the Earth*, 1990: 12-13).

During the meeting sponsored by the government, the World Bank, IDB, USAID, and the AHE worked in conjunction with the SRN and the Ministry of Public Health to develop an action plan for the south that could be translated into legislation. CODESUR was the result and designed to link enterprise in the south to central authorities, consistent with

---

[51] The parties whose representatives attended the 1988 meeting were the Party of Innovation and Unity (PINU), Enrique Aguilar Paz; Christian Democratic Party, Efrain Diaz Arrivillaga, and representatives from the Liberal Party, Guillermo Sevilla; and National Party, Roberto Martinez Lozano.

neoliberal decentralisation efforts, and was connected to the neoliberal-oriented NDP. CODESUR was expected to work with the existing governing bodies in each of the southern departments, Choluteca and Valle, the head of the Unit for Regional Planning of SECPLAN, and the heads of regional state institutions with jurisdiction in the south (*La Tribuna*, 15 April 1988).

Turcios was given lead authority over the commission, which augmented his power to represent the interests of the people in the south, illustrating the strengthening of civil society. However, to oversee the agreed-upon economic development strategy for the region, he was placed in an awkward position. The strategy was consistent with neoliberal reform and included promoting aquaculture development as outlined in the NDP. Turcios had previously articulated opposition towards aquaculture. In exchange for the position to head CODESUR, he agreed to support the agenda and indicated that he would vote in favour of the resolutions being considered for passage in the National Congress (*La Tribuna*, 15 April 1988).

At the second meeting held in April 1989, civil society actors agreed on several points and suggested that the government take action on each. Of these, the first priority was preventing mangrove degradation, in line with the marine and coastal resource management plan proposed by CODDEFFAGOLF in conjunction with COHDEFOR on 25 November 1988 (*El Tiempo*, 3 April 1989). FECOPRUH issued a joint statement in Choluteca in March and published the position in *El Tiempo* in April 1989, exemplifying their efforts to provide information to the public.

Following the meeting, all four candidates released statements claiming that conservation should be a priority, signaling the government's intention to address the issue. However, the boldest objectives were put forth by Diaz Arrivillaga, the Christian Democratic Party member, who argued, along with the UNDP, for the creation and approval of the first General Environmental Law and the creation of a National Environmental Institute under the President of the Republic, nascent forms of environmental governance (*La Tribuna*, 3 June 1989).

FECOPRUH's connections to CODDEFFAGOLF and the AHE illustrate the diversity of actors involved in politicising environment and development issues in the context of economic transformation in southern Honduras. They also highlight the importance of assessing the links of movements such as CODDEFFAGOLF with other actors in civil society. The environmental issues addressed by these actors symbolised the effects of past and present transformations, leading to increased attention on the management of forested areas. In response to the new debate, the government convened the first meeting for the management of forested areas in 1989.

First Meeting for the Management of Forested Areas 1989

The Honduran Government called the meeting in Jicaro Galan, Valle, from 30 January to 3 February 1989. In the south, an estimated 48 participants released a statement following the meeting. The participants represented 14 different institutions, including government representatives, individuals from the private sector and four international organisations. The published statement recognised that there were legal and institutional conflicts that affected the management of forested areas (*La Tribuna*, 17 February 1989). The inclusion of language recognising institutional weaknesses within the government was a triumph for CODDEFFAGOLF and the Honduran environmental movement, as the statement legitimised issues that they had been raising regarding poor management of forest resources.

The statement recognised that the GOF was affected by the degradation of natural resources within designated protected forest zones, the cause of which was blamed on competing and powerful interests in the region that had a lack of knowledge of protected forest zones and prevented the government from taking action. The political attention led the signatories to encourage the government to create an institution to manage the protected forest zones, take responsibility for the development of legal aspects of their management, and consolidate the system to address the issue of mangrove deforestation (*La Tribuna*, 17 February 1989). It was, however, more than a decade before a management regime was actually approved.

The meeting called on the government, the private sector, the armed forces and the general public to participate in the conservation of protected zones, and was one of the first times that co-management was proposed as an option. They also encouraged the Honduran government to consider participating in international agreements for the protection and conservation of the country's natural resources including CITES, World Heritage Sites, and the RAMSAR Convention on the Conservation of Wetlands.

These aforementioned institutions were visible expressions of an increased attention towards global environmental issues, starting with the UN's Conference on the Human

Environment held in Stockholm, Sweden in 1972 and intensifying after the Brundtland Commission Report of 1987, *Our Common Future*. Civil society actors also began to call on their governments to support these international environmental regimes. State institutions were encouraged by the participants to coordinate a strategy to conserve protected forest zones, since different agencies had overlapping authorities in relation to the wetlands. These agencies were also encouraged to work more closely with NGOs to support the management in alignment with Honduran national priorities directed towards neoliberal reform (*La Tribuna*, 17 February 1989).

Due to the issues raised in relation to aquaculture and fisheries resources, meetings with civil society representatives and the private sector were also initiated to revise the 1959 Fisheries Law. Here, the government agreed that it needed to address a broader set of issues and begin consultation to discuss the management and control of human activities in the region. The impacts of economic transformation appeared to be recognised, as well as the need to reform institutions to be able to address the impacts effectively. Civil society organisations, such as CODDEFFAGOLF, and the private sector were invited to discuss their interests in conjunction with artisan fishermen, the shrimp industry, salt producers and other actors using mangrove wood. The goal was to reach agreement on controlling the use of the resources (*La Tribuna*, 15 March 1989). These efforts culminated in the passage of a declaration in 1990 that created the first wildlife areas and refuges in the south around the region of San Bernardo and Punta Condega.

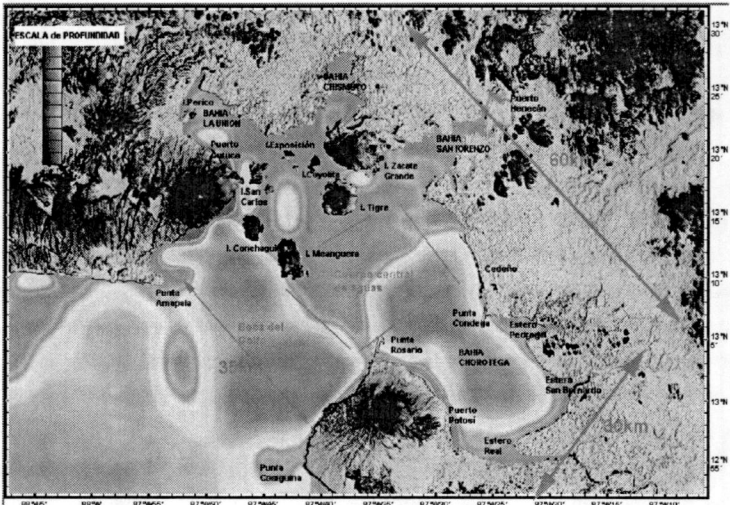

State Territoriality: Declaration of Wildlife Areas and Refuges inthe Gulf of Fonseca 1990

As the government was confronted by civil society actors, it was increasingly pressured to demarcate clearly the areas in which various human activities could take place within the wetlands. Due to CODDEFFAGOLF and AHE pressure, the government declared some of the areas as protected, adding credibility and legitimacy to their campaigns, leading to further international support and influencing the direction of neoliberal futures in the south. In terms of a technique of power, territoriality is an efficient way of maintaining centralised control over peripheral spaces by defining them in terms of state and local interests. The state's power in relation to a space can be strengthened through processes of territoriality by (1) classification of a space; (2) creating boundaries, such as a wildlife area or refuge, and (3) enforcing the boundaries of that space (Johnston *et al.*, 2000: 824). Consequently, the government's response demonstrated an increase in state territoriality of the physical space around the GOF as it sought to order the classification of the wetland's use.

As neoliberal policies were implemented more assertively in southern Honduras, the private sector acquired more power in relation to the coastal wetlands. The state's interests in promoting economic development had caused it to disregard the environmental impacts associated with various activities such as aquaculture. The repercussions of this early neglect became visible as resistance and contestation to the state and private-industry practices emerged, with the overall objective of influencing their future actions in relation to the space around the GOF.

As environment and development issues grabbed attention, there was a congruent rise in state territoriality of the physical space of the GOF. It arose in the form of protected areas and other management regimes in the south alongside the rural coastal poor's diminished control of wetland resources. Although physical spaces were demarcated for specific uses, the state often failed to comply with its own rules, leaving the wildlife areas and refuges as empty promises. Consequently, the disconnection between the definition of these newly created spaces and their actual use led CODDEFFAGOLF to deploy more assertive strategies to contest the actions of the state. CODDEFFAGOLF continued to contest the socio-economic and environmental impacts as it sought to acquire greater access to and control over these spaces and seek government and private sector compliance within the established legal framework. In some cases, the organisation took direct control over some physical spaces through protest actions or enforcement of their accord.

As aquaculture continued to expand and pressure for its limitation increased, the government made conciliatory gestures through the development of new legislation to protect the region's resources. By December 1990, SRN took action and released the Declaration of Wildlife Areas and Refuges in the GOF through Resolution 041-90, signed on 11 October 1990. Passing the resolution was consistent with RENARE's new position to protect the mangrove ecosystem in the south, and represented state territoriality of the region and recognition of human impacts on the wetlands.

The government declaration overseen by RENARE stated that shrimp-farming businesses which had received concessions to establish on lands in the GOF would be inspected by the authorities. The purpose was to ensure that they were within the legal boundaries for their activities. At the time, it was estimated that shrimp farms covered approximately 5,200 ha, nearly 2,000 ha more than in 1988 when CODDEFFAGOLF was formed to stop expansion. It was estimated that shrimp farming could cover nearly 20,000 ha by 1995 if it continued at the current rate (*Newsmagazine of the Friends of the Earth*, 1990: 12/13). RENARE gave existing businesses 90 days from 4 October 1990 to establish the official limits of their operations in accordance with the concessions given. Furthermore, no new operations would be permitted in the areas designated as wildlife refuges.

The resolution established wildlife refuges and protected areas at La Alemania, Guapinol, El Quebrachal, Montecristo, El Tionostal, and Jicarito del Golfo de Fonseca, and made infrastructure development illegal in these areas. Each area was selected due to its proximity to the emerging shrimp industry, some of which were to become central battlegrounds as CODDEFFAGOLF defended their protected status. In the law, artisan fishermen were permitted to continue using them as fishing grounds, and the state was permitted to conduct research as necessary. The resolution was signed by German Zelaya Meza, Director of RENARE, and Mario Nufio Gamero, Minister of Natural Resources on behalf of President Rafael Leonardo Callejas (CODDEFFAGOLF *Boletin Informativo* No. 2, Ultima Hora, 1990: 8).

Summary

Within two years of CODDEFFAGOLF's formation, it successfully politicised the conflicts over access to and control of the mangrove wetlands in southern Honduras. Three objectives were achieved as a result of the successful politicisation of its claims against the state and the aquaculture industry: first, the organisation placed the environmental debate in the south on the Presidential candidates' agendas; second, greater attention was drawn to the socio-economic and environmental issues surrounding aquaculture expansion; and third, it led to the state taking actions to assess the impacts on mangrove forests, subsequently leading to the first meeting on forested areas and the creation of the first wildlife areas and refuges in the south. The latter was discussed in terms of state territoriality in order to illustrate that, although the movement's objectives were achieved, access and control over the space still lay in the hands of the state. Lack of enforcement of the protected areas demonstrated that private interests were the real winner in terms of access.

In Chapter 9, I emphasise how and why CODDEFFAGOLF's claims and actions shifted to the aquaculture industry's impacts on fishermen's livelihoods as two competing images of what *ought* to happen in relation to the use, conservation, and management of the wetlands played out. Three claims against the industry continued to be emphasised by CODDEFFAGOLF and various tactics used to advance them. First, the construction of shrimp ponds was leading to mangrove deforestation. Second, aquaculture expansion led to a decline in the fisheries due to the extensive by-catch associated with the capture of wild postlarvae, adversely affecting fishermen's catch. Finally, the industry's expansion was diminishing fishermen's access to their traditional fishing grounds in the seasonal lagoons[52] and estuaries. These three claims were identified as the most dominant discursive positions articulated by the organisation throughout the period.

---

[52] Seasonal lagoons form during the rainy season when the land is inundated as a result of natural flooding and are highly productive areas for fishermen. However, they are also an ideal location to construct shrimp ponds, as the water can flow naturally into the area when the tide comes in or the water could easily be pumped from the estuary into the ponds, illustrating the *'natural agency'* of the environment.

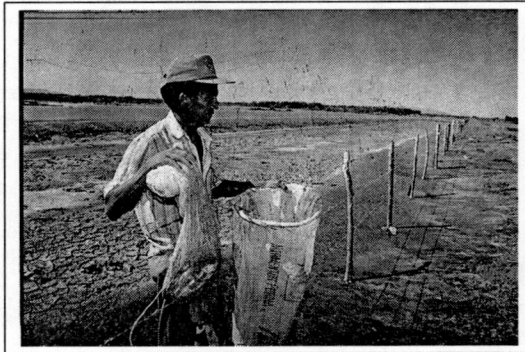

FIGURE 8.2: DIMINISHED ACCESS TO TRADITIONAL FISHING GROUNDS
Source: CODDEFFAGOLF

CODDEFFAGOLF successfully gained international support from a number of NGOs, which further facilitated the development of its strategies and tactics to contest aquaculture expansion. Consequently, the formation of transnational environmental political networks is central to Chapter 9. At the local level, CODDEFFAGOLF continued to focus on artisan fishermen's livelihoods as access to the seasonal lagoons was increasingly diminished. Furthermore, as competition within the local fisheries market increased due to the introduction of aquaculture, CODDEFFAGOLF focused on articulating their constituents' interests widely. Combined with their international campaign, which emphasised mangrove deforestation, the issues gained attention at a variety of levels beyond the local and national as CODDEFFAGOLF further politicised the issues.

Chapter 9

CODDEFFAGOLF:
The Formation of Transnational Environmental
Political Networks

Introduction

In Chapter 9, I argue that conflict over natural resources in the GOF occurred between advocates of neoliberal policy reforms, and those who adopted environmental ideals as a framework to contest the effects of these reforms. Increasingly, CODDEFFAGOLF advocated the protection of the mangrove wetlands and resisted neoliberal advocates on behalf of a faction of the artisan fishing community's interests. By highlighting mangrove deforestation and the impacts of aquaculture on the wetlands, CODDEFFAGOLF also aligned its agenda with a number of international ENGOs that sought to prevent deforestation and unsustainable agricultural practices. The rise of contestation to aquaculture was congruent with the emergence and strengthening of the global environmental movement.

I explain the various strategies and tactics used by CODDEFFAGOLF, the trends in the global environmental movement and why it was advantageous for local and international movements to link. I highlight the characteristics of Friends of the Earth (FOE) and WWF to illustrate the differences within the international environmental movement and how each campaigned for the use of different tactics for the same ends. Combined, each increased the power of the others to protect the mangrove wetlands. Examples of direct action, environmental education, the strengthening of media relations, the politicisation of violence, and the results of the campaigns are addressed. Conversely, I illustrate how the actions of these nascent networks led the government to establish new institutions such as the National Commission for Environment and Development (CONAMA) to address environmental issues following neoliberal reform efforts.

CODDEFFAGOLF Further Politicises Aquaculture's Impacts

The story that unfolds narrates the clash of two competing images of the future socio-economic and environmental reality in relation to the wetlands. Aquaculture development and environmental conservation became the two dominant issues as numerous actors on both sides of the argument organised to defend their positions strategically. In turn, the social, economic and environmental issues were increasingly politicised within the public arena as actors released statements and took actions consistent with the 'reality' they were attempting to produce.

The passage of the Declaration of Wildlife Areas and Refuges in 1990 and the threat of future expansion, estimated to be 1,000 ha per annum, signaled a shift in CODDEFFAGOLF's strategy. The organisation increasingly directed attention to the shrimp industry and mangrove deforestation. In the early 1990s, CODDEFFAGOLF's discourse focused directly on the impacts of aquaculture development on fishermen's livelihoods. In conjunction with multiple ENGOs they fought for conservation of natural resources. This discursive shift polarised attitudes towards shrimp farming in the south. Conversely, USAID and the government consistently emphasised economic development and food security. The division between the two camps increased throughout the 1990s. The culmination of the conflict encouraged CODDEFFAGOLF to strengthen its position through the formation of transnational environmental political networks and, subsequently, elevate its campaign from the local to the global level.

There were two factors that facilitated CODDEFFAGOLF's ability to strengthen links to international actors. First, the global environmental movement had achieved a higher profile during the 1980s. However, the success of international ENGOs was often dependent on links to local environmental movements that had emerged world-wide. The international ENGOs had the financing, members and political visibility to raise local concerns. Local environmental movements had concrete issues and local knowledge of the situation and provided ENGOs with the empirical evidence and substance needed to politicise the issues effectively. In other words, each needed the other and added legitimacy to their campaigns.

Second, deforestation in the Central American region was on the agenda of most of the international ENGOs, such as WWF, FOE and Greenpeace, during the 1980s. Subsequently, mangrove deforestation fitted within the thematic areas of their campaigns, aligning their interests and, subsequently, their discursive positions. I demonstrate that alignment occurred concurrently, as each adopted aspects of the others' discourses to strengthen their positions. New tactics and strategies were also devised and concerted as they sought to politicise the issues further. The result was the early formation of transnational environmental political networks focused on stopping aquaculture expansion.

CODDEFFAGOLF's Strategies and Tactics and the Formation of Transnational Environmental Political Networks

CODDEFFAGOLF was increasingly concerned that the Honduran government would seek to relocate local populations where aquaculture expansion was planned as they had under previous agrarian reform efforts. Inhabitants had no title to the coastal lowlands, it was technically state-owned land, and thus they could be forced to leave if the government chose to remove them. With the backing of FOE and WWF, CODDEFFAGOLF decided to intervene by taking direct action against the industry, while pursuing an outreach campaign to highlight the fishermen's plight and the impacts on the mangrove wetlands. Three strategies were enacted: non-violent protests were held; a media campaign was developed; and plans to develop an environmental education programme (EEP) commenced, targeting teachers and children. FOE supported the direct action and media campaign while WWF supported the EEP. The overarching goal was to persuade the government to agree on a management plan, halt industry expansion, and reduce the impacts resulting from by-catch that were linked to the capture of wild postlarvae.

The links between CODDEFFAGOLF and a number of ENGOs were consistent with trends within the global environmental movement that had begun in the 1970s. The environmental movements in a number of industrialised countries began to connect with local social, environmental and indigenous movements in the developing world, and by the 1980s increasingly gained international attention.

As local environmental organisations sprang up in Honduras, they increasingly relied on relations to larger international ENGOs, such as WWF and FOE[53], to acquire funding and generate international support. WWF's transition from a small organisation concentrating on problems such as endangered species and habitat destruction, into a multi-faceted international institution in the 1980s was indicative of the globalising trend within the environmental movement. Its strategy evolved further in the 1990s and

---

[53] It is important to note that although FOE and WWF are both considered ENGOs, their ideological views to environmentalism are neither consistent nor homogenous. At the international level they pursue strategies and tactics that vary.

aimed to decentralise their decision-making process and increase cooperation with local organisations, as did FOE. WWF's work with CODDEFFAGOLF was a partial fulfillment of this strategy in the Central American region (www.wwf.org). It is worth mentioning since there is an obvious correlation between the strengthening of the global environmental movement as capitalist transformation and globalising processes were becoming more rapid, particularly after the end of the Cold War.

FOE provided a small grant to support CODDEFFAGOLF's campaign. Founded in 1971, FOE characterises itself as the world's largest grassroots environmental network and its motto is: mobilise, resist, and change[54]. Their main strategy is to politicise the issues in order to influence government policies, favouring changes aligned with social and environmental justice. In order to accomplish their objectives, strengthening ties in the developing world were pertinent (www.foei.org).

For FOE, links with CODDEFFAGOLF were advantageous as it allowed them to broaden their campaign against deforestation in Central America and highlight factors believed to contribute to mangrove deforestation in the region. FOE began using CODDEFFAGOLF's campaign in its April/May 1990 newsletter as one example of its efforts in the Central American region. It attacked the environmental and social impacts of US policy in Central America, especially as the debates on deforestation[55] came to the fore in the 1980s. In their newsletter, an article was published called *Whale-Sized Fight Over Shrimp* (*FOE*, 1990) in which the organisation put forth CODDEFFAGOLF's claims (FOE, 1990: 12/13). Such publications were used by ENGOs to disseminate information during the later 1980s as it was the primary means to influence public perception and, therefore, the discursive debates surrounding mangrove deforestation.

In order to politicise environmental issues and generate media attention, FOE backed direct action by CODDEFFAGOLF in January 1990 when supporters occupied a 350 ha area in which the mangrove forests were going to be cleared for the construction of a shrimp pond (FOE, 1990: 13). Future non-violent protests were planned due to the organisation's success in gaining media attention through this method[56].

---

[54] Subsequently, in order to mobilise, FOE has traditionally been highly dependent on links with local environmental groups to target its priority issues, one of which was deforestation in the 1980s. To accomplish its objectives, FOE worked with various groups to resist government and corporate actions that resulted in socio-economic and environmental impacts.

[55] For more on deforestation in Central America see Weinberg (1991).

[56] I would like to note that CODDEFFAGOLF had held at least seven similar protests in the Departments of Choluteca and Valle since its founding in 1988. If someone wishes to research each individual protest, I will gladly place them in contact with Jorge Varela. However, my thesis

Environmental Education and Outreach

In conjunction with WWF, CODDEFFAGOLF increased its outreach and environmental education efforts and began targeting teachers, journalists, the armed forces, and individuals from the private sector in the south. Efforts to 'educate' the public should be seen not only as a technique of power to build support for the campaign, but as a wise strategy to obtain access to the public in order to influence the discursive terrain by seeking to change public perceptions.[57]

The outreach campaign culminated in July 1990, with training called *El Maestro y Su Aporte a la Conservación y al Desarrollo*. It was held in both Nacoame and Alianza and had around 90 participants. CODDEFFAGOLF explicitly stated that its purpose was to begin changing attitudes (discourses) and behaviours (actions) in relation to wetland conservation. Individuals from RENARE, UNAH and CODDEFFAGOLF provided training that covered basic ecology, problems affecting the flora and fauna of the GOF, consequences of poor management of the watersheds, and an overview of environmental problems at the regional, national, and international levels as they collectively sought to influence public perceptions (CODDEFFAGOLF 1990a: 6).

CODDEFFAGOLF and its partners sought to influence public opinion to gain more momentum for their campaign within the political public arena. The strategy was motivated by their ultimate goal to obtain greater access to and control over the physical geographic space that they were trying to protect. As a result, the goals associated with specific material outcomes in relation to a physical space can coincide with efforts to gain influence discursively in order to build support for specific outcomes in relation to it.

A number of methods were pursued over the years to influence public perceptions on various scales. WWF and FOE disseminated information via newsletters and publications to internationalise the issue of mangrove deforestation in southern Honduras (CODDEFFAGOLF, 1991a: 5). WWF also supported more exchanges between local

does not provide a play-by-play account of each act of contestation. My intention is to reveal examples of various CODDEFFAGOLF strategies throughout the socio-historical trajectory of the conflict.
[57] Environmental education activities were funded by WWF, which had supported CODDEFFAGOLF's efforts since 1988, when it was founded. I do not analyse the effectiveness of these programmes, which is what would have to have been done to quantify changes in local discourses. My point was to illustrate that environmental education was a strategy of power used by the environmental movement to change people's ideas in relation to how the wetlands *ought* to be used, managed and conserved. Further research is needed to determine whether or not those efforts were effective and why, or why not.

movement leaders worldwide and invited Mauricio Alvarado, President of the Central Directive of CODDEFFAGOLF, to attend an international meeting on forested areas with the purpose of exchanging information and sharing experiences with other conservation professionals, increasingly internationalising the issues (CODDEFFAGOLF, 1991a: 3).

These early exchanges and experiences were particularly important for strengthening transnational environmental networks which enabled CODDEFFAGOLF to advance its objectives and acquire the necessary financial support. Furthermore, interaction between ENGOs worked towards discursive alignment as actors established strategies to contest environmental degradation worldwide. For example, WWF was able to showcase CODDEFFAGOLF's efforts as its own in alignment with its strategy. One of WWF's goals was to demonstrate that it was working closely with local actors while strengthening networks between organisations.

Throughout 1991, CODDEFFAGOLF expanded its network of regional and international contacts, leading to additional support from the Latin American Network for Tropical Forests (Red Latinoamericana de Bosques Tropicales LBT), International Tropical Timber Organisation (ITTO) and World Neighbours among others. CODDEFFAGOLF also hosted the Inter-American Foundation (IAF), an independent US institution funding grassroots development projects throughout Latin America with funds from US President Kennedy's Alliance for Progress. The IAF was also interested in drawing international attention to the problems in southern Honduras by making a television documentary focused on how local communities were organising to defend the natural environment. CODDEFFAGOLF was selected as a case study, providing additional means to get their message to the public (CODDEFFAGOLF, 1991a: 5). One means by which CODDEFFAGOLF was able to defend coastal resources more effectively was by scaling up the Honduran environmental movement nationally.

CODDEFFAGOLF Scales up the Honduran Environmental Movement Nationally

CODDEFFAGOLF sought to capitalise quickly on the international attention it had gained to build further support. On 15 June 1991, CODDEFFAGOLF began working with the wider environmental movement in Honduras to constitute the Federación de Organizaciones no Gubernamentales Ambientalistas de Honduras (Federation of Honduran Non-Governmental Environmental Organisations, FEDAMBIENTE). Whereas the government supported Fundación Vida, FEDAMBIENTE was established independently. Its principal objective was to consolidate the collective power of several NGOs and their members to fight powerful business interests they believed to be posing a threat to the country's natural resources. Initially, FEDAMBIENTE had 24 participating NGOs, nine of which acted as the central directive (http://www3.undp.org/rc/forums/mgr/sdnpca/msg01272.html).

FEDAMBIENTE was essentially an extension of CODDEFFAGOLF, yet expanded the power of a number of other environmental groups with a shared interest in conservation and development issues. The AHE was no longer considered representative of the Honduran environmental movement after it faced corruption charges following Jaime Gustillo's departure (Varela, 2006b). FEDAMBIENTE was founded to fill the role played by the US-funded AHE. Varela was appointed as its first President, which increased CODDEFFAGOLF's public profile. It also placed him in a more advantageous position to speak on behalf of the Honduran environmental movement, and increased his power to influence the wider debates taking place in the context of neoliberal transformation (CODDEFFAGOLF, 1991b: 7).

The Campaign's Outcomes

The collective efforts and tactics used were successful in convincing the government to discuss the development of a management plan for the GOF. On 9 May 1990, CODDEFFAGOLF called a roundtable conference to discuss the environmental issues affecting the south. Authorities with RENARE, ANDAH, FPX and USAID attended and agreed that there was a need to elaborate terms of reference for the creation of GOF management plan. The roundtable was one of the first large meetings that brought together actors with varying interests to discuss, more seriously the creation of a management plan for both environment and development issues in the south.

During the forum, the group agreed to set aside areas for different uses such as forest reserves, national parks and fishing areas. Particular attention focused on the areas around El Jicarito, close to Guapinol, a fishing village closely aligned with CODDEFFAGOLF and on the coast near Punta Condega. The area was near shrimp farms, and its protection was planned to benefit the environment, the fishing community and the industry. ANDAH had two reasons for supporting these efforts which aligned with their previous position; they argued that mangrove forests served as the primary habitat for the wild shrimp postlarvae needed to stock the ponds, and they acted as a natural source of protection for shrimp ponds against storm surges.

The result was the creation of the National Commission for Environment and Development, (Comision Nacional del Medio Ambiente y Desarrollo CONAMA) by President Callejas as a Technical Commission with civil society, government and private sector representatives (CODDEFFAGOLF 1990: 4). Jorge Varela was selected to be one of the representatives on the Commission, which agreed to meet in Choluteca, 29 and 30 September CODDEFFAGOLF 1990: 8). In preparation for the meeting, a number of Honduran environmental organisations were brought together to prioritise regional and national environmental issues to be advanced through CONAMA (CODDEFFAGOLF, 1990a: 6).

The second roundtable precipitated the creation of Fundación VIDA on 5 June 1990.[58] The organisation was a non-profit entity established to represent the Honduran environmental movement as a whole. CONAMA, Fundación VIDA, and Varela's position on the Commission gave CODDEFFAGOLF greater influence over national debates and provided the organisation with the space to continue scaling up environment and development issues relevant to their interests. Fundación VIDA also represented the successful scaling up of the Honduran environmental movement.

The creation of this organisation was important for three reasons. First, it signified the consolidation and strengthening of the Honduran environmental movement. Second, it led to more coordinated actions as the burgeoning environmental movement increasingly contested the impacts related to neoliberal reform throughout the entire Central American region. Third, it established a forum for collective action and prioritisation of the issues at the national level. In turn, the priorities could be advanced through CONOMA, effectively providing the movement with a more unified voice. Conversely, CONOMA represented the Callejas administration's need to address environment and development issues as it aggressively pursued neoliberal reform. The government's actions were also partially a response to the international community's efforts in preparation for the UN Conference on Environment and Development (UNCED) scheduled to occur in 1992.

---

[58] The agreed upon mission of the organisation is to contribute to the conservation and sustainable management of natural resources, to secure financing, administer and execute projects, and create strategies focused on environment and development issues at the national, regional and municipal levels (www.fundacionvida.org).

Creation of the National Commission for Environment and Development

The creation of CONAMA symbolised the transition toward new forms of environmental governance under neoliberal reform, as the interests of the environmental movement came into the consciousness of the neoliberal elite. The state sought to exert its authority in relation to the environment and Honduran society in the run-up to UNCED. Presided over by Franklin Bertrand Anduray, designated by the President of the Republic, CONAMA included representatives from more than 70 organisations and included ecologists, labourers, and government and private institutions (*La Tribuna*, 2 March 1992). A number of the organisations that exerted pressure on the government on environmental issues throughout the 1980s led the government to consider the passage of a General Environmental Law.

International inter-governmental organizations and ENGOs used their political power to lobby the government in support of the law. For example, the Honduran government was encouraged by the UNDP, under the guidance of Paolo Uberti, to pass the legislation since Honduras was one of the only countries in the region that did not yet have one. The UNDP also wanted the Honduran government to prepare for the UNCED scheduled to take place in Rio de Janeiro, Brasil in 1992. CONAMA's purpose was to provide civil society organisations with an opportunity to provide input into the creation of the General Environmental Law and the UNCED process by contributing to the completion of the '*Honduran National Report on Environment and Development*[59] which would be presented by the government during the conference (CODDEFFAGOLF, 1991b: 7). FOE, WWF, and other environmental organisations decided to pursue further actions to continue applying pressure on the Honduran government to adopt the General Environmental Law. Collectively, they internationalised the campaign, and worked to strengthen media relations and draw increased attention to the links between aquaculture expansion and mangrove degradation. CODDEFFAGOLF's strategies of resistance were increasingly diversified as they continued to resist the negative effects of the neoliberal reform agenda.

CODDEFFAGOLF Strengthens Media Relations

In addition to developing stronger links with national and international actors interested in the conservation and protection of the environment, CODDEFFAGOLF also successfully increased its contacts with the Honduran media to disseminate stories pertaining to its

---

[59] Translated from Spanish [Informe Nacional de Honduras sobre Medio Ambiente y Desarrollo]

cause. The organisation's efforts resulted in a strong relationship with Dario Guzman, a journalist working for *La Tribuna*. In August 1990, Guzman wrote an article entitled 'Shrimp Fever is Killing the Gulf of Fonseca: Shrimp Farmers Exploit Land that Exceeds the 10,000 ha Technically Recommended'. The article highlighted the claims made by CODDEFFAGOLF against the industry, publicly blaming it for diminishing access to the livelihoods of the fishermen it represented by expanding operations in the seasonal lagoons. CODDEFFAGOLF substantiated its claim by pointing to concessions granted to over 10,000 ha of land for shrimp pond construction around the Negro, Sampile, and Choluteca rivers along the estuaries where they flow into the GOF. However, the issuance of concessions did not mean that development had actually taken place.

In 1990, it was estimated that there were at least 60 large farms and about 30 small-to-medium-scale producers operating in the region (*La Tribuna*, 9 August 1990). At this point, the government had issued concessions to more than 30,000 ha of land. The amount conceded was 10,000 ha more than had originally been estimated as available for aquaculture development (*La Tribuna*, 9 August 1990). However, this number was inaccurate as it took into account the total number of hectares conceded by all agencies with authority to issue concessions. In some cases, agencies had been issuing overlapping concessions making these numbers appear higher. In other cases, a single agency had issued overlapping concessions unknowingly.[60] The government's previous efforts to encourage existing shrimp farms to delimit the boundaries for the areas that had been conceded was one step the government had already taken to begin determining where concessions overlapped.

CODDEFFAGOLF, Aquaculture and Mangrove Degradation

CODDEFFAGOLF and the AHE also claimed that nearly 1,000 ha of mangrove forests were being destroyed every year, affecting populations of international migratory birds, reptiles, mammals and marine species. This claim contradicted the industry's statements that shrimp farm expansion had taken place primarily on salt flats where mangrove forests were not prevalent, but where the seasonal lagoons formed during the rainy season. Tactically, the approach used to politicise the issues appeared to be based on linking two sets of apparently unrelated claims; illustrating how 'facts' are used selectively and strategically in a conflict scenario.

The industry immediately denied CODDEFFAGOLF's claims and stated that the mangroves were important for a diversity of flora and fauna, important to the industry,

---

[60] Personal interview with Leonel Guillen, San Lorenzo, Honduras 2004.

and future generations (*La Tribuna*, 9 August 1990). ANDAH's objective was to counter the discourse generated by CODDEFFAGOLF. The uncertainty over land availability for the production of shrimp and future aquaculture expansion led ANDAH and investors in the industry to demand once again that the government clearly demarcate the areas in which shrimp could be produced (*La Tribuna*, 9 August 1990).

By agreeing on these areas, the industry gained greater legitimacy for the location of its development and would also be able to place the onus on government officials responsible for designating those spaces if claims against the industry arose. Clearly demarcating the areas that had been conceded to different investors would also assist in preventing conflicts between shrimp businesses with overlapping claims to the same space.

In USAID's Environmental Profile 1989, the agency recognised that "the laws related to these resources are not abided by or are obstructed" and "the government has not imposed limits on the renting of lands, so it is the same to obtain 7,000 ha or 5,000 ha of land"[61] (*La Tribuna*, 9 August 1990). Of course, uncertainty reduced the security of the concessions issued, and made future investments and expansion less certain, increasing risk for private investors counter to the neoliberal agenda. This issue was USAID's concern. Consequently, the government, in conjunction with USAID, regional institutions and international organizations, responded to the concerns articulated by the environmental movement and the expanding shrimp industry. An effort was initiated to define the areas available for shrimp production to increase the security of the concessions issued legally and to identify individuals who were developing operations on land that had been unlawfully conceded. Title to the land, or at least legal arrangements around the land, was a core aspect of neoliberal agrarian reform, thus becoming a priority.

Summary

As the concerns of the AHE and CODDEFFAGOLF gained greater national and international attention, the issues of unclear boundaries for aquaculture expansion, the impacts of the use of drag-nets for the capture of post-larvae on fisheries resources, enclosure of common pool resources, and mangrove deforestation became increasingly important. However, there was a noticeable rise in claims linking the industry and

---

[61] Translated from Spanish: [las leyes que se relacionan con estos recursos no se cumplen o son mediatizadas" y "no ha existido, por parte del gobierno, limites al arrendamiento de tierras, ya que da lo mismo obtener siete o cinco mil hectáreas de tierra]

mangrove deforestation. Furthermore, CODDEFFAGOLF was able to successfully politicise the issues by linking the industry and mangrove deforestation. In conjunction with a number of international NGOs the dominant regional and international discourse around southern Honduras was the industry's impacts on the wetlands, illustrating the power of nascent transnational environmental political networks. In Chapter 10 I argue that the state responded to CODDEFFAGOLF's calls for action by seeking to revise the 1959 Fisheries Law leading to the creation of the General Directorate of Fishing and Aquaculture Law in 1991 and increased attention on addressing issues in the fisheries sector.

Chapter 10
Expansion of the Fisheries and Aquaculture Sectors

Introduction

In Chapter 10 I argue that the politicisation of the fisheries resource issues in the GOF increasingly led IFIs, the state and USAID to invest in the fisheries sector to acquire support within segments of the artisan fishing community while also developing the necessary labour force for future aquaculture expansion. I argue that the attention directed towards the artisan fishing community was a technique of power used to enroll fishermen in the neoliberal reform process. Consequently, legal changes that followed exacerbated the conflicts as the fisheries sector was transformed further.

Regional Capacity Building for Fisheries and Aquaculture Expansion

The changes in the rural labour market that led to rivalries between artisan fishermen prompted the government, the OLDEPESCA and other international organisations to focus more attention on the aquaculture and fisheries sector throughout the 1990s. The attention directed towards the artisan fishing community by 1991 focused on the needs of those marginalised from the early efforts to develop the sector. The opposition to aquaculture practices and its impacts was the impetus.

The past programmes supporting the artisan fishing community were largely directed toward developing an appropriate labour force for aquaculture expansion by training artisan fishermen to capture wild postlarvae, run hatcheries and sell fisheries products. They also worked to increase fisheries catch and strengthen the market to improve the region's fishermen's livelihoods. Vicente Borjas had stated that the fisheries sector was a principal priority for the Honduran government, because it generated employment and foreign currency for the national economy, primarily through aquaculture development (*La Tribuna*, 26 May 1989). The section of the artisan fishing community that did not receive the benefits associated with these earlier programmes contested industry practices. Consequently, new efforts to incorporate artisan fishermen into the wider Honduran economy began to focus on communities with the largest populations of fishermen and those that were most affected by the expanding industry including, Cedeño, San Lorenzo, El Tulito, and San Jeronimo (*La Tribuna*, 26 May 1989).

The attention directed towards the artisan fishing community was partially to allay fears related to future expansion in particular, the enclosure of the seasonal lagoons, and impacts on fisheries due to by-catch related to the capture of wild postlarvaes. By the end of 1990, RENARE commissioned a study to designate the areas where aquaculture expansion was to take place. German Zelaya, Director of RENARE, announced that the study was completed to identify areas for shrimp cultivation, fishing, tourism, and forestry conservation as the government attempted to assess multiple uses in the region as the conflicts between various interest groups became more visible (*La Tribuna*, 17 December 1990). Simultaneously, the government, USAID and several others convened the first Central American Aquaculture Symposium to bring together various actors on a biennial basis to address concerns, share experiences, and exchange information and research results with others involved in the industry throughout the region.

Use of Drag-Nets for the Capture of Wild Postlarvaes Banned

Efforts to create DIGEPESCA coincided with a ban on the use of drag-nets by larvae gatherers. It also coincided with government efforts to shift the industry's practices from using wild-caught to laboratory-produced postlarvae in the ponds. Both were aligned with neoliberal reform efforts as the first sought to reduce waste and inefficiency, and the second to reduce demand for rural labour by increasing productivity through technological improvements, thereby increasing market efficiency. Supporting the transition to laboratory-produced postlarvae would lead to vertical integration of the industry, provide legal justification for the shift of practices, and reduce the number of postlarvae purchased from those artisan fishermen who supplied the industry. However, the transition took some time and the industry continued to depend on wild postlarvae for several years.

DIGEPESCA was tasked with taking responsibility for the supervision and control of drag-net use to complement the January 1990 resolution passed by RENARE regulating their use. Since RENARE was dissolved by the creation of DIGEPESCA, the responsibility rested with this new agency. CODDEFFAGOLF had been claiming that these nets were affecting fisheries resources due to their small mesh sizes, thought to contribute to higher levels of by-catch (CODDEFFAGOLF, 1991c: 6).

The ban on drag-nets did not, however, stop the industry's use of wild postlarvae to stock its ponds; it only outlawed the fishermen's use of this specific type of net thought to be contributing to by-catch. As a result, fishing for postlarvae remained important to the industry. The changes associated with implementation of these laws were one of the main reasons for further divisions and conflicts within the artisan fishing community, between segments of the artisan fishing community, and between a segment of the artisan fishing community and the industry.

A number of artisan fishermen simply ignored the ban and continued using the nets. The fishermen represented by CODDEFFAGOLF were opposed to these fishermen. CODDEFFAGOLF sometimes took actions to prevent the use of the nets. Furthermore, the commercial shrimp industry continued to purchase postlarvae caught by fishermen using nets, which undermined the efficacy of the ban. Eulalio Alvarez, a fisherman in the region for over 25 years, raised concerns about the ban on drag-nets. He stated that the resolution was going to make life harder for smaller-scale fishermen in the GOF, who primarily used the banned nets (*La Tribuna*, 30 April 1991). His statement was indicative of the concerns associated with how fisheries reforms were affecting the sector.

At the time, the government lacked information regarding the size of local fishing communities, the types of equipment that they used, how they interacted with the fisheries population or who was fishing when, where, why, or how. This complicated efforts to train and educate fishermen. However, the resolution was intended primarily to target postlarvae collectors who relied on the drag-nets. CODDEFFAGOLF agreed to take the lead and train local larvae gatherers to raise public awareness of the new legislation and the effects on the fisheries, leading to four training courses for fishermen in April 1990 in San Lorenzo, Cedeño, El Tulito, and San Jeronimo, some of the communities affected most by the economic transformations taking place (CODDEFFAGOLF, 1991c: 6).

Legally, the fishermen represented by CODDEFFAGOLF attained their objective to ban the use of this net; CODDEFFAGOLF's interest in stopping industry expansion to protect the wetlands failed as it ensued, which also meant that their articulated goal to ensure artisan fishermen's access to their traditional fishing grounds in the estuaries and seasonal lagoons was compromised. The industry continued to expand, and efforts to consolidate and integrate various aspects of it were apparent during the first Central American Aquaculture Symposium held in 1991.

Central American Aquaculture Symposium

Government efforts to focus on fisheries and aquaculture-related resources culminated in the first Central American Aquaculture Symposium held 24–26 April 1991 in Tegucigalpa. Representatives from the SRN, President Callejas' administration, the aquaculture industry, the US, Central America governments and a number of other countries gathered to discuss current research associated with aquaculture development. The conference was organised by FPX and ANDAH, and included presentations, technical sessions, commercial exhibitions and tours of the shrimp farms.

At the time of the conference, it was estimated that there were 18 large farms operating in the region including GGMSB, Sifar de Honduras, Cultivos Marinos, Cultivos del Mar, Aguas Marinas Chismullo, Agua Especies in addition to a number of small and medium-scale operators that occupied 25–100 ha. The largest shrimp farms, especially GGMSB, were targeted by CODDEFFAGOLF. They opposed USAID's goal to support expansion. ANDAH argued that the aquaculture industry in Honduras had a competitive advantage in the global market and should continue to be supported (*La Tribuna*, 26 April 1991).

John Sanbrailo, USAID Director, was the keynote speaker during the symposium. He made three points. He emphasised USAID's role in aquaculture development and committed the agency to its future development through the provision of investment incentives to encourage private enterprises, with the objective of increasing foreign exchange earnings, a neoliberal objective. He announced a five-year project to further develop NTAX to back this claim. Furthermore, he reaffirmed support for ANDAH's objective to expand the industry by 1,000 ha per annum to achieve the goal of having 15,000 ha under production within the next five years. He estimated that the industry could generate over $100 million and provide 15,000 people with employment.

Finally, he supported the goal of transition from using wild-caught postlarvae to using laboratory-produced larvaes. He stated that USAID would provide appropriate technologies, technical assistance and training to achieve this goal. He claimed laboratory-produced postlarvae would secure a more consistent supply for the industry and reduce impacts on wild postlarvae in the estuaries. DIGEPESCA thus announced that it would begin work on a resolution to establish closed seasons for the capture of wild postlarvae to reduce impacts on the fisheries, resulting in the subsequent passage of Resolution No. 019-91 of 11 November, 1991 (*La Tribuna*, 26 April 1991).

Sanbrailo's third point had the most serious implications for certain sections of local fishing communities; the establishment of laboratories to produce postlarvae would exclude some artisan fishermen from the commodity chain as the industry sought to become vertically integrated. CODDEFFAGOLF and ANDAH were the beneficiaries as it was a goal that both organisations advocated conjointly. The creation of closed seasons also aligned with CODDEFFAGOLF's interests, as it would reduce the impact of the capture of wild postlarvae on other fisheries resources. The ban on drag-nets would potentially reduce by-catch and serve to protect the fisheries population. The government had taken obvious steps to address the concerns of both the industry and CODDEFFAGOLF, but emphasised its objective to continue with expansion. The photo below depicts GGMSB's operations in 1989.

Hernan Piñedo Bardales, the President of FPX, concurred with USAID's position, their funding body, and released a joint public statement indicating that there was a total area of 30,000 ha of salt flats available for the production of shrimp, half of which could be put under production in the next five. The new estimate was based on a study completed by Ruben Guevara, Vice Minister of Natural Resources. However, Guevara stated that there were several obstacles confronting the expansion of the industry, including the lack of well-defined policies and the growing conservation movement (*La Tribuna*, 26 April 1991). The environmental movement was thus articulated as a continued threat to the neoliberal reform process, which included the promotion and expansion of aquaculture as an NTAX.

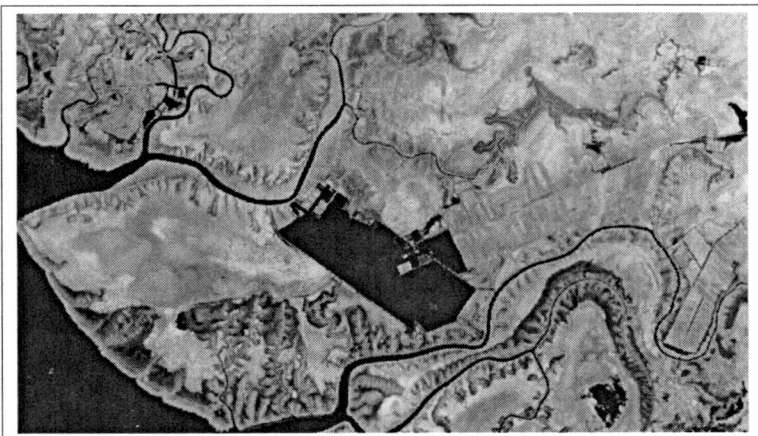

FIGURE **10.1:** CONSTRUCTION OF THE SECOND PHASE OF **GGMSB'**S EXPANSION ON THE
SALT FLATS OF **S**AN **B**ERNARDO (DARK AREA) OF **P**UNTA **G**UATALES IN **1989.**
Source: NOAA

ANDAH and the Government Respond to Claims of Mangrove Deforestation

The government and ANDAH initiated efforts to defend themselves against CODDEFFAGOLF's claims. They emphasised, along with Richard Pretto Malca, Panama's National Director of Agricultural Extension, that Central American shrimp farming had integrated rural lives economically by satisfying four principal necessities: production for domestic food supply (food security discourse); generation of foreign exchange; generation of investment; and employment (neoliberal economic development discourse) (*La Tribuna* 26 April 1991).

During the symposium, ANDAH bought an advertisement in *La Tribuna* with the statement: "Let us use our natural resources rationally" (*La Tribuna*, 26 April 1991).[62] Essentially, this statement linked natural resource use, rationality and the industry, which was important given the attention drawn to its negative aspects by the ENGOs. The emphasis on rational resource use, economic development and job creation are examples of language used repetitively throughout the process of neoliberal transformation. These concepts are difficult to oppose as they appeal to the wider public's interests and they are clear and unambiguous.

Jorge Varela applauded AID and ANDAH for indicating that they would work towards the rational development of the industry. However, he stated that it was a shame that the participants at the symposium did not discuss the recent destruction of the natural lagoons in Guameru or mangrove deforestation in La Vaca de El Tulito or Pasadero de Miguel, all areas where the industry was expanding. A technical consultant with the IUCN stated that "One of the most evident problems is the divorce between what they [ANDAH] say and what they practise"[63] (CODDEFFAGOLF, 1991b, Editorial: 1-3). He thus argued that ANDAH was inconsistent. Although they claimed to be pursuing rational development, the technical consultant drew attention to the industry's destruction of the seasonal lagoons. The battle over the discursive terrain to influence public opinion

---

[62] Translated from Spanish: [Hagamos un uso racional de nuestros recursos naturales]

[63] Translated from Spanish: [Uno de los problemas mas evidentes es el divorcio entre lo que se dice y lo que se practica]

continued. As the links between the industry and mangrove degradation were debated, the conflicts associated with the fisheries continued.

Government Enforces the Law Banning Drag-Nets and CODDEFFAGOLF Denounces the Industry

Ironically, on the last day of the conference, government officials and military authorities went into Guapinol, Municipality of Marcovia (one of the communities closely aligned with CODDEFFAGOLF), to conduct an inspection of the fishermen, with the intention of decommissioning the illegal drag-nets. During the operation they found 40 nets that had apparently been stolen, 23 supposedly from GGMSB and two from CULCAMAR. At the time, it was unknown from where the others were taken. Two fishermen were arrested, Donatilo Cruz Flores (Fiscal General of CODDEFFAGOLF) and Abelisaro Alvarez (President of the Section for CODDEFFAGOLF in Guapinol) by the Dirección Nacional de Investigaciónes (National Directorate of Investigations, DNI) and incarcerated in Choluteca. It was thought that CODDEFFAGOLF was taking direct action to halt the capture of the wild postlarvae by stealing the nets used by the *larveros* (larvae gatherers). CODDEFFAGOLF argued that the government and the industry were pursuing a strategy of intimidation (CODDEFFAGOLF, 1991b, Denuncia Publica, San Lorenzo: 5).

CODDEFFAGOLF denounced the arrests and released an inflammatory public statement: "For the moment there has been no bloodshed, the postlarvae collectors and the artisan fishermen are brothers, but if there is bloodshed, it will be the fault of the aristocratic 'boss' and the government"[64] – the agents of neoliberalism (CODDEFFAGOLF, 1991b, Denuncia Publica, San Lorenzo: 5). This statement directly attacked both the industry and the government, alluding to the possibility of future violent action. The statement also recognised the unfolding division within the artisan fishing community, between those who caught wild postlarvaes and those who fished for adult shrimp and other species. In their public denouncement, CODDEFFAGOLF emphasised that the problem on the south coast should not be perceived as a conflict between CODDEFFAGOLF and the shrimp farmers, but as a national problem that needed urgent attention, an indirect reference to the Callejas administration (CODDEFFAGOLF, 1991b, Denuncia Publica, San Lorenzo: 5).

---

[64] Translated from Spanish: [Hasta el momento no ha corrido la sangre, pues tanto los larveros como los pescadores artesanales somos hermanos... pero si llegara a correr, la culpa sera del aristocrata "gerente" y del gobierno.]

Less than a week later, CODDEFFAGOLF responded by mobilising artisan fishermen and staged a protest in Tegucigalpa to criticise the government and USAID for allowing shrimp aquaculture expansion to continue. The central aim of the protest was to put forth a formal public complaint that aquaculture expansion was resulting in the destruction of the south's coastal resources. The protest occurred on *Dia del Trabajo* (Labor Day), as CODDEFFAGOLF aligned its efforts with the labour movement in Honduras and the *Plataforma de Lucha,* or Platform of Struggle. The *Plataforma de Lucha* was an anti-neoliberal policy campaign driven primarily by the labour movement in opposition to President Callejas' economic reform agenda, which they thought would lower the cost of labour. Of course, CODDEFFAGOLF was concerned about the environmental impacts, so the two, along with numerous other civil society actors, pressurised the government to make policy changes (CODDEFFAGOLF, 1991b, Editorial: 2-3).

CODDEFFAGOLF's links to the labour movement and the *Plataforma de Lucha*, established in 1989 when Callejas came to power, signified the official amalgamation of opposition to neoliberalism and the burgeoning of the Honduran popular movement in the 1990s. The wider movement included community groups, unions of peasants and workers', and a number of environmental organisations. The opposition to neoliberal policies coalesced, leading to resistance and an articulation of alternatives to "neoliberalism and free market integration into the global economy" (Robinson, 2003: 132). The battle fought in the south was between two competing social imaginaries in relation to the coastal wetlands.

CODDEFFAGOLF takes Direct Actions against the Larvae Gatherers

At the local level, CODDEFFAGOLF began pursuing more assertive strategies to enforce the ban on the capture of wild postlarvaes. On 24 July 1991 a number of CODDEFFAGOLF-affiliated fishermen were arrested in their base of Amapala, for attacking one of GGMSB's operations in La Brea, Department of Valle. It was uncertain whether the action taken by CODDEFFAGOLF was in retaliation to the earlier arrest of two of its members caught with stolen nets in Guapinol, or the continuation of a new strategy to intervene directly and stop the industry's practices. Regardless, CODDEFFAGOLF's opposition became increasingly direct as neoliberal reform ensued.

Three newspapers reported that somewhere between 13 and 25 fishermen had been detained by the Public Security Forces (FSP). Those detained were accused of being illegally armed, breaking and entering, destroying private property and robbery. Silvaño Osorto, a local fisherman from La Brea, stated that he and several other fishermen who supplied GGMSB with wild-caught postlarvae were the victims of armed bands formed by CODDEFFAGOLF. He was fishing for wild postlarvae in Tembladera estuary when he and his companions were surrounded by about 20 men armed with pistols and automatic assault rifles (AK-47s), who claimed to have authority over the coastal resources. The detail of this last point is worth mentioning as the industry was later blamed for being the direct cause of violence in southern Honduras. Osorto stated that they took the equipment (apparently to Amapala), released their catch and warned them to stop fishing or they would be attacked again (*El Heraldo*, 29 July 1991). Actions of this nature occurred more frequently as the divisions within the artisan fishing community broke out.

CODDEFFAGOLF's members were also accused of attacking postlarvae collectors in San Pedregal, a community near the shrimp farms located in San Bernardo, in close proximity to the San Bernardo River that supplies water for the shrimp farms in the region. The industry claimed that CODDEFFAGOLF had based its operations from the Naval Base in Amapala (where the armed men were detained) to intimidate postlarvae collectors and monopolise the postlarvae-collection market. ANDAH called on the President to recognise the socio-economic development role that the industry played in the south and revisit CODDEFFAGOLF's legal status (*El Heraldo*, 29 July 1991).

Within days of the detention of its members, CODDEFFAGOLF led a protest. It was one of its more effective strategies of resistance and included 600 to 800 locals, mostly artisan fishermen opposed to the capture of wild shrimp postlarvae. The protestors demanded the release of their fellow fishermen after taking control of the only bridge that led into Choluteca from the east. They demanded that the government find a solution to the problems between the industry and the fishermen (*El Tiempo*, 5 August 1991).

ANDAH Issues a Formal Legal Complaint Against CODDEFFAGOLF

The industry immediately sought to exploit the situation and undermine CODDEFFAGOLF's public status. On 29 July, ANDAH wrote to President Callejas, denouncing CODDEFFAGOLF's actions in *El Heraldo*. In their statement, ANDAH's board of directors made several allegations against CODDEFFAGOLF, suggesting that it had illegally organised armed gangs to threaten the *larveros* (larvae gatherers) working for the industry, and to steal equipment and materials from installations. CODDEFFAGOLF's

President, Mauricio Alvarado, was also criticised for actions taken in the area around San Lorenzo on 7 July 1991. ANDAH claimed that the same group set fire to installations of Aquacultura Fonseca and attempted to attack mechanical equipment owned by Hondufarm, attacked the vehicles transporting the wild postlarvae for Sea Farms of Honduras, and attempted to assassinate executives, business owners and workers associated with the shrimp industry. ANDAH stated that CODDEFFAGOLF was provoking rivalries between artisan fishermen who supplied wild postlarvae to the industry and fishermen who captured adult shrimp to sell in the local fish markets (*El Heraldo*, 29 July, 1991).

The government also opposed the actions of CODDEFFAGOLF. The Vice Minister of the SRN, Ruben Guevara, stated that CODDEFFAGOLF had overstepped the line. He argued that this was the first time that they had captured members of CODDEFFAGOLF acting illegally, but that it had committed several illicit acts against the industry in the past. He stated that no legal action had been taken because the industry preferred to engage in dialogue rather than relying on the legal process. German Zelaya, former Director of RENARE which was eliminated and became DIGEPESCA when the agency was created in the early 1990s, stated that he felt CODDEFFAGOLF took the law into its own hands solely for the purpose of trying to achieve their own interests, and that they had bypassed the authority of the SRN to address their various concerns (*La Tribuna* and *El Heraldo*, 2 August 1991; also in *La Prensa*, 31 July 1991).[65]

Several CODDEFFAGOLF members admitted to participating in the operations, but denied robbery even though Donatilo Cruz Flores, a member of the board of directors, had been caught in Guapinol with 40 fishing nets in his home, some of which were the property of GGMSB. CODDEFFFAGOLF, however, claimed that the actions were retaliatory, due to the industry's refusal to negotiate a settlement to continued environmental destruction of the southern zone. They placed particular emphasis on the impact of capture of wild postlarvae on other marine species and justified their actions by claiming they intended to fight for the preservation of the marine flora and fauna. CODDEFFAGOLF demanded that two shrimp-fishing seasons be established to accommodate both groups of fishermen, and that the industry reforest mangroves and stop the drainage of four lagoons intended for expansion in the San Bernardo region

---

[65] *La Tribuna*, 2 August 1991 "ANDAH" Repudia Violencia de CODDEFFAGOLF Contra Pescadores page 9. The same article was also printed in *El Heraldo*, August 2 1991, Repudian Violencia de CODDEFFAGOLF Contra Pescadores, 43 and a shorter version in *La Prensa*, 31 July 1991 "Comite de Defensa del Golfo No Coopera con RENARE".

near San Pedregal (*El Heraldo*, 29 July; *Honduras This Week*, 3 August; and *La Tribuna*, 29 July 1991).

On 1 and 2 August, negotiations were held between CODDEFFAGOLF and ANDAH and mediated by a Catholic priest, Father Lopez Tuero. ANDAH agreed to the release of the detainees and to suspend the judicial process. In return, CODDEFFAGOLF agreed to stop pressuring ANDAH. The fishermen were released on 5 August 1991 after the agreement was negotiated (*El Tiempo*, 5 August 1991).

Did Aquaculture Development Lead to the Post-Larvae Black-Market? CODDEFFAGOLF Accused of Acting as *Coyotes*

CODDEFFAGOLF's actions seemed to have an effect on the neoliberal agenda as investors became concerned about the conflict around the industry in the region. According to Mario Nufio Gamero with the SRN, CODDEFFAGOLF's actions discouraged the expansion of operations. Some investors suspended their decision to come into the region, whilst others decided to wait until the corresponding authorities could provide guarantees to the industry that the region could be secured. Subsequently, he met with German Zelaya, Director of DIGEPESCA, and Raul Aguero Neda, the Director of the Consejo Hondureno de la Empresa Privada (Honduran Council for Private Business, COHEP), to decide how to secure the region for the industry's continued expansion (*La Tribuna*, 31 July 1991). Some of the issues with which they were most concerned were the assertiveness of the tactics and the increase in black-market activities.

Although the dispute was successfully mediated, a number of claims against CODDEFFAGOLF appeared in the press. The conflict was revealed as a dispute over access to and control over fisheries resources and the socio-economic and environmental impacts wrought by economic change. The changes associated with the introduction of the industry into the south was not only a direct cause of conflict between groups of artisan fishermen, but was also related to an increase in black-market activities in all three of the countries surrounding the GOF, exacerbating regional transboundary resource conflicts.

Black-market activities connected to the fisheries increasingly led to confrontations between naval forces and fishermen in each country. Ultimately, they escalated into unresolved transboundary natural resource conflicts between Honduras, Nicaragua, and El Salvador – conflicts that still persist. Consequently, they also generated instability, confrontation and wider conflict around the entire fisheries in the GOF. These factors taken together complicated efforts to move towards shared resource management in the GOF and became an impediment to regional economic integration as territorial disputes continued. In turn, it appeared that as neoliberal reform occurred throughout the 1990s, other black-market activities, such as drug smuggling and arms trafficking, became prevalent in the GOF.[66] Some groups may have been involved in smuggling all three across borders and CODDEFFAGOLF was accused of acting as an intermediary, or as *coyotes*[67] and smugglers. Felix Diaz Torres, President of the Association of Shrimp Postlarvae Fishermen of the Gulf of Fonseca[68], stated that CODDEFFAGOLF was trying to monopolise or corner the market so that local fishermen had to go through them to get their product to market.

The fishermen associated with some of the cooperatives established by the government in the 1970s and 1980s who had been trained as larvae gatherers confronted challenges in terms of getting their product to market by the early 1990s. Their challenges can be attributed partly to the industry's efforts to organise and train *larveros* to work directly for the industry to ensure a steady supply of wild-caught postlarvae for production. Consequently, some of the older cooperatives were affected by the changes in the market, leading some to pursue black-market activities and others to identify alternative places to sell their product. The result was further divisions within the artisan fishing community as the labour market changed (*La Tribuna* and *El Heraldo*, 2 August 1991). For example, those who did not have contracts with the industry sought alternative avenues for their product which sometimes involved selling it in Nicaragua or El Salvador. In relation to the black-market, many fishermen, regardless of their affiliation with CODDEFFAGOLF, were now accused of stealing the shrimp from the ponds and smuggling it across borders. Possibly, this resulted in violence between the guards protecting the industry's ponds and the fishermen. However, there is no evidence to verify the connection between theft of shrimp, the industry and violence; the connection

---

[66] This statement is based upon the repetitive viewing of newspaper articles pertaining to these issues. However, since my dissertation is not about black-markets in the GOF, drug trafficking, gun smuggling, or the illegal import and export of fisheries resources, and therefore I am not going to address them further.

[67] In Spanish, *coyotes* generally refer to individuals who act as intermediaries to smuggle illegal immigrants into the US, but is also often used in relation to the smuggling of drugs, weapons or other types of contraband. In this case, *coyote* refers to CODDEFFAGOLF as an intermediary in the fisheries market associated with price fixing and the smuggling of the produce across borders to avoid taxes, most likely into El Salvador and Nicaragua.

[68] It is worth noting that the Association of Shrimp Postlarvae Fishermen was established with support from ANDAH and government funds as cooperatives were being established in the 1980s.

was only alluded to in the newspapers, but was used later by CODDEFFAGOLF to politicise industry practices by connecting it to violence (see Chapter 13).

The claims that CODDEFFAGOLF intended to act as *coyotes* was supported by Juan Alvarez, ex-Director of CODDEFFAGOLF in Guapinol. He stated that Donatilo Cruz, the organisation's Fiscal Officer, wanted to monopolise the fisheries market in the GOF. Postlarvae caught in Honduras could easily be smuggled into Nicaragua or El Salvador and sold to the shrimp industry in those countries. Alaverez claimed that CODDEFFAGOLF was exploiting local fishermen by paying them low salaries and reaching agreements with the *coyotes* in Tegucigalpa, Nicaragua, and El Salvador to fix prices and assist with getting the contraband into the market (*La Tribuna* and *El Heraldo*, 2 August, 1991). Whether or not these claims were true is unknown. Nevertheless, negative perceptions of CODDEFFAGOLF were more apparent within the public arena as these stories were disseminated through the media.

Similar claims were made by other local fishermen, such as Luis Nuñez, who argued that "we have rejected their offers [to get the product to market] because we do not want them to take away our work that we depend on for our families' wellbeing, as well the CODDEFFAGOLF people intend to become the *coyotes* of fishing by marginalising us from the direct sale of our products"[69]. Another fisherman, Jose Escalante from the community of San Bernardo, argued that CODDEFFAGOLF's actions were intended to monopolise the fisheries market in the GOF, and control the purchase from the postlarvae collectors and the price of fish on the local market (*El Heraldo*, 29 July 1991).[70] His statement suggested that there was some type of illegal activity taking place within the fisheries market. In order to create stability and provide a secure environment for future investment in the region, the government worked with CODDEFFAGOLF and ANDAH to establish a commission to oversee natural resource use in the region.

---

[69] Direct translation from Spanish: "les hemos rechazado sus ofertas, porque no queremos que nos quiten el trabajo de lo que viven nuestras familias, pues los de CODDEFFAGOLF, tienen la intención de quedar como coyotes de la pesca, al apartarnos de la venta directa de nuestros productos".
[70] "En el Golfo de Fonseca: Bandas Ecológicas Siembran el Terror": "In the Gulf of Fonseca: Ecological Gangs Plant the seeds of Terror". Increasingly, CODDEFFAGOLF and others began to be labeled as eco-terrorists and radical environmentalists. The title of this article was one of the first to put forth this image through the media yet had been articulated by government and industry officials prior to the publication in the newspaper.

The government successfully negotiated an institutional remedy between ANDAH and CODDEFFAGOLF that included representatives from the armed forces (to ensure security and enforce the law), Governors of the Departments of Choluteca and Valle, municipal representatives in the south, and cooperatives representing the postlarvae collectors. The result was the creation of the Commission for the Vigilance, Verification, and Control of the Problems of the GOF (CVC) (CODDEFFAGOLF, Boletin No. 5, July 1991, Pescadores y Camaroneros del Golfo de Fonseca Inician Dialogo, 2). The CVC was one of the first collective environmental governance mechanisms established by a collective group of actors in the south, and it arose in the context of the neoliberal changes taking place.

The government also began working on a resolution to establish a closed season on the collection of postlarvae and adult shrimp from 15 December 1991 through 15 August 1992 on the littoral Pacific coast. The National Congress passed this law through Resolution No. 019-91 of 11 November 1991, but never actually implemented it due to transboundary resource conflicts with Nicaragua and El Salvador. Since fishermen from both countries were purportedly coming into Honduran territorial waters to fish, the closed season would have negatively impacted Honduran fishermen but not those illegally entering Honduran territory to fish. The government, therefore, decided not to enforce the ban (*La Tribuna*, 30 December 1991).

Creation of the Asociación de Recolectores de Mariscos del Sur (Association of Southern Shellfish Harvesters, ARMASUR)

In August, shortly after the escalation of the conflict, CODDEFFAGOLF shifted its strategy and began to consolidate artisan fishermen, some of whom supplied the industry with postlarvaes and were from the communities of Guapinol, Punta Raton, San Jose de las Conchas, and Cedeño, all of which were close to industry operations. It is uncertain whether the postlarvae gatherers targeted by CODDEFFAGOLF were those unsuccessful in their efforts to sell their product to the industry or if they were those more involved in the black-market activities associated with the sale of postlarvae in Nicaragua and El Salvador. The point is that the transboundary resource issues impeded the effective management of the natural resources.

CODDEFFAGOLF's strategy was quite astute given that the shift towards laboratory-produced larvae would affect the collectors' livelihoods as the industry sought to vertically integrate. Although some postlarvae collectors were employed directly by the commercial industry, others were contractors. If the transition to laboratory-produced postlarvae were successful, it would minimise opportunities for postlarvae collectors to obtain contracts. CODDEFFAGOLF sought to use the potential impact on their livelihoods to strengthen its base and gain further support in artisan fishing communities, although it supported the industry's and government's positions in relation to increasing the use of laboratory-produced postlarvae.

The Regional Director of the Honduran Institute for Cooperatives (IHDECOOP)[71] convened a meeting in Cedeño in conjunction with CODDEFFAGOLF in August 1991 to establish ARMASUR. The new cooperative was created to counter the Association of Shrimp Postlarvae Fishermen of the Gulf of Fonseca, founded and supported by ANDAH. The new group was created as an affiliate of CODDEFFAGOLF and intended to identify alternative markets and increase earnings for the organisation (CODDEFFAGOLF, 1991b: 5).

The action was consistent with CODDEFFAGOLF's objectives to consolidate its power, and with the claims made by local fishermen that the organisation wanted greater control as intermediaries in the fisheries market. Finally, it provided CODDEFFAGOLF with additional representation before the CVC. ARMASUR was in an ideal position to provide an alternative perspective from within the community of postlarvae collectors, a position mirroring that of CODDEFFAGOLF, as it became a local extension of the organisation, as FEDAMBIENTE had become a national extension and representative body to achieve their interests.

---

[71] This government agency was created through the Cooperatives Law, Ley de Cooperativas de Honduras Decree 65-87 passed in 1987. The institute was created to support cooperative development in relation to economic sectors.

FIGURE 10.2: LOCAL FISHERMAN
Source: Photograph No. 41 in Zonificacion de los Bosques de
Mangle del Golfo de Fonseca, Honduras, C.A. 2002.

Summary

In this chapter, I began by discussing government and regional organisations' efforts to
further develop the fisheries sector in order to capitalise it while seeking to enlist local
artisan fishermen into their efforts to expand shrimp farming and allay the concerns of
those opposed. The government also began discussing the establishment of multiple
use areas in order to clarify access to different spaces within the wetlands. As these
efforts were pursued, the government also began working towards revision of the 1959
Fisheries Law to update it in accord with national interests in this period. The state took
action in relation to a ban on drag-nets which was an interest of both CODDEFFAGOLF
and ANDAH, although the reasons for those interests diverged. I stated two reasons for
the ban on drag-nets and then discussed issues involving oversight and enforcement.
The ban ultimately led to further dissension within the artisan fishing community,
perpetuating conflicts amongst individuals associated with it. Finally, I illustrated how
the government's lack of knowledge in relation to the fisheries community and the
potential effects of legal changes were partially responsible for the exacerbation of the
conflicts that had emerged. In Chapter 11, I illustrate how the resulting conflicts
encouraged the government to take steps to mitigate the negative environmental
impacts of aquaculture in the south by developing a better understanding of the issues
and passing laws to give the government the necessary authority to take action.

Chapter 11

Neoliberal and Anti-Neoliberal Coalitions: Internationalisation of the Aquaculture and Environment Debate

Introduction

In this chapter, I show that although efforts were made to address the social and environmental impacts of neoliberal reform, aquaculture expansion continued throughout the mid-1990s. As Callejas confronted the backlash within civil society, the government began to take a more active role in relation to environmental issues locally. Government agencies called for state support to strengthen their capacities to address the socio-economic and environmental issues affecting the south. I illustrate how these issues were scaled-up to the regional level as new institutions emerged to deal with environment and development issues in Central America.

As environmental governance was increasingly emphasised, CODDEFFAGOLF's formal legal claims against the government and the industry led state agencies in the region to take action as the negative impacts of aquaculture on mangrove forests became apparent. The passage of the first General Environmental Law signified the Honduran government's recognition of the wider impacts of neoliberal reforms. In conjunction with El Salvador and Nicaragua, the Honduran government contracted a private organisation called Tropical Resources and Development (TRD to issue a report on mangrove loss in the region of the Gulf of Fonseca. The TRD report was one of the first to recognise officially the role of disputes within the fisheries sector as a source of conflict. The extent of mangrove loss due to aquaculture expansion continued to be contested leading to further conflicts, which intensified as the government took steps to pass new environmental legislation. The characteristics of the industry are discussed, demonstrating the rapidity of shrimp farm expansion during this period.

The Reinas administration, elected on an anti-neoliberal platform, emphasised enforcement, regulation, issuance of licenses, and the need for environmental impact evaluations to be completed. The state agreed to apply pressure to enforce newly created laws but in alignment with neoliberal reform, with a specific focus on land titling and tenure issues in the south. Although the state had taken several steps to change its position and actions, violations of the law and the state's inability to enforce a moratorium led to industry violations precipitating further protests and the sailing of the Greenpeace ship, Moby Dick, into the GOF in protest. The campaign initiated by CODDEFFAGOLF and Greenpeace was rapidly scaled up during this period and led to the signing of the Choluteca Declaration with other environmental organisations as global civil society begun contesting the impacts of aquaculture development worldwide.

FIGURE 11.1: **M**OBY **D**ICK *SAILING INTO THE* **G**ULF *OF* **F**ONSECA
Source: La Tribuna, 1994.

ENGOs and local movements worldwide collectively organised to protest the expansion of aquaculture, resulting in the formation of the Industrial Shrimp Action Network (ISA-net). This organisation represented the augmentation of transnational environmental political movements' power and the elevation of local conflicts into the international arena, resulting in the global aquaculture and environment debate. I illustrate how this organisation used the media, conducting presentations at international conferences, and lobbying governments, to achieve their interests. The dominant discourse of the global movement was that shrimp farming was destroying mangrove forests and local livelihoods. Their goal was a global boycott and a moratorium on shrimp farming, to encourage consumers to stop buying shrimp, and establish an independent certification council to assess the commercial industry's practices through criteria based on the social and environmental effects of aquaculture.

Industry Expansion in the early-1990s

Despite the conflicts around aquaculture and fisheries-related resources, the industry continued to plan its expansion, representing the continuance of the neoliberal project in the south, supported by President Callejas. By the autumn of 1992, it was estimated that there were approximately 25 larger businesses dedicated to the production of shrimp in the south, 6 packing plants, and 6 factories dedicated to the production of ice. According to ANDAH, 22,000 ha of land were still available for the cultivation of shrimp (La Tribuna, 26 October 1992).

President Callejas decided to travel to the region to reiterate his support for the industry as pressure from environmental groups continued to be applied (La Tribuna, 26 October 1992). He attended the first shrimp harvest by the company Criadoeros Marinos S.A. (CRIMASA) located in El Faro and El Triunfo, Choluteca. Callejas was accompanied by a number of government and military officials and made a statement that the industry could generate up to $2 billion worth of shrimp by 2000 if efforts to expand were successful. He discussed infrastructure needs, identified the construction of roads and electricity of being of prime importance, and reiterated his wish to see the neoliberal agenda continued. Callejas asked ANDAH to revisit the agreements in place with the government to reassess the options for expanding the shrimp industry further (La Tribuna, 8 December 1992). However, the increased need to direct attention towards environmental issues in the context of neoliberal transformations in Honduras culminated in 1993 with the passage of Honduras' first General Environment Law.

General Environmental Law 1993

The Honduran National Congress passed the General Environmental Law in 1993, following UNCED in 1992 (La Gaceta, 30 June 1993). The law created the first Ministry of Environment (SEDA, which became SERNA in 1995) and was symbolic of the increased government attention to environmental governance. It also represented the convergence of two social imaginaries – neoliberalism and environmentalism – that were not necessarily incompatible. The manner in which environmental law and conservation were enacted from this point forward illustrates the co-optation of environmental discourses into the neoliberal project. Subsequently, it led to what has been termed 'neoliberal environmental governance', that emphasises market-based approaches to environmental conservation, apparent in the later part of the 1990s and into the twenty-first century.

As the Honduran environmental movement had gained momentum throughout the country, the environment increasingly became an important part of the Callejas administration's neoliberal agenda. In the south, Jorge Arevalo Carcamo, an extension agent with COHDEFOR, wrote in *La Tribuna* that there were a number of measures that the government needed to take to address environmental degradation. He referenced the legal agreement 0800-90 that created Fundación Vida, which was established to fund projects led by NGOs and encouraged the strengthening of civil society (*La Tribuna*, 25 June 1993). Arevalo argued that the government needed to:

1. support NGOs so that they could address environmental issues more actively due to a lack of state capacity in this area;
2. strengthen the National Counsel of Protected Areas (CONAPH)[72] (the newly created agency responsible for protected areas) as the LMDSA was implemented;
3. create an independent commission to determine how contiguous areas of habitat could be kept intact in the areas designated as a part of *Paseo Pantera;*
4. work with the IDB to strengthen emerging regional environmental institutions such as the CCAD;
5. encourage the Presidents of the Central American region to collaborate more closely on environmental issues; and
6. make the role of the state oriented more towards the conservation of natural resources and work with the public and private sectors to confront the challenges now that the General Environmental Law had been passed (*La Tribuna*, 25 June 1993).

The General Environmental Law affected shrimp farming practices in two ways. First, it required environmental impact assessments to be completed (Article 5 and 2) and overseen by the National System for Environmental Impact Evaluations (SINEA) (Article 11(d)). Second, it gave the Ministry authority to authorise concessions and permission for commercial enterprises to operate (Article 11 (m), 3).

In line with Honduran efforts to focus increased attention on environmental issues, governments of other Latin American countries in the region began advocating, through the CCAD, the strengthening of the System for Regional Integration's (SICA) focus on the impacts associated with the capitalist transformations taking place throughout the

---

[72] Acuerdo 0875-93 Creación del Consejo Nacional de Áreas Protegidas La Gaceta No. 27,079 del 25 de Junio de 1993

region. CCAD was created with the support of USAID, which was increasingly realising the problems of the political and economic reforms that it had been pushing under the neoliberal agenda as transboundary resource conflicts intensified.

USAID believed that establishing closer political and economic ties through trade and development would create stability and, therefore lead to greater cooperation on natural resource issues. Subsequently, whilst individual governments strengthened their environmental policies, USAID focused on regional institution building. Their efforts were directed towards focusing on shared resource issues in Central America according to ecological rather than the political boundaries. The *Paseo Pantera*, later the Meso-American Biological Corridor (MBC) was indicative of the trend (personal interview with Martin Schwartz with the University of Zamorano, formerly with CARE and USAID, 2005).

Although a number of powers were transferred to the newly created Ministry, AFE-COHDEFOR remained responsible for overseeing forest resources whether for extractive or conservation related purposes. However, AFE-COHDEFOR, along with other government agencies, were increasingly encouraged to take a more active role to protect the environment as locals, regional politicians, and environmental groups contested its practices. With the passage of the General Environmental Law, COHDEFOR remained responsible for the management and administration of public forests, including national and *ejidal* lands, as well as private lands making it central to the implementation of the LMDSA (see Chapter 4) as neoliberal agrarian reform efforts were pursued. As a result, COHDEFOR began to reassess its responsibilities in relation to the LMDSA and the management of protected areas and subsequently, developed a new legal framework to manage natural resources in line with neoliberal objectives (*La Tribuna*, 25 March 1993).

The emphasis on environmental governance in the context of the implementation of neoliberal economic policies was exemplified by the statement of Reyes Chirinos, with COHDEFOR, that the new environmental legislation prompted his organisation to change their policies to clarify its environmental responsibilities in relation to ecotourism and land tenure issues, as well as the management and administration of national parks and wildlife refuges within the context of agrarian reform under the LMDSA. He also stated that it would investigate claims of environmental degradation and attempt to more actively impose fines for the degradation of forest resources as environmental governance and strengthening the rule of law became increasingly important (*La Tribuna*, 25 March 1993).

Arevalo, COHDEFOR's lead extension agent in the south, admitted that the aquaculture industry had not previously taken environmental impacts into consideration as it expanded. It concluded that an estimated 30 per cent of mangrove forests in the south had been destroyed. COHDEFOR announced that it would impose fines and penalties more actively, but claimed that it was debilitated due to a lack of political will at the national level, and the powerful interests of the shrimp and salt industries and opposition among mangrove wood-cutters and fishermen. The shrimp industry opposed COHDEFOR's stance and argued that they were bringing in over $45 million in foreign exchange revenues that benefited the country (*La Tribuna*, 10 September 1993).[73] Arevalo argued that if the government and the industry did not complete a study to determine the capacity of the remaining 15,000 ha available for shrimp farming or evaluate the environmental impacts, "we will stop the development and the killing of this green gold", aligning himself with CODDEFFAGOLF's position (*La Tribuna*, 10 September 1993).

Tropical Resources and Development Study

In the same year, the TRD released the results of a study that it had been contracted to complete for the countries surrounding the GOF. It was important because one of CODDEFFAGOLF's claims was that the shrimp industry was destroying mangrove forests, the discourse put forth by Arevalo. The TRD analysis was intended to identify how much mangrove forest loss could be attributed to industry practices. The study cited COHDEFOR's 1987 report that stated that the total area of mangrove forests intact in 1973 when the first shrimp pilot project commenced was 30,697 ha. It concluded that by 1982 the total area of mangrove forests had declined to 28,776 ha, equaling a total loss of 1,921 ha or 6 per cent of the total area. The TRD assessment, completed in 1992 and disseminated in 1993, estimated that the total area of mangrove forests had declined to 23,937 ha, indicating an additional loss of 4,839 ha or 17 per cent of the total area between 1982 and 1992. 2,131 ha of this loss were directly attributed to shrimp pond construction. 2,708 ha were attributed to local use of the resource, estimated to be 250–350 ha per annum. The report also estimated that 2,174 ha of salt flats had been occupied for pond construction between 1982 and 1992 (Vergne et al., 1993).

---

[73] *La Tribuna* "Camaroneras Subestiman Multas" by Jorge Arevalo Carcamo, Departmento Fomento y Extension de COHDEFOR.

TABLE 11.1: LOST COVERAGE (IN HA) OF THE MANGROVE FOREST IN THE GOF

Category	1973	1982	1992
Mangrove	30,697	28,776	23,937

Source: Vergne et al., 1993.

Observations indicated that the majority of the commercial shrimp farms were those that utilised the salt flats as they had the political and economic connections to acquire concessions to these spaces, whereas local artisan shrimp farmers were likely to have cleared mangrove forests without obtaining concessions. It was noted that the limiting factor for artisan shrimp farmers to access land for production was not the cost, which was only the equivalent of $5 per ha, but connections with the right people in the government to acquire the concessions. Land speculation was fairly common and lucrative. In fact, the TRD report estimated that the total area available for possible production, 31,419 ha of land, had already been conceded, mostly salt flats which encouraged the black-market and land speculation (Vergne et al., 1993). Those who controlled the concessions would sublease them or sell them on the black market to the highest bidders. For example, Rigoberto Duarte, the ex-President of ANDAH, was accused of selling land he controlled to Agro-Internacional, a North American company wanting to invest in aquaculture in the Department of Choluteca. Elias Asfura, a presidential candidate for the National Party, was also accused of the same in the Department of Valle (La Prensa, 15 November 1993).

TRD estimated that the shrimp industry, heavily reliant on the capture of wild postlarvae to stock the ponds, was using 1.3 to 3.5 billion postlarvae each year to stock the ponds (Vergne et al., 1993: 51). Claims made by CODDEFFAGOLF and sections of the artisan fishing community persisted regarding the impacts of larvae gathering on fisheries resources and, therefore, their earnings, as well as diminished access to the estuaries and seasonal lagoons to fish. CODDEFFAGOLF raised issues in relation to the government's lack of support for small-scale producers. Specifically, CODDEFFAGOLF was referring to artisan shrimp farmers and fisheries cooperatives whom had wanted to produce on a small scale but were not provided with the financing and technical assistance given to the commercial farms. Reiterating some of the priorities already agreed upon through the CVC, TRD recommended that the government create a management plan, take advantage of USAID's CRSP, specify the types of nets and equipment that could be used to fish, and establish closed seasons in relation to the capture of various species of fish, adult shrimp, postlarvaes, bivalves and molluscs (Vergne *et al.*, 1993).

CODDEFFAGOLF Discourse Emphasises Protection of the Wetlands

The discursive struggle associated with the impacts of aquaculture expansion on the mangrove forests escalated towards the end of 1993 as industry practices continued on the same trajectory. The TRD report also provided CODDEFFAGOLF with ammunition to counter the claims of the industry. With newfound support from Greenpeace, it joined a nascent global movement to boycott the purchase of cultivated shrimp from Honduras since expansion continued to occur unabated – a new strategy of resistance targeting the consumer market, the heart of neoliberalism. The links to Greenpeace signified a turning point in the conflict and the early global aquaculture and environment debate as the issues increasingly transcended the south, Honduras, and the Central American region.

CODDEFFAGOLF's members marched on Tegucigalpa to announce the boycott and claimed that the industry was committing 'genocide' and 'ecocide' as it increasingly attempted to link the industry to violent actions, human rights abuses and environmental degradation. Varela stated that five fishermen had been assassinated by industry security and demanded that the government bring together the CVC (comprised of industry representatives), CODDEFFAGOLF and government agencies to assess the problem and seek solutions (*La Prensa*, 15 November 1993).

On the basis of the TRD study, CODDEFFAGOLF also justified the boycott on the grounds that the shrimp industry was contributing to the massive destruction of the mangrove forests. They stated that, in 1973, there were approximately 30,000 ha of mangrove forests within the territorial boundaries of Honduras and that now there were only 24,000 hectares since *fiebre del camaron cultivado* (fever to cultivate shrimp) commenced in 1984. These arguments were also based on the TRD study. None of the loss CODDEFFAGOLF estimated in their claim was attributed to local use (*La Tribuna*, 26 November 1993).

CODDEFFAGOLF publicly claimed that aquaculture had developed more than 28,000 ha of land and claimed that 6,000 ha were once mangrove forests and seasonal lagoons used as traditional fishing grounds during the rainy season. CODDEFFAGOLF emphasised the impacts on the habitat of international migratory birds and numerous species of marine flora and fauna and claimed that the areas most greatly affected were around the rivers Sampile, Tinto o Negro and Choluteca all of which were located in the San Bernardo region. The blame was directed at foreign investors, a number of politicians, and military personnel investing in the south (*La Tribuna*, 26 November 1993).

ANDAH argued that they had developed only 13,000 ha of arid land, plains, and salt flats, with little impact on the mangrove forests. Shrimp farmers had never utilised the entire area conceded and in 1993 had placed only about half of the total area conceded under production, but had a goal to expand by an additional 1,000 ha per annum on the salt flats to achieve Callejas' objective to increase production. ANDAH admitted that a boycott could result in considerable damage: an estimated $231,000 a year. ANDAH was adamant that aquaculture expansion should be allowed to take place on the salt flats near the rivers, estuaries, seasonal lagoons, and near the rivers that flowed into the GOF, and stated that they would lobby the government to demarcate the boundaries clearly for various uses. Uncertain legal boundaries for various human activities in the region persisted, complicating efforts to manage the resources (*La Tribuna*, 26 November 1993).

The increased public pressure, spearheaded by CODDEFFAGOLF, led the Minister of Environment (SEDA), Carlos Medina, to announce that the shrimp industry in the wetlands of the GOF would not be permitted until environmental impact studies (EIS) had been completed in accord with Articles 2 and 5 of the General Environmental Law. However, he also stated that "The shrimpers agree with protecting the mangroves in accord with the General Environmental Law because they want to avoid the decline of the industry due to environmental degradation, as has happened in Ecuador, Taiwan, and Indonesia" (*La Tribuna*, 27 December 1993). Medina had recently authorised nine of thirteen businesses that had applied for permission to expand to carry out the expansion on the salt flats. He stated that there were approximately 25,000 ha available for the cultivation of shrimp on the salt flats and that only 11,500 ha were currently being used for production (*La Tribuna*, 27 December 1993). Consequently, government actions continued to contradict its rhetoric.

As aquaculture expansion continued overlapping legal jurisdictions and unclear roles continued to complicate efforts to manage marine and coastal resources. Although the passage of the General Environmental Law in 1993 made it illegal to concede protected areas for the cultivation of shrimp DIGEPESCA had issued concessions to expand into protected areas (*El Periodico*, 25 January 1994). COHDEFOR was eventually provided with more power and management authority to govern forest resources in March 1994 with the passage of the Incentives for Forestation, Reforestation, and the Protection of Forests Law (Ley de Incentivos a la Forestacion Reforestacion y a la Proteccion del Bosque[74], 29 March, 1994, *La Gaceta*, Number 27,311, Decree Number 163-93). The law required COHDEFOR was create incentives to incorporate the private sector into government efforts to promote agro-forestry, reforestation, and protection for forested areas. Promoting private sector involvement in conservation efforts underpinned the nascent transition to neoliberal forms of environmental governance, which emphasised market-based solutions to environmental problems and a wider role for the private sector in resource management efforts.

The extent to which CODDEFFAGOLF, the Honduran environmental movement and international ENGOs encouraged the shaping of this law is not entirely known. However, it has been viewed as a successful outcome of their collective efforts. It represented the government's increased recognition of the importance of taking into account environmental factors in relation to development processes and action taken to address the demands of the environmental movement since the 1980s. However, it was apparent that the state also attempted to take into consideration private sector demands by devolving authority to various agencies to engage it in resource management efforts.

The wider movement under which CODDEFFAGOLF pursued its objectives, the Platforma de Lucha, achieved greater success when President Roberto Reina came to power. President Reina's campaign was based on an anti-neoliberal agenda, backed by the Plataforma de Lucha as well as CODDEFFAGOLF (Green, 1995: 229). The efforts of Varela had been recognsed in March 1994 when CODDEFFAGOLF was awarded WWF's J. Paul Getty 500 Prize of $50,000, in recognition of its efforts to promote and complement national policies designed to prevent the destruction of the country's natural resources (*La Tribuna*, 10 March 1994). To continue applying pressure, Greenpeace's Central American Coordinator, Lorenzo Cardenal called on the governments of the region to make efforts towards a Sustainable Development Alliance (Letter to *Ecocomentarios* in *La Tribuna*, 13 October 1994, submitted by Elmer Lopez Rodriguez, Coordinator for the Biodiversity Campaign, Greenpeace Central America).

---

[74] Ley de Incentivos a la Forestación Reforestación y a la Protección del Bosque. This law serves as the legal basis for the creation of PROMANGLE in conjunction with the International Tropical Timber Organisation (ITTO) later in the period.

State Response

As pressure continued to mount, the Reinas' new administration, recognising the widespread concern about its previous policies and the lack of implementing and enforcing them, issued statements that it would direct more effort towards environmental issues. Reinas acknowledged the need to continue developing appropriate legal instruments as well as institutions to implement recently enacted laws. The government claimed that they would assess the impact of existing and emerging industries in relation to the environment. It was stated that the creation of the Secretariat of Environment in the 1990s, the passing of the country's first General Environmental Law and the National System for the Regulation and Evaluation of Environmental Impacts (SINEIA)[75] were efforts towards these ends as the government sought to strengthen environmental governance. It was also stated that strengthening these types of institutions were shared goals among the Presidents of the Central American States through the CCAD (*La Tribuna*, 27 October 1994).

The government emphasised that SINEIA would play an increasingly important role in the regional efforts within Honduras. SINEA had been given legal authority in the General Environmental Law to secure plans, policies, programmes and projects in relation to industrial installations. It was also responsible for oversight of any public or private activities susceptible to contaminating or degrading the environment. The law provided the agency with the authority to ensure that new industries completed an environmental impact evaluation to determine the potential effects of their activities. It also required them to identify ways to mitigate these effects, and apply for environmental licences before commencing operations. It was made clear that licences would not be issued, including to the shrimp industry, unless an Environmental Impact Evaluation (EIE) had been completed (*La Tribuna*, 27 October 1994).

ANDAH continued to emphasise that there were still 20,000 ha of salt flats that could be used for shrimp farming.[76] It claimed that only 5,725 ha were currently under production for aquaculture, generating about 6 million lbs of shrimp a year. The shrimp produced were worth $25 million, making it the third largest export according to the Central Bank of Honduras. In turn, the industry continued to emphasise the indirect benefits to the

---

[75] Reglamento Sistema Nacional de Evaluación de Impacto Ambiental.
[76] La distribución del Golfo de Fonseca de Honduras en relación con su uso es la siguiente: "Uso del Territorio del Golfo de Fonseca Hondureño", 1994 Uso Área (ha) Porcentaje Bosque de mangle 43 678 61,2; Camaroneras 22 113 31,0; Playones salitrales 3 000 4,2; Salineras 1 842 2,6; Otros 693 1,0 Total Golfo Fonseca Honduras 71 326 100,0 Source: INCAE: "La Industria del Camarón en Honduras, Análisis de Sostenibilidad".

families of 15,000 people directly employed with the industry, 70 per cent seasonally and 30 per cent permanently, with up to 90,000 receiving indirect benefits associated with those employed by the industry, continuing to put forth the discourse of economic development (La Tribuna, 5 August 1994). It was estimated that by 1995 nearly 3,000 artisan fishermen were employed to supply the industry with wild postlarvae (El Heraldo, 6 November 2002). Subsequently, ANDAH argued that the Reinas administration should continue to seek expansion of aquaculture in the south due to the economic benefits it was generating for the country.

Fishermen represented by CODDEFFAGOLF claimed that they had not seen any of the benefits associated with the industry and as a result of its expansion had been forced to fish in the open waters of the GOF, sometimes within the territorial waters of Nicaragua and El Salvador (see Chapter 10). They claimed that this put their lives at risk since it was illegal for them to fish in the territorial waters of other countries. They also argued that it was putting their livelihoods in danger due to the risk of having their boats and fishing gear confiscated if caught by Nicaruagan or Salvadoreñan authorities[77] (La Tribuna, 5 August 1994). These claims revealed increased concerns associated with transboundary resource management and recognised the power of the narrative of territorial disputes, a concern for all three countries surrounding the GOF. ANDAH retorted, claiming it wanted to work with the government to take action in several areas to:

1. designate specific areas for shrimp farming to prevent negative ecological impacts on marine resources;
2. designate areas of salt flats that could be leased or conceded for the production of shrimp but with open access to the estuaries so that fishermen were able to fish in these locations;
3. delimit areas as reserves for the protection of the flora and fauna; and
4. establish areas for artisan fishing by drafting a joint agreement between ANDAH, the government, FPX, and CODDEFFAGOLF (La Tribuna, 5 August 1994).

ANDAH stated that their long-term goal was to work with the government to develop management plans for the entire GOF in conjunction with El Salvador and Nicaragua in

---

[77] In 1996 El Heraldo reported that fishermen from El Salvador were captured in Honduran territorial waters. More frequently, articles began to appear in the papers demonstrating an increase in transboundary resource conflicts as the pressure on resource use increased. The article used as an example here was titled: "Capturados mas Pescadores Salvadoreños en el Golfo de Fonseca"

order to address the transboundary resource issues (*La Tribuna*, 5 August 1994). Nevertheless, CODDEFFAGOLF's campaign to stop industry expansion continued, leading to a government moratorium on shrimp farming which signified another turning point in the conflict.

Moratorium Placed on Shrimp Farm Expansion

CODDEFFAGOLF, in conjunction with increased government support under the Reinas administration, persuaded the Secretariat of the Environment (SEDA was changed to SERNA in 1995), to place a moratorium on aquaculture expansion through a formal decree in January 1995. Utilising the new powers provided in the General Environmental Law, SERNA suspended the industry's environmental licenses and concessions to the land for twelve months (Natural Resources, Secretariat of Environment, Agreement Number 0041-95, *La Gaceta* No. 27,554, 6 January 1995).

ANDAH immediately filed a formal complaint opposing the moratorium and released a public statement that articulated its opposition to the unilateral decision of the government to declare it, and stated that it prevented further growth of the aquaculture sector. They called on the Honduran Foundation for Social Investment (FHIS),[78] created under the Callejas administration, to mitigate the negative impacts of neoliberal reform and to create a rural compensation fund with the goal of supporting local environmental projects and programmes rather than ceasing industry growth (*La Tribuna*, 30 January 1995).

The government's new Environmental Prosecutor, Clarissa Vega, expressed the need to complete an analysis of the current environmental laws because certain aspects of legislation were contradictory and duplicated roles (*El Heraldo*, 18 February, 1995). The National Congress's Commission on Natural Resources argued that tenure insecurity was one of the main factors leading to deforestation in the country and, therefore land titling should be a priority in accord with neoliberal agrarian reform. The Commission thus called on COHDEFOR to annually register property located on public lands (national, *ejidal* and in protected areas). They also called for a review of all lands registered since 1972,[79] the last time major agrarian reform efforts were undertaken.

---

[78] Fundación Hondureña de Inversiones Sociales

[79] 1972 because this is when the last major efforts for land reform/titling had taken place.

The National Congress also called on COHDEFOR to establish closed seasons on the capture of wild shrimp postlarvae and other species, especially in the GOF from May to December, to guarantee their sustainable reproduction. Finally, the Commission called on COHDEFOR to establish, within the next 60 days, management plans designed to permit, declare, maintain and administer public forests. The increased attention towards governance of the natural resources was another indication that the state was identifying means to abate the negative impacts of rapid neoliberal reform under the Callejas administration (*El Heraldo*, 5 May 1995).

Despite this rhetoric, economic development continued to supersede environmental concerns. The moratorium was all but ignored by businesses within the shrimp industry as it was rarely enforced. Moreover, the government continued to extend concessions and environmental licences to the industry for expansion, illustrating inconsistency between the government's articulated policy and its actions on the ground. CODDEFFAGOLF and others decided to continue pressuring the government and elevated the campaign against shrimp farming.

The campaign culminated in April 1996. CODDEFFAGOLF organised 3,000 people to march on the capital. During the protest, Greenpeace sailed its ship, *Moby Dick*, into the GOF to protest, in conjunction with CODDEFFAGOLF, the expanding industry on the same grounds it had in the past (*El Heraldo*, 11 April 1996). Greenpeace called on the government to enforce the ban on aquaculture expansion and begin addressing the effects on the wetlands. The campaign coincided with the first Shrimp Tribunal held by the United Nations Commission for Sustainable Development (UNCSD) in New York, April 1996. Coordinated by the Natural Resources Defense Council (NRDC) people were invited from Africa, Latin America and Southeast Asia to state their cases before the UN Commission (*Latin American Newsletters: Caribbean and Central American Report*, 16 May 1996).

In June 1996, the government ceased recognition of any tenure arrangements associated with shrimp farming, as well as environmental licences or permits issued for activities inside the protected areas prior to the passage of the moratorium. It also stated that AFE-COHDEFOR would not accept any solicitations for environmental licences or permits in protected areas (*La Gaceta*, 20 August, 1996).

FIGURE 11.2: GGMSB'S OPERATIONS IN 1997. PHASES 2 AND 3 OF GGMSB
Source: NOAA

Greenpeace and CODDEFFAGOLF announced a global boycott on the consumption of shrimp, targeting consumers in the US and Europe, while arguing for a global moratorium on production. Honduras was selected as the location from which to launch the campaign, maintaining pressure on the government to enforce their moratorium (*Latin American Newsletters: Caribbean and Central American Report*, 3 October 1996).

In conjunction with Greenpeace, CODDEFFAGOLF successfully convened a meeting on 15 16 October 1996 to encourage the passage of a global moratorium on aquaculture expansion and to launch the boycott. This boycott was intended to target the consumer and thus the export market promoted under neoliberal reform efforts. Saul Montufar, CODDEFFAGOLF, stated that "Environmentalists are not opposed to the shrimp industry but to its current operating methods" (*Inter-Press Service*, 18 October 1996). The meeting concluded with the signing of the Choluteca Declaration, to which over 20 environmental organisations from Latin America, Europe and Asia were signatories (Appendix V). The intention was to send the declaration to the United Nations and multi-lateral lending institutions to encourage the curtailment of aid to developing countries pursuing aquaculture expansion. Shrimp farming practices in India, Ecuador, Honduras, Nicaragua, Thailand, Brazil and Bangladesh were also highlighted in the document. The group agreed to follow up by presenting their claims at the February 1997 World Aquaculture Society Conference and holding an NGO Planning Forum on Shrimp Aquaculture in Santa Barbara, CA the following year 14-17 October, 1997.

The collective efforts of CODDEFFAGOLF and Greenpeace forced INA and the National Congress to amend the Agricultural Modernisation Law to address land tenure issues thought to contribute to environmental degradation. Once passed, the executive agreement clarified the tenure arrangements on national lands. It dealt specifically with communities and populations dependent on forests resources. The amendment of the law finally recognised the rights of people occupying state forested lands permitting them to continue occupying it. Chapter VI, Section I of the law also authorised AFE-COHDEFOR to oversee the development of management plans for protected areas on state lands and to enforce those plans once devised. Any new concessions or permits for extractive activities within these spaces were to be overseen by this agency (*La Gaceta*, 25 March 1997). The passage of this amendment was important because it recognised the right of those living in coastal communities in southern Honduras, while also allowing them to continue residing in current and future protected areas. The moratorium on shrimp farming was extended for another year on 23 October 1997 following the NGO Planning Forum on Shrimp Aquaculture held in Santa Barbara 14-17 October 1997.

The global aquaculture and environment debate seemed to be at its zenith as anti-neoliberal coalitions emerged in opposition to the shrimp industry. These coalitions increasingly expounded alternative visions for the future and argued for social and environmental factors to be taken into greater consideration at a variety of institutional levels. Fragmented movements came together to contest the negative effects of capitalist transformations, taking place in line with neoliberalisation. Localised movements banded together with larger organisations, although their interests may have varied, and stood on common ground in opposition to the corporations and governments behind globalisation leading to the formation of an anti-neoliberal coalition and the strengthening of transnational environmental political networks.

The Formation of Anti-Neoliberal Coalitions: The Industrial Shrimp Action Network (ISA-net)

In contrast to past *campesino* movements, CODDEFFAGOLF had received the backing of an extensive global network that included the public, the press and international organisations consisting of environmental and social activists. Representatives from 14 nations and several organisations – WWF, Greenpeace International, Environmental Defense Fund (EDF), Mangrove Action Project (MAP), and the NRDC – came together for the purpose of discussing their position on shrimp aquaculture practices worldwide. Their efforts built on the success of the Choluteca Declaration,[80] signed in Honduras the previous year leading to the formation of new coalitions that opposed the effects of neoliberalism.

FIGURE 11.3: PROTESTS AGAINST SHRIMP FARM EXPANSION, NEAR PUNTA CONDEGA
[Translation: No more shrimp farms in the Gulf of Fonseca]. *Source: El Periodico*, 23 July, 1997

The major success of the NGO Planning Forum, coordinated by a number of actors worldwide after signing the Choluteca Declaration, was the formation of the Industrial Shrimp Action Network (ISA-net) on 16 October 1997 – World Food Day – appropriately timed given that food security was generally used as the justification worldwide for the promotion of shrimp farming. The stated mission of the organisation was to "oppose the expansion of destructive industrial shrimp farming with such consequences as impoverishment and displacement of local communities, degradation of mangrove forests and other coastal and inland ecosystems, loss of agricultural land, pollution, and the loss of cultural and biological diversity."[81]

---

[80] CODDEFFAGOLF also signed the Declaration of Mombasa, February 1998 in Kenya and most recently the Declaration of Guayaquil singed in Ecuador the week of 16 November 1998. (CODDEFFAGOLF, Boletín Informativo No. 46, "Declaración de Guayaquil, Hecha por los Miembros de la Red Internacional de Acción Sobre la Industria Camaronera", September/December 1998: 13).
[81] www.shrimpaction.com, ISA-net website).

ISA-net became one of the first formal transnational environmental political networks opposed to industrial shrimp farming. It was created to lobby international agencies and governments to improve policies associated with natural resource management, and encourage consumers to stop buying shrimp, as it targeted the cornerstone of the neoliberal market, the consumer. The organisation also agreed to provide information and analysis to the UNCSD Shrimp Tribunal, first held in April 1996.[82]

One of the group's first actions was to attend the December 1997 FAO's Technical Consultation on Shrimp in Bangkok, to raise their collective concerns.[83] WWF stated that they would work with the World Bank and the FAO to explore environmental certification standards for aquaculture. They also began working on a position statement in relation to aquaculture that they planned to release in December 1998 (See Appendix VII). Each of these efforts was designed to influence public perception at the international level as the campaign against the shrimp industry escalated into the current global aquaculture and environment debate.

Domingo 15 de febrero de 1998

La camaricultura ha tenido un auge en los últimos años en la zona Sur del país.

FIG. 11.4: MANGROVE WETLANDS NEAR PUNTA CONDEGA

[Translation: Shrimp farming has grown over the years in the southern zone of the country.]
Source: La Tribuna, 15 February 1998

Global Aquaculture Alliance – A Neoliberal Coalition

The international campaign against shrimp farming led the industry to organise on a global scale to defend its interests against those contesting industry practices at the

---

[82] (http://www.ramsar.org/about/about_shrimp_action.htm)
[83] (http://www.ramsar.org/about/about_shrimp_action.htm).

local, national and international levels. The result was the creation of the Global Aquaculture Alliance (GAA) and, eventually, the Aquaculture Certification Council (ACC). Marketing, lobbying governments and demonising the environmental movement as 'radical' underpinned their campaign as they sought to delegitimise it within the public arena. Consequently, the industry's discourse shifted, and socially and environmentally produced shrimp were advocated.

It was apparent that the increased pressure environmental organisations were exacting on the state and private sectors led to change within a number of commercial businesses. The threat led the industry to counter the formation of ISA-net by creating the GAA. The environmental movement's efforts to create certification programmes to provide the consumer with a means to make environmentally and socially responsible decisions led the industry to establish its own certification process: co-opting the process to discursively define what was meant by 'socially' and 'environmentally' responsible shrimp. The global aquaculture and environment debate now had two distinct camps at the international level, ISA-net and the GAA. Ideologically one advocated environmentalism and the other advocated aquaculture, a material manifestation of neoliberal reform.

The GAA was formed by a small group of aquaculture industry leaders in the same year that ISA-net was formed. George Chamberlain, President of the GAA, owns Black Tiger Aquaculture in Malaysia and discussed establishing the organisation in conjunction with several industry leaders including Jim Heerin, Chairman of Sea Farms International (GGMSB). Discussions began between 1995 and 1996 in Tegucigalpa as a result of the moratorium placed on aquaculture in Honduras, India and other locations around the world (personal interview with Susan Chamberlain, Office Manager GAA, Tegucigalpa, Honduras, 25 August 2004).

ANDAH was one of the founding members, as well as Sea Joy of Honduras, Grupo Deli, and Sea Farms International (represented by Jim Heerin who was also a Board member of the GAA), three of the largest producers in southern Honduras. Based in St. Louis, Missouri, USA, the GAA claims to be an international, nonprofit trade association dedicated to advancing environmentally and socially responsible aquaculture. It argues that aquaculture is the only sustainable means of increasing seafood supply to meet growing food needs, incorporating food security discourse.[84]

---

[84] www.gaaliance.org/

The GAA, along with the National Fisheries Institute (NFI) lobbies governments and publicises aquaculture's role in providing food security. It developed an effective public relations campaign and organised lobbying efforts to influence perceptions in relation to aquaculture. It campaigns on behalf of the commercial industry in international fora, assists its members in advocacy to national governments, and provides technical information to its members to improve practices, mirroring some of the same strategies as the environmental movement.[85] Chamberlain stated that this was critical since "there are several hundred local and international NGOs worldwide contesting the industry, they are very well organised, and know how to utilise each other's strengths to achieve their goals" (personal interview with George Chamberlain, President of the GAA, Tegucigalpa, Honduras, 25 August 2004). He also argued that "the industry needed to take a more defensive role due to pressure from Greenpeace, WWF, Environmental Defense, and others which were trying to force them to change their practices" (personal interview with George Chamberlain, President of the GAA, Tegucigalpa, Honduras, 25 August 2004).

Since product certification had become one of the primary means by which environmental organisations sought to hold corporations accountable for their practices, the founding members of the GAA also decided to establish their own certification body. Subsequently, the ACC, based in Kirkland, Washington, USA was established as an international NGO that offers 'process' certification for shrimp production facilities with a primary orientation towards seafood buyers.

The co-optation of certification processes increasingly became common practice for a number of industries producing commodities such as coffee and shrimp. Like most of the industry-established programmes, Chamberlain argued that the ACC was a third-party certification programme (personal interview with George Chamberlain, President of the GAA, Tegucigalpa, Honduras, 25 August 2004). However, some of the GAA board members also served on the ACC board, which undermined its third-party independent status.

The ACC was established by the GAA in the interests of its members to certify social, environmental and food safety standards at aquaculture facilities throughout the world. Furthermore, the ACC bills itself as an organisation that independently and objectively

---

[85] www.gaaliance.org/

certifies the industry.[86] The certification programme was indicative of further private-sector involvement in environmental governance within the commercial industry as it attempted to establish what was meant by 'social' and 'environmental' standards. It also indicated that the industry accepted that actions in relation to social and environmental issues had to be taken into account and, at minimum, publicly addressed.

During the 2004 Central American Aquaculture Symposium, William Moore, President of the ACC, argued that "there is a shortage of good independent information about the environmental friendliness of aquaculture products; the ACC was created to fulfill this role".[87] The increased international pressure from the environmental movement forced the aquaculture industry to organise and incorporate social and environmental priorities into their discourse

Led by Jim Heerin as of 2006, the ACC's goal was to apply the GAA's Best Aquaculture Practices (BAP) agreed on by its members in a certification system. The ACC argued that by implementing its BAP standards programme its members could better meet the demands of the growing global market for wholesome seafood produced in an environmentally and socially responsible manner.[88] In Honduras, GGMSB was one of the first producers to agree to establish an environmental and social programme. These programmes primarily targeted communities near the larger shrimp farms or where the company's employees lived. The establishment of the programme in 1997 occurred in conjunction with the creation of the GAA. GGMSB's chairman played a key role in founding both (personal interview Dominique Gautier, Environmental Coordinator for Sea Farms International, April 2005).

As the industry and CODDEFFAGOLF worked to create opposing images of the local-level reality, the issues were debated more extensively in international fora as each competed to dominate discursively. As the debates at the international level became more heated, the issues continued to play out at the local level. The environmental movement continued to push its agenda so far that aquaculture businesses had to change their practices. ENGOs continued to hold the industry and the state accountable in relation to the governance of marine and coastal resources. However, the weakness of the nation-state and a lack of sufficient resources made it difficult to enforce compliance, but the efforts to target consumers resulted in the commercial industry

---

[86] http://www.aquaculturecertification.org/
[87] Presentation by William Moore, President of the ACC, Tegucigalpa, Honduras, 25 August 2004
[88] http://www.aquaculturecertification.org/

taking an active role, incorporating socially and environmentally responsible criteria into its operations.

The nation-states of the Central American (CA) region, in conjunction with ENGOs, took specific actions concurrently as the industry began to shift its practices. They agreed to define the CA region by ecological boundaries (e.g. MBC, similar to the *Paseo Pantera*) that transcended political boundaries, although management regimes remained in the sovereign realm of the state. Co-management agreements reached between the three countries were often discourse on paper and used as political tools rather than being viable, integrated coastal zone management (ICZM) programmes. Most SICA-CCAD funds were directed towards 'environment and development' issues, with 'development' emphasised more than 'environment'. Regardless, CCAD was the regional institution responsible for addressing shared-resource issues, but its power was limited due to wider geo-political strategic interests in the region that could be overridden by any nation-state's president or, in El Salvador, the Foreign Affairs Office[89]. For example, territorial disputes between the three countries that share the GOF overrode collective resource management issues. Consequently, transboundary resource management was too difficult an objective as nascent national and regional environmental institutions emerged. However, their emergence coincided with regional political and economic integration and neoliberal reform efforts in each country.

CODDEFFAGOLF and ANDAH Cooperate to Establish the GOF as part of Meso-American Biological Corridor

As the moratorium on aquaculture in Honduras was about to expire in August 1998, CODDEFFAGOLF announced that it would take its case against the Honduran government and the industry before the Central American Water Tribunal.[90] Varela stated that he had been working with the government and the industry in an attempt to reach agreement on the management of protected areas, but decided to bring the

---

[89] Further research needs to address what incentives create possibilities for international cooperation at a regional institutional level. My thesis is not about this topic. However, it is worth noting that the complexity of institutional relations in the Central American region, and economic ties to the US and the EU, possibly provide the largest incentives for CA nation-states to comply with programmes funded to establish habitat corridors such as the *Paseo Pantera* or the MBC. Trade and environment are important objectives of USAID, often discussed in terms of the development of green markets. These initiatives can be attributed partially to the efforts of the environmental movement locally, regionally and globally to put pressure on regional IGOs and IFIs to change their practices.

[90] The Central American Water Tribunal was created in line with the International Water Tribunal. It was one of the first actions of the Latin American Water Tribunal, in its attempts to create a parallel legal system. This enables diverse sectors of civil society to use their organisational skills, and to prosecute those responsible for damaging or otherwise abusing water resources and aquatic environments in the region (http://www.waternunc.com/gb/CAWT.htm)

issues to the Tribunal to increase pressure because of government inaction (*La Tribuna*, 18 August 1998).

As pressure mounted, ANDAH decided to support CODDEFFAGOLF's efforts to designate the area around the GOF as a part of the MBC and incorporate the entire region into the SINAPH, in support of state territorialisation of the space around the GOF. The creation of the Tribunal, the elevation of the claims into extra-judicial fora, and the call for the designation of protected areas marked a turning point in the conflict as it continued to be scaled-up. The government agreed to look into designating protected areas in accord with Honduran and regional environmental legislation. The state's gesture was considered to be a significant step for both CODDEFFAGOLF and ANDAH and demonstrated the latter's increased interest in taking into consideration both environmental and social issues (*La Tribuna*, 9 October 1998). The imperative to do so increased dramatically by the end of October 1998 when Hurricane Mitch devastated the country and the Central American region (discussed in the next chapter).

Summary

The almost complete absence of any relation between past policies directed towards the conservation of the natural environment and macro-economic and sector policies, heavily promoted by the Honduran government, the US and IFIs under the guise of neoliberal reform, made it increasingly difficult to protect the natural resource base and local livelihoods. The multiple examples used in Part III illustrate that, although neoliberal legal reforms gained pace throughout the 1990s, they occurred in congruence with social and environmental legal reforms intended to address the downside of reform efforts.

It was apparent that by the 1990s the ENGOs, along with other social movements emerging throughout Honduras, were able to place the conservation and protection of the country's natural resources firmly on the political agenda. Subsequently, the pressure generated from within civil society at the local, national and international levels led to the passage of the first General Environmental Law, the revamping of the 1959 Fisheries Law, and the creation of the National System of Protected Areas. Although civil society actions influenced the state's discourse and actions their contestations did not necessarily follow from one neoliberal policy to another. However, I have illustrated that the increased strength in civil society to contest the actions of the state ran parallel to the implementation of these policies.

The changes in the legal framework were representative of the power that actors within civil society were able to exert in order to realign the state's interests to their own. As I have illustrated, in a number of cases where policies did exist to protect the environment, they were rarely applied, and often disregarded or, in some cases, entirely ignored by the government and the private sector. I attributed this aspect of state failure to the weak presence of the government within the region. I argued that there were very few institutions capable of dealing with the wide array of issues that had emerged as neoliberal reform was undertaken.

Thus far, I have demonstrated that as neoliberal reform gained pace in Honduras and throughout the region in the 1990s, civil society actors organised collectively to advance their interests while also resisting the 'real' and perceived impacts of past and present macro-economic and political changes occurring. CODDEFFAGOLF, founded in 1988, was an example. As issues in the south continued to become more prominent and politicised, CODDEFFAGOLF and ANDAH continued to strengthen their links to a number of other national and international institutions, scaling up the conflict, referred to as the global aquaculture and environment debate. The creation of ISA-net and the GAA were used as examples. Consequently, as neoliberal reforms ensued, the state and regional institutions began to focus more attention on the social and environmental impacts associated with national and regional political and economic reforms.

In Part IV, I begin with Hurricane Mitch, as it had a huge impact on the southern coast of Honduras as well as the communities that lie on or near the coast. I argue that it created an opportunity to question past development practices and the neoliberal agenda. Although comment had been growing about the downside of neoliberal reform, such questions increased post-Mitch. Consequently, the 1990s saw further strengthening of the environmental movement to contest the reconstruction process. Specifically, CODDEFFAGOLF resisted efforts of the state to rebuild the industry and promote aquaculture further after Hurricane Mitch. The state was perceived to have implemented policies more favourable to the industry after Ricardo Maduro assumed the Presidency (NPH). In Chapter 13, I illustrate that CODDEFFAGOLF pursued an even more assertive strategy to protect the wetlands during this period.

Part IV

Post Hurricane Mitch: 1998 – 2006: Towards Neoliberal Environmental Governance

Chapter 12

Reconstruction and the Convergence of Neoliberal and Environmental Discourse

Introduction

In Chapter 12, I argue that neoliberal and environmental imaginaries converged more rapidly following Hurricane Mitch as the country sought to rebuild. I substantiate this argument by addressing the debates surrounding conservation and neoliberal reform. I begin by describing Hurricane Mitch and how international and civil society actors, the state, and the private sector began to negotiate the future of the country after it. Two social imaginaries dominated the debates, one that argued for the continuance and strengthening of neoliberal reform through reconstruction, and the other focusing attention on the need for better forms of environmental governance. 'Environmental governance' refers to the process or act of governing natural resources, often by promoting their conservation, use or management. Examples of new forms of environmental governance that emerged after the hurricane are provided in Chapter 13. Here, I focus on the debates that preceded the transition.

The socio-economic and environmental impacts of Hurricane Mitch nationwide are discussed to establish the severity of the crisis confronted and why the debates became so pronounced. Specific attention is then directed towards the impacts of the hurricane in relation to southern Honduras. Although there were several issues that could have been addressed, I have chosen to focus on the effects related to the fisheries sector, primarily the aquaculture industry and artisan fisheries, since they were the most relevant to the conflict assessed.

As development dollars, $2 billion or more, began to flow into the country after Mitch, the future of neoliberal reform was increasingly contested as numerous actors competed to influence how these funds would be spent. I argue that the hurricane's impact was significant enough that it changed the course of the environment and development debates in Honduras, exemplified by the IDB's, and the state's, call for strengthening regional environmental institutions. However, I demonstrate that aquaculture, and the agro-export sector in general was prioritised due to their importance in generating foreign exchange revenue, an important aspect of neoliberal policy. Consequently, the state and international actors had to pursue their objectives within the context of increased attention to environmental factors, which were, at least discursively, incorporated into the *Master Plan for Reconstruction and National Transformation*, the state's neoliberal blueprint for the future of Honduras. Finally, I argue that the post-1998 period had more to do with extra-national factors, some rooted in the same types of neoliberal reforms taking place in the US and Europe. Later, this point is illustrated by directing attention towards the Central American Free Trade Agreement (CAFTA) as the US sought to restructure the entire region in line with its neoliberal agenda.

As CODDEFFAGOLF and civil society actors sought to create a different future for the country, the focus on the agro-export sector and aquaculture was heavily criticised. The devastation the hurricane wrought on the country was blamed on past neoliberal reform efforts. As differing approaches to reconstruction were negotiated and contested within the public arena the administration of President Flores Facuse, Liberal Party (1998–2002), continued to pursue a number of initiatives that demonstrated the state's interest in environmental protection. Although the impetus to establish more assertive environmental policies was apparent, a number of changes based on neoliberal ideology were rapidly pushed through the Honduran National Congress as a part of the reconstruction efforts. However, policies related to both neoliberal and environmental ideologies became more explicit as the social imaginaries related to both converged, leading to a new era defined as neoliberal environmental governance.

The actions of civil society actors, the challenges confronted by shrimp businesses after the hurricane, and the pressure on the state to incorporate both social and environmental issues into the reconstruction plans, led the Flores administration to establish protected areas around the GOF, declare it as a part of the MBC, and as RAMSAR site 1,000, *The Convention on Wetlands of International Importance especially as Waterfowl Habitat* Environmental audits and state enforcement of environmental laws increased in accord with these efforts. Although demonstrable success was illustrated by the Flores administration in terms of seeking better forms of environmental governance, the shrimp industry in the south continued to be a primary beneficiary.

The efforts of the Flores administration were significantly undermined when President Maduro (NPH) came to power, leading CODDEFFAGOLF to target more assertively the companies in which he had invested. As Maduro's administration sought to rebuild the ponds where the largest producers were operating in the region of San Bernardo, protests and confrontations with GGMSB and El Faro increased quite significantly. CODDEFFAGOLF's role is addressed more substantively in Chapter 13. Furthermore, the IFC, Central American Bank for Economic Integration (Banco Centroamericano de Integracion Economica BCIE), and USAID investments in the larger operations drove a number of small and medium-scale operators out of business. The challenges the industry confronted during this period appeared insurmountable for some, influencing the degree to which the commercial industry was able to focus on social and environmental issues after Hurricane Mitch.

Hurricane Mitch

On 30 October 1998, class 5 Hurricane Mitch came in from the Caribbean causing widespread damage throughout Central America. The situation arising in the wake of the hurricane led a number of civil society actors to propose new options, alternatives that considered the socio-economic and environmental impacts related to previous neoliberal reform efforts. Most government officials and international organisations accepted that the severity of the human and economic impact was not merely the result of the hurricane itself, but also were exacerbated by past disregard for anthropogenic changes in the environment. They were attributed to the over-exploitation of the country's natural resources as neoliberal economic reform was embraced throughout the 1980s and the 1990s (Skidmore and Smith, 2001: 341).

Honduras bore the brunt of the hurricane's impact; as it sat over the country for nearly three days devastating huge swaths of the landscape, including the capital, Tegucigalpa. It resulted in serious losses to life and property and damage to homes, roads, schools, and hospitals, as well as the tourism, fishing, and aquaculture and agriculture industries. As a result, the hurricane destroyed over 60 per cent of the national infrastructure, 70 per cent of agricultural output, 25 per cent of the educational infrastructure, and severely debilitated the commercial and banking sectors, affecting 70 per cent of the economy (*El Heraldo*, 13 November 1998). The impact on agricultural output – 60 per cent of the country's exports – was particularly noticeable; with losses totaling more than $800 million (*El Heraldo*, 24 June 1999). In a country with over 45 per cent of its population living in poverty, the severity of the crisis was difficult to comprehend (*El*

Heraldo, 24 July 2003). The neoliberal reform efforts that had been pursued for more than a decade seemed to have been reversed overnight.

TABLE 12.1: HURRICANE MITCH: NUMBER OF DEAD, MISSING AND INJURED AS OF 14 NOVEMBER 1988

Country	No. dead	No. missing	No. injured
Honduras	7,000	12,000	1,932,482 (32% of the population)
Nicaragua	3,800	5,856	1,000,505 (23% of the population)
El Salvador	239	137	58,788 (1% of the population)
Guatemala	197	250	77,900 (0.7% of the population)
Total	11,236	18,243	3,069,675

Source: CODDEFFAGOLF Boletin Informativo No. 46 September–December, 1998: 3.

The south was particularly devastated as the downstream recipient of massive flooding. For example, in Valle more than 32,611 homes were damaged, 1,187 houses destroyed, 39 people were killed, 13 disappeared, and 6,130 *manzanas* of land used for cultivation were lost (*El Heraldo*, 13 November 1998). CODDEFFAGOLF calculated that the fisheries in the GOF faced reparations of nearly $8 million due to the destruction of boats, motors, fishing gear, basic infrastructure and access to fishing grounds in the lagoons, estuaries and mangroves.[91] Punta Raton, Guapinol, and Cedeño were the worst affected areas. Over 200 families were made homeless in Punta Raton and over 3,000 fishermen had to leave their homes in the areas around Guapinol and Cedeño (*El Heraldo*, 5 December 1998). It was estimated that the total number of people displaced exceeded 10,000 with over 100,000 people unemployed in the south; the situation was considered an acute crisis in a region that was already Honduras' poorest and most denuded (Boyer and Pell, 1999: 2). CODDEFFAGOLF requested that the Minister of Environment, Elvin Santos, and the Minister of Agriculture, Pedro Sevilla, evaluate the impact on agro-exports and fisheries, and include them in national reconstruction plans (CODDEFFAGOLF, Boletin Informativo No. 46, *Invitan a la SAG y Medio Ambiente a Evaluar Danos a los Pescadores*, September/December 1998: 6).

---

[91] "Golfo de Fonseca: Honduras Ha Aprovechado Mejor los Recursos Costeros" (*El Heraldo*, 10 September 2003). The region of the Pacific is rich in species such as *robalo, pargo, corvina*, and *mero*, all staples of the Honduran diet on the southern coast. It was estimated that there were about 20,000 Honduran artisan fishermen in the GOF by 2003, according to Miguel Suazo of DIGEPESCA.

The aquaculture industry was decimated by the hurricane as 6,000 ha of ponds were washed away. Losses were estimated at $50 million for the entire industry, including 600 tons of shrimp. The hurricane wiped out most of the coastal infrastructure making transportation in the south difficult (*El Heraldo*, 24 June 1999). Flooding of rivers and the influx of storm sewage to the coast inundated most low-lying areas, affecting coastal communities and the shrimp industry which lost substantial production and capital stock. Floods also washed away entire shrimp farms and eroded beaches, river banks and hillsides. In some cases, the hurricane changed the course of rivers. Floods inundated the region with fresh water, changing the salinity levels in the estuaries. Of primary concern were the potential impact of the changes on the wild postlarvae and, subsequently, the livelihoods of artisan fishermen (*Times-Piscayne*, 4 February 1999).

The contribution of fishermen to the overall economy was considered minimal and restoring the aquaculture industry's operations to pre-Mitch levels over-rode most local concerns associated with the impacts of Mitch in the south. Some people attributed the government's disinterest to the relative isolation of fishing communities and the smaller populations associated with them.[92] The perception of the state's actions were consistent with the government's emphasis on rebuilding the agro-export sector, as it was the key to generating needed foreign exchange revenue (*El Heraldo*, 5 December 1998). Consequently, reconstruction of the aquaculture sector became one of the main priorities for the government in the region, as it was the largest agro-export industry in southern Honduras by 1998 and influenced the future trajectory of the conflict between 1998 and 2006.

Given the severity of the disaster, the response to Mitch was rapid. Local grassroots organisations and national and international actors were forced to collaborate as financial aid and relief flowed into the country to finance the reconstruction efforts. However, numerous actors had differing priorities in regard to reconstruction and the future of the shrimp industry in southern Honduras. In accord with neoliberal reform, the US, the EU, Japan and a number of other countries quickly came to Honduras' aid, providing technical and financial support as the country prepared for the reconstruction. The Honduran government was able to secure $2.7 billion in loans and grants during the International Summit for Central American Reconstruction held in Stockholm, Sweden in

---

[92] A fisherman named Adilberto Hernandez noticed that the government was interested not in the impacts on local artisan fishermen, but on the larger-scale productive sectors in the south such as aquaculture and agricultural-based activities.

1999. However, the costs of reconstruction were estimated to be $5.2 billion, nearly twice as much as the secured loans (Boyer and Pell, 1999: 5).

For some, the hurricane provided an opportunity to change historical practices and renegotiate the terms of neoliberal futures. The President of the IDB, Enrique Iglesias, stated that reconstruction created the space necessary to remake the fiscal and economic structures of the nations affected by Mitch and also an opportunity to develop and strengthen environmental institutions in order to address the issues (*El Heraldo*, 13 November 1998). It seemed that the need for environmental governance in the neoliberal era was increasingly recognised due to the impact of the hurricane. The event brought numerous actors together in response to the natural disaster and led to positive steps to protect the wetlands, local livelihoods, and the agriculture and aquaculture industries in the south.

Post-Hurricane Mitch Reconstruction – Government, Shrimp Industry, and International Actors

The role of the government, the commercial shrimp industry, and international actors in the post-Hurricane Mitch context was particularly important. The hurricane raised key questions about past policy practices and led to greater consideration of social and environmental issues as civil society actors successfully placed their interests on the reconstruction agenda. CODDEFFAGOLF continued to defend its interests ardently and push its environment and development agenda in the south, while the state and the commercial industry sought to recover from the hurricane's impacts as quickly as possible.

By April 1999, the government released its *Master Plan for Reconstruction and National Transformation* which included a list of projects totaling more than $4 billion. Two citizen organisations that emerged after the hurricane – Citizen Forum and Interforos – had lobbied Flores to ensure that the finances secured included the following priorities: citizen participation, decentralisation, transparency, sustainable development, environmental protection and combating poverty, none of which they felt were in the existing plan, nor addressed by past neoliberal reform efforts. The Citizen Forum decided to formulate an alternative plan, which the President recognised after pressure from donor groups and numerous civil society actors, resulting in a statement that outlined joint priorities. The statement committed the government to aid transparency, decentralisation, and combating environmental and social problems (Jeffrey, 1999: 4-5).

The hurricane changed the course of the debates about aquaculture development and environmental conservation in southern Honduras. The discussions surrounding commercial aquaculture shifted towards reconstruction as the government and IFIs rapidly invested in the agro-export sector, one of the main sources of foreign exchange for the country. Environmentalists saw the event as an opportunity to restore previously denuded wetlands. These two competing perceptions of what should happen after Mitch dominated the discursive struggle between 1998 and 2006. Regardless of their disagreements, both the industry and the environmental movement had a greater awareness of their vulnerability to natural hazards and took steps to mitigate future impacts, leading to collaborative efforts towards ICZM.

In the south, the expansion of the aquaculture industry, intensification of agricultural activities in the coastal lowlands, and unprecedented deforestation had contributed to the continued deterioration of the natural environment, perpetuating conflicts between numerous social actors as the country tried to reconstruct. Those opposed to neoliberal reform in the past argued that the culprit was Honduras' export-oriented development model that was heavily promoted by the US and IFIs over the previous two decades.

The agro-export model was criticised for having led to the over-exploitation of natural resources. Neoliberal agrarian reform in particular was denunciated for contributing to massive deforestation, the clear-cutting of lands for agricultural development and the removal of mangrove forests for aquaculture development in the south (Boyer and Pell, 1999: 5). Increased agricultural activity over the previous decades was congruent with higher levels of use of fertilisers, chemicals and pesticides, leading to contamination of soils and waters in the south after the hurricane. Consequently, the macro-economic and political changes since 1973 were argued as the cause of environmental degradation and deteriorating social conditions in the south, providing greater legitimacy for CODDEFFAGOLF and other civil society actors to contest more recent neoliberal policies as reform efforts continued.

The government's main priority for reconstruction in the south was to rehabilitate the agricultural and aquaculture sectors, and repair damaged infrastructure impeding the full functioning of the agro-export sector, the base of the Honduran economy. Although he was not considered a strong advocate of neoliberal policies, President Flores had three post-Hurricane Mitch priorities for economic development that he had already been emphasising prior to the hurricane: tourism, expanding the *maquilas* and strengthening the agro-export sector (Jeffrey, 1999: 3).

President Flores also recognised that Hurricane Mitch provided an opportunity for the state, local and international actors to emphasise environmental protection as neoliberal reform ensued. Under the Flores administration, the state took a more active role in the creation of protected areas, while trying to ensure compliance and enforcement of laws. Whereas the period between 1988 and 1998 was defined by the emergence of new policies to protect the environment, after Hurricane Mitch state efforts to implement and enforce these policies became more important.

Consequently, although the hurricane increased the pace of neoliberal reform efforts in Honduras, it can also be attributed as a factor leading to an increased attention to nature–society relations. Subsequently, new efforts were pursued by the Honduran government that led to the strengthening of a number of national and regional institutions focused on environment and development issues in the GOF, and throughout the entire Central American region.

Hurricane Mitch immediately generated the political will for several pending decisions – pertaining to both neoliberal economic reform and environmental conservation – to be pushed rapidly through the Honduran National Congress. The strengthening of environmental governance therefore became a priority under the reform efforts between 1998 and 2006, leading to a new era. However, CODDEFFAGOLF was not solely responsible for government actions taken to protect the environment after Hurricane Mitch. It was only one organisation, among many, that influenced government policies after the hurricane – the disaster provided the impetus. After Hurricane Mitch there was an obvious convergence of interests between the state, the private sector and civil society actors, representing an increased interest in environmental governance as neoliberal reform continued.

The Politics of Post-Hurricane Mitch Reconstruction: Southern Honduras

Regardless of the commitment to the environment, it was apparent that economic elites used the hurricane to justify privatisation efforts and rebuild the agro-export sector quickly by providing more concessions to foreign corporations and the private sector nationwide (Jeffrey, 1999: 1-2). Less than two months after the hurricane, the Honduran

National Congress passed long-awaited laws on concessions in agro-investment, mining and tourism advocated as early as the Callejas administration.[93]

One of the most contentious efforts the government pursued was the revision of Constitutional Article 107, which gave foreign investors the right to purchase land within 25 miles of the country's borders, including coastal lands. When ratified, it allowed foreigners to have access to and control of up to 26,000 km² of national territory (CODDEFFAGOLF Boletin Informativo No. 47, January/February 1999: 9). Although the number of hectares acquired by foreigners since Hurricane Mitch is not clear, efforts to promote foreign investment in the land is apparent (see Chapter 4; neoliberal land reform, Thorpe, 2002).

At the time, up to 86 per cent of Hondurans still lacked legal title to the spaces where they lived and worked, leading to widespread resistance to this policy change (*El Heraldo*, 29 July 2003). CODDEFFAGOLF's primary concern was with land rights throughout the country, but it was also concerned that private investors would acquire control over the coastal wetlands before they were designated as protected areas. Consequently, it contested this constitutional revision among other policies. Along with the increased pace of neoliberal reform efforts, efforts to protect the environment also occurred more rapidly as CODDEFFAGOLF worked in collaboration with other social actors to influence the future of land use in the south.

The political momentum that the Flores administration established to address environmental issues dissipated when the National Party Candidate, Ricardo Maduro, was elected President in 2001. There was a noticeable increase in industry violations of the new environmental laws. Regardless of the political differences between Flores and Maduro, the reconstruction of the aquaculture industry was a priority for both.

CODDEFFAGOLF and international ENGOs shifted their attention to protesting the largest companies operating in the south, particularly GGMSB and Hondufarms after 2001. It was an effective tactic on the part of CODDEFFAGOLF as President Maduro was an investor in both companies: by targeting them, political pressure was applied at the highest levels of government.

---

[93] US Commercial Service, Report Title: Honduras Country Commercial Guide 2002: 5

Although Maduro had pledged to carry on the Flores administration's legacy and further enforce environmental protection laws, complaints against the government and its representative agencies in the south increased significantly during his administration, exacerbating environmental conflicts throughout the country and in the south (*Economist*, Intelligence Unit 'Riskwire', 29 October 2003). These problems can be partially attributed to the Maduro administration's apparent lack of concern with illegal actions as it sought to reconstruct the industry or, perhaps CODDEFFAGOLF became more successful at highlighting them.

The Aquaculture Industry: Post-Hurricane Mitch Reconstruction

Since Hurricane Mitch, the shrimp industry has been rebuilt and entrepreneurs have confronted a number of challenges. Aside from dealing with public protests and increased pressure from international NGOs and foreign governments in regard to its social and environmental practices, the industry has been confronted with a number of extra-national factors outside its control (*El Heraldo*, 24 June 1999).

Before the hurricane there were approximately 18,500 ha in shrimp production; by 2001 this had declined to 12,500 ha. The farms that have remained idle post-Hurricane Mitch are mostly small and medium operations. Reconstruction was slow; neoliberal legal and economic changes in the US and the EU, coupled with the threat of disease, a decline in the price of shrimp on the global market, environmental factors, and lack of infrastructure affected the industry's position, ultimately leading to a number of the small and medium-scale businesses ceasing operations by 2003. The larger businesses had huge sums of financial support from the international community to rebuild and faced less of a challenge.

In 1999, ANDAH had stated that they understood the need to protect certain areas, but but suggested that too much protection might affect the industry, prohibit new expansion and limit activities associated with the capture of wild postlarvae in the region, which could affect the livelihoods of local fishermen. At the time, 200 shrimp farms were operating in the region (CODDEFFAGOLF Boletin Informativo No. 51 September/October 1999 "Editorial": 2).[94] While the challenges at the local level were serious, the road to reconstruction was strewn with a number of international political and economic challenges that also confronted the industry.

---

[94] CODEEFFAGOLF took ANDAH's statement from Acuicultura en Honduras Ed. #1 ANDAH, Honduras C.A.: 18-19 Jose Antonio Fuentes and the Director of Biodiverstiy for SERNA wrote it in June 1999.

Extra-National Factors and the Challenges Faced by the Aquaculture Industry during Reconstruction

Following Hurricane Mitch, the World Bank's IFC[95] announced, along with Carlos Lara Watson, General Manager of GGMSB, that they were investing $6 million into the company to reactivate production to pre-Hurricane Mitch levels in order to rebuild the agro-export sector.[96] The IFC was already a shareholder of the company due to its initial investments in the industry in the early 1980s, when Arias and others initiated production (*La Tribuna*, 24 June 1999; *El Heraldo*, 24 June 1999).

The total cost of reconstruction was an estimated $18 million. The BCIE agreed to match IFC investment by $6 million, and the industry committed $6 million. The IFC funds were provided to repair infrastructure such as roads, electricity and water pumps, and to reconstruct buildings and expand production of an additional 951 ha of salt flats which were highly contested.[97] Funds were also provided to rebuild and improve the laboratory in Cedeño, the ice plant in Choluteca and the packing plant in San Lorenzo. New World Bank and IFC loan conditions stipulated that environmental management and recovery were required of companies in which they invested (*La Tribuna* and *El Heraldo*, 24 June 1999).

As shrimp industry reconstruction ensued, the actors behind it began to re-establish operations in areas where the ponds had been washed away. Most of their attention was directed to reconstructing the areas that had already been used for previous production. As the industry sought to rebuild, it faced a number of exogenous pressures, which ultimately led to the cessation of the social and environmental programmes they had initiated in 1997. Increased pressure from CODDEFFAGOLF to stop reconstruction did not make it any easier for the commercial industry. However, it was the external factors that weakened the industry to the point where it had to eliminate its social and environmental programmes in the midst of further expansion.

---

[95] The agreement was signed between Peter Woicke, Executive Vice President of the IFC and the President of GGMSB, Jorge Bueso Arias, who was also one of the original investors in shrimp farming in the south.

[96] GGMSB had taken steps to start a social and environmental programme after the formation of the GAA in 1997. Following the provision of this loan the company's efforts in these areas were more heavily promoted.

[97] : *El Heraldo*, "Camaroneros Denuncian Improvisación del Gobierno en Rehabilitación Vial", 6 October 2000. In 2000, shrimp farmers criticised the government for not taking measures quickly enough to reestablish vital infrastructure in the south, which they claimed was preventing the export of an additional 200,000 pounds of shrimp, or $1 million worth each year. Roads, ice facilities, processing plants, etc. were vital to industry productivity.

An increase in global production and, therefore, competition, especially in China, Taiwan, and Vietnam, as well as a number of environmental factors, such as disease, led to a 40 per cent decline in the price of shrimp by 2002, affecting producers (*El Heraldo*, 24 February 2003). The industry's weakened position forced it to explore methods of adding value to its products prior to export to improve its competitive position. The industry also became more interested in expanding current operations, according to Alberto Zelaya, President of ANDAH (*El Heraldo*, 11 April 2003). Businesses had already conceded 37,012 ha for the production of shrimp, and only an estimated 12,500 ha were under production (*El Heraldo*, 4 April 2003). Figure 14.1 depicts the number of hectares used for shrimp farming and the amount produced in tons.

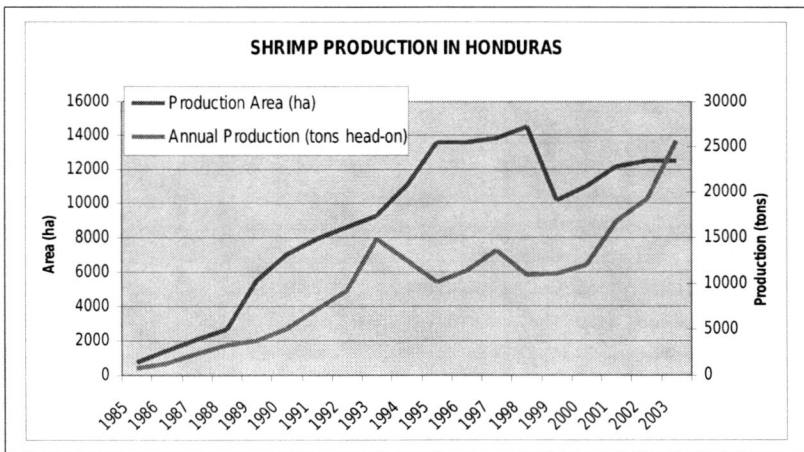

SHRIMP PRODUCTION IN HONDURAS

2003
Source: ANDAH, 2005

The threats to the industry were exacerbated further when the US passed an 'anti-dumping' law, following a round of WTO talks in Uruguay, placing restrictions on imports into the US from any country exporting items that comprised more than 3 per cent of the market. Given that the industry in southern Honduras exported 70 per cent of its product to the US, Zelaya called on the government of Honduras and the US to suspend the application of the anti-dumping law in relation to exports of shrimp from Honduras (*El Heraldo*, 4 November 2003).

In 2004, the International Trade Commission (ITC)[98] unanimously agreed that the US shrimp industry had been affected by imports from six countries, leading to the imposition of tariffs on them (*Washington Post*, 25 February 2004). Honduras was excluded for two reasons. GGMSB is a part of Sea Farms International (SFI), owned by US foreign investors, and accounted for over 50 per cent of production in southern Honduras, 70 per cent of which was exported to the US. In turn, the US had been investing extensive resources into reconstructing the shrimp industry, most of which were directed to GGMSB's operations between 1999 and 2001.[99] In other words, imposing restrictions on Honduran shrimp exports into the US would have been counter to the interests of the US government and private businesses.

TABLE 12.2: DISTRIBUTION OF LANDS OF ANDAH AFFILIATES IN OPERATION IN 1997, PRIOR TO HURRICANE MITCH

Size of Range (ha)	Number of Farms	% Total	Area	% Total
0–10 (extensive)	52	64	228.5	4
10–100 (semi-intensive)	16	20	533.0	9
100–200	4	5	536.0	9
200–300	2	2	450.0	7
300–400	3	4	985.0	16
400–500	2	2	912.0	15
500 or more	2	2	2569.0	41
Total	81	—	6213.5	—

Source: Ministry of Natural Resources, Honduras; 1997

Despite the enormous challenges confronted by the industry, in July 2003 ANDAH reported that the commercial industry produced 131.9 million pounds of shrimp (*El Heraldo*, 17 July 2003). In the same year, the Central Bank of Honduras (BCH) indicated that there were 14 laboratories supplying the industry with postlarvae, eight processing and packing plants, and 252 shrimp farms located in the Departments of Choluteca and Valle registered with the government. Of the 252 shrimp farms, 86 were affiliated with ANDAH. The impacts associated with Hurricane Mitch and the continued collapse in price

[98] The ITC is a US federal agency within the Department of Commerce (DOC) that analyses the impacts of imports on US industries.

[99] The goal of this project was to develop an information and technology transfer programme to educate and change the attitudes, perceptions and practices of resource users, resource managers and the general public to the sustainable use of coastal and marine resources in the Gulf of Fonseca region of Honduras and Nicaragua (Chaparro, R. *et al.*, "Providing Educational Support to Mitigate Future Disasters by Developing an Informed Citizenry", 25 February 2002).

on the international market forced 150 small and medium-scale producers to abandon their operations, according to the President of ANDAH. He stated that there were only 40 viable businesses in the GOF, 25 of which dominated the industry (*La Tribuna*, 18 September 2003). The region of San Bernardo comprised 83% of all shrimp farming practices in the south. Furthermore, 80% of the aquaculture industry was controlled by three companies; GGMSB (SFI), El Faro and Grupo Deli. In 2003, SFI controlled 50% of the Honduran market and continues to be the largest group represented by ANDAH (personal interview with Joaquin Romero, GGMSB/SFI, April 2005).

The satellite image below depicts GGMSB as well as a number of other commercial industry operations in the region of San Bernardo in 2003.

FIGURE. **12.2: GGMSB'S** OPERATIONS IN **2003** IN SAN BARNARDO AND OTHER SHRIMP OPERATIONS IN THE REGION. *Source: NOAA*

The reconstruction of GGMSB and other shrimp farms in the region was completed in 2003; nearly five years after Hurricane Mitch devastated the country. The original mangrove forests that had developed along the estuaries were preserved, according to GGMSB.

Most of the small farms were family-run or small cooperatives that were never able to reactivate production due to a lack of capital or insurance, and most did not have title to their property to use as collateral for a loan. Furthermore, they were seldom the recipients of development assistance in the reconstruction efforts. Most efforts focused on the larger farms since they produced the highest volumes and generated the largest amount of foreign exchange revenue. It was estimated that the hurricane's effects led to over a million pounds of shrimp being illegally exported into El Salvador where the value of the product was higher due to the 'dollarisation' of the country and fewer taxes imposed on the product (*La Tribuna*, 18 September 2003). Consequently, after Hurricane Mitch there was an increase in black-market activities in the south as formal market exchanges were impeded due to the effects of the hurricane.

The Chairman of SFI, Jim Heerin, argued that the anti-dumping action in the US was a factor that helped to increase prices, therefore diminishing the severity of the financial impacts on the industry. Restrictions on imports of shrimp from elsewhere led to more demand from suppliers like GGMSB, increasing its profits (personal interview with Jim Heerin, Chairman Sea Farms International, 17 March 2004). However, it was not enough to bring industry profits back to their pre-Hurricane Mitch levels. By July 2005, the company decided to eliminate its social and environmental programmes, along with all other non-productive assets, due to the financial situation that had unfolded since Mitch. The future direction of the company has been highly contested and plagued by infighting between the American and Honduran investors on the Board. The CEO, Ralph Parkman one of the first people to produce shrimp outside of Punta Raton near the *Estero Los Barracones* in the early 1980s was forced to step down in July 2005 (personal interview with Dr. Daniel Meyer following the 14 July 2005 SFI Board Meeting in Miami, interview August 2005).

Summary

The reconstruction of the agro-export sector and progress towards implementation of neoliberal reform objectives gained pace, although past reform efforts were increasingly questioned as the possible cause of the severity of the devastation wrought on the country. The rapid rebuilding of the aquaculture industry served only to exacerbate the issues. Consequently, the conflict between CODDEFFAGOLF and the commercial shrimp industry intensified during this period, as each sought to redefine the physical-spatial reality in southern Honduras in line with their interests and is discussed in Chapter 13.

CODDEFFAGOLF in particular saw the disaster as an opportunity to implement the conservation goals for which they had been fighting over the previous decade. The industry and the state were concerned about restoring productivity to pre-Mitch levels to regenerate lost foreign exchange revenue. However, neoliberal visions of the future had begun to consider both the environment and conservation as the two worldviews converged between 1998 and 2006.

The more recent history in the south is discussed as an historical trend towards neoliberal environmental governance. The process of increased attention to the socio-economic and environmental impacts related to past practices coincided with efforts to integrate the governance institutions of the Central American region politically, economically, and environmentally. Subsequently, throughout much of Central America, the institutions and organisations responsible for the governance of the marine and coastal environment have gone through rapid transformations in the context of globalisation.

Although there was an increase in concern for the environment at the national and regional levels, I illustrate how the institutions behind neoliberal reform have, in some cases, co-opted the environment as discourse, consistent with the findings of Dore. Dore has argued that the international development establishment throughout the 1990s has begun to wrap itself in the discourse of environmentalism to further legitimise the neoliberal agenda (Dore, in Collinson, 1997: 11). Neoliberalism has reframed conservation debates using the logic of the market, linking environmental governance to trade, and devolving systems of governance as the state seeks to produce new subjects of economic development and conservation in the midst of rapid political-economic change.

Here, I argue that, as the state sought to reproduce new images of the physical-spatial reality in the south, the neoliberal futures it pursued remained contested. Both CODDEFFAGOLF and other civil society actors further politicised the social and environmental impacts connected to the aquaculture industry. CODDEFFAGOLF's power relative to the state and the private sector had increased following the democratic transformation in Honduras, which placed it in a better position to negotiate the future of the use, management and conservation of the marine and coastal environments of southern Honduras. The organisation was strengthened further by the state's slow response to the hurricane, which delegitimised its capacity to act as the sole agent of power over the marine and coastal environments. Ultimately, the hurricane led to new debates concerning ICZM and transboundary resource management.

Chapter 13

Violence and Protected Areas

Introduction

In the post-Hurricane Mitch context, CODDEFFAGOLF symbolised the emergence of new powers embedded within civil society; assisting with the transition from government to governance as local actors created new spaces within the public arena to negotiate and contest the future of neoliberalisation in the south and throughout the country. Following Hurricane Mitch, the voices of the local people and civil society actors worldwide were increasingly heard, leading the state to make swift decisions (discussed in Chapter 12) that had implications for the environment, society and politics in southern Honduras. Although the Flores and Maduro administrations' approaches differed, the transition towards novel forms of environmental governance ensued through the creation of protected areas and transboundary management regimes. CODDEFFAGOLF continued to apply the necessary pressure to sustain attention towards both environment and development issues around the GOF.

I begin Chapter 13 by discussing CODDEFFAGOLF's characteristics, interests and actions in the post-Mitch context. I argue that its campaign against the aquaculture industry focused on the largest commercial businesses in the south. In order to politicise the impacts related to state and industry actions, the organisation espoused a narrative that linked violence to the industry's practices and mangrove degradation. Both narratives were articulated more broadly as ENGOs and others opposed to neoliberal reforms in the region sought to influence consumer decisions. I unravel these claims and demonstrate that local perceptions of the existence of conflict in the south and the purported violence attributed to the industry were inconsistent with the discourses that linked commercial aquaculture, violence, and mangrove degradation. However, through these over-politicised representations CODDEFFAGOLF was able to bring increased international attention to the issue of environmental protection in southern Honduras, resulting in the creation of protected areas. Once established, the organisation worked to ensure that the government and the private sector complied with the legal basis of protected area designations. The state increasingly made an effort to work with civil society actors and the private sector to develop a new aquaculture law that incorporated both social and environmental concerns into fisheries and aquaculture sector development.

I conclude Chapter 13 by arguing that the local politics of resistance to neoliberalism have coincided with a populist resurgence throughout Latin America. The impacts of past development processes in relation to the environment and the labour market led numerous organisations alongside CODDEFFAGOLF to contest further neoliberalisation. Consequently, USAID, IFIs and the private sector have been forced to incorporate social and environmental concerns into NDPs and, more recently, CAFTA, which has an environment and labour component incorporated into the final text. The result has been the institutionalisation of various forms of environmental governance into the neoliberal objective to promote free trade.

CODDEFFAGOLF Post-Hurricane Mitch

By 1999, CODDEFFAGOLF claimed to represent 10,000 artisan fishermen in the GOF: the rural poor, salt producers, teachers and primary school students. According to DIGEPESCA's Miguel Suazo, in 2003 there were an estimated 20,000 artisan fishermen, almost 3,000 of whom continued to supply the industry with wild-caught postlarvae to stock their ponds (*El Heraldo* 10 September 2003). Financial and human resources had been invested to reconstruct the aquaculture sector post hurricane Mitch, while very few went to issues affecting local fishermen's livelihoods. Most of the reconstruction aid designated for the south went to the largest shrimp farms, especially GGMSB, Hondufarms, El Faro and others, which were the targets of most of CODDEFFAGOLF's protests during the period between 1998 and 2006.

CODDEFFAGOLF continued to highlight the inconsistencies between the state's actions, articulated policy discourses and Honduran environmental law; exemplified by the government's continued provision of environmental licences and permits for the industry to expand and lack of legal enforcement, while rhetorically advocating for environmental conservation. CODDEFFAGOLF successfully targeted government agencies and the industry, further documenting their impacts on the coastal wetlands of the GOF while demanding that laws be enforced and complied with in an effort to hold the government and the industry accountable.

CODDEFFAGOLF's strategies and tactics continued to include formal public complaints, demonstrations, lobbying and attending local, national, and international conferences to raise awareness of the issues affecting the south. Internationally, CODDEFFAGOLF developed stronger alliances with other ENGOs and further strengthened the transnational environmental political networks opposed to global aquaculture expansion (see Chapter 11). In order to attack the industry, CODDEFFAGOLF and actors such as the Environmental Justice Foundation (EJF) and Greenpeace garnered international attention by putting forth an image linking the industry with violence and mangrove deforestation. Despite the fact that neither of these narratives lined up with local perceptions, this tactic was one of the means by which the industry's practices were further politicised during this period, escalating the conflict as actors continued to contest shrimp farming's impacts on society and the environment.

Violence and the Environment

Theories that link violent conflict and the environment have recently come to the fore in academic debates (Peluso and Watts 2001). These debates generally fall into two categories. The first is the environmental security literature (Homer-Dixon 1993 and Baechler 2006) and concerns itself with violence resulting from the degradation of the natural environment. This literature tends to focus on the relations between environmental scarcity and degradation and its contribution to large-scale violence and is often state-centric. The second focuses on the relationships between local level violence and the socio-historical contexts where they occur, as well as macro level processes that affect power dynamics; this approach lies within the tradition of political ecology.

Political ecologists critique the former approach as being neo-Malthusian by directing its lens towards environmentally determinist explanations of violence while ignoring political and social variables (Hecht 1985 in Peluso and Watts 2001). Political ecology of violence theory reverses the direction of the relationship between the environment and violence, broadly focusing on violence against people and the environment (Peluso and Watts 2001). In the following sections, I demonstrate that the concept of violence has been politicised in this conflict scenario. My results reveal facets that need to be further explored when theorizing about violent conflict and the environment.

The discourse linking the shrimp industry and violence appeared more frequently after 2001 as CODDEFFAGOLF sought to politicise the industry as being in violation of human rights laws. The claims generated by CODDEFFAGOLF were disseminated broadly through a number of transnational environmental political networks that sought to create an image of the industry as the cause of violent conflict in southern Honduras. For example, Dina Morel, Assistant Director of CODDEFFAGOLF, began to link the history of industry development publicly to the murders of 12 artisan fishermen in the GOF and the destruction of their access to traditional fishing grounds (*El Heraldo*, 6 November 2002). The deaths that Morel linked to the industry have never been proven in court and all records associated to most of the cases were apparently washed away during Hurricane Mitch.

Later, in 2003 the EJF completed a report entitled *Smash & Grab: Conflict Corruption and Human Rights Abuses in the Shrimp Farming Industry*, which provided a list of fishermen who had purportedly been murdered by the industry in southern Honduras, and claimed that the industry had repeatedly threatened Varela's life. In 2001, Stonich and Vandergeest (2001) published *Violence, Environment, and Industrial Shrimp Farming* in which they claimed that the industry was the direct cause of violence in southern Honduras.

The EJF report, Stonich and Vandergeest's (2001) publication and issues raised in other publications were investigated in a BBC documentary entitled *Price of Prawns*, made by Jeremy Bristow (Senior Producer of the Natural History Unit), and shown in December 2004. The publications and reports linking the industry and violence do not mention the industry's position in most cases. However, the BBC documentary did try to capture its views, albeit in 28 seconds:

> *09:48 Jeremy Bristow (Senior Producer BBC Natural History Unit)*
>
> Have you ever made any threats to Jorge Varela - to his life or to his property?
>
> *09:54 Hector Corrales (General Manager of Grupo Granjas Marinas San Bernardo (GGMSB))*
>
> No. No. That's not the way we do business, Jeremy. Is there a benefit in threatening him – or, threatening his life or his property? No, absolutely not.
>
> *10:15 Jeremy Bristow*
>
> Jorge Varela does not produce any evidence of the threats against him. But there has been violence and crime on the farms. According to the police there's been 500 cases of theft and 70 violent incidents in the last six years.[100]

The BBC's investigation into the matter revealed that the industry had reported over 500 cases of theft that had led to violent confrontations in some instances. Subsequently, I made an enquiry into local perceptions of the industry, conflict and

---

[100] BBC "*Price of Prawns*" Transcript, December 2004.

violence. Local respondents who participated were asked one question relevant to violence and the environment in southern Honduras: "Do you know of any conflicts (fights, arguments, disputes, or legal demands) in relation to the use, access, administration, conservation or control of the mangrove wetlands or the lands surrounding it?" The questionnaire intentionally excluded any reference to the actors in the south.

Of 364 respondents, 31 per cent (114/364) stated that they were aware of conflicts surrounding the mangrove wetlands. Of these, 38 per cent (43/114) associated the conflicts with the shrimp industry. However, it is uncertain whether or not fear of reprisal prevented respondents from answering accurately, making this aspect of my survey particularly uncertain. Regardless of the uncertainty, Figure 13.1 shows the percentage of respondents that perceived any type of conflicts around the wetlands as very violent, somewhat violent, on occasions violent, not very violent, and not violent at all. Given that only 12 per cent of the 364 respondents perceived the industry as linked to conflict, the results contradicted claims made that the shrimp industry's practices had been the direct cause of conflict and violence in southern Honduras, at least as perceived by the rural coastal poor. Naturally, this finding raises serious questions in relation to Stonich's claim.

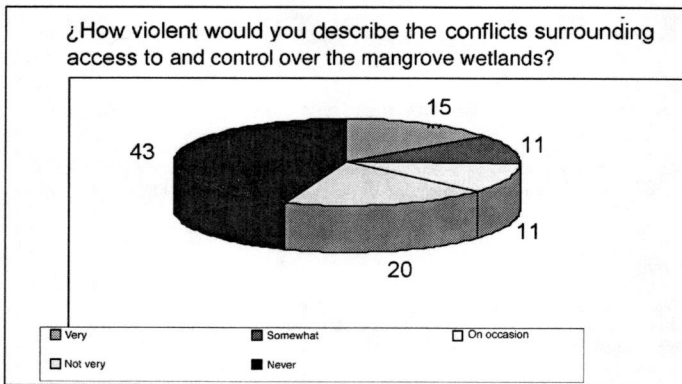

¿How violent would you describe the conflicts surrounding access to and control over the mangrove wetlands?

15

11

11

20

43

Very   Somewhat   On occasion   Not very   Never

FIGURE 13.1: PERCEPTIONS OF THE HOUSEHOLD RESPONDENTS IN RELATION TO HOW VIOLENT THEY PERCEIVED THE CONFLICTS SURROUNDING THE MANGROVE WETLANDS
Source: Wilburn, Socioeconomic Analysis: July 2005

The Purported Roots of Violence

The statements produced are indicative of how events and their interpretations were used strategically by actors opposing aquaculture. Most of the publications produced in relation to violence and the environment fail to mention incidences of theft that the industry has had to address. ANDAH estimated that 20 per cent of the shrimp produced are stolen by organised groups of artisan fishermen, taken to Nicaragua and El Salvador and sold on the black-market, exacerbating disputes between the three countries at the national level. El Faro estimated that their company alone was losing shrimp valued at $1,650,000 annually as a result of theft (*El Heraldo*, 5 November 2004).

Shrimp theft has been a problem since the 1980s; consequently, ANDAH always aimed to ensure facility security. However, the purported scale and degree on which it was now occurring was having a significant impact on the industry. In the past, ANDAH had demanded that the government take measures to protect the industry. Due to the government's inability to provide protection, the industry hired armed guards to protect the perimeter of their ponds. It was thought that clashes between those stealing the shrimp and the industry guards may have been a factor leading to the deaths, and why 70 violent incidents had been reported over the last six years in conjunction with 500 reported cases of theft (BBC, 2004). Despite the incongruence between the claims and local perceptions about violence the discourses suggesting the industry was the direct cause of violence and mangrove degradation succeeded in bringing further international attention to the situation in southern Honduras; and thus, to increase pressure on the government to acquiesce to CODDEFFAGOLF's demands.

Mangrove Degradation and the Industry

During the reconstruction efforts, CODDEFFAGOLF's campaign also focused more directly on the aquaculture industry's impact on the wetlands. CODDEFFAGOLF continued to publicly critique the government and ANDAH for the degradation of the mangroves. CODDEFFAGOLF's formal public complaints coincided with direct actions such as public protests, marches, or blockades. Formal complaints and public statements were released between cycles of protests in order to sustain public attention through the media. Attacks became more specific as CODDEFFAGOLF sought to castigate specific companies and individuals in violation of the laws. To apply continued pressure, CODDEFFAGOLF provided information to various NGOs through nascent networks such as ISA-net.

In order to develop a clearer understanding of local perceptions of mangrove use and degradation, I turned towards the communities in the south to assess individual positions regarding the impacts upon them. Respondents were asked if they thought the mangrove forests had changed over previous years, and via an open-ended question, the reasons for the changes. Of the total who responded, 32 per cent attributed some of the changes to shrimp farming practices (117/362), whereas 78 per cent (283/362) attributed most of the changes to local use of mangrove wood for fuel, construction, and other subsistence use activities (Wilburn, 2005). The perceived impact of local use was therefore greater than impacts attributed to the industry's practices. These results were not surprising when analysed against the rest of the data, which revealed that 67 per cent (240/356) of household respondents use mangrove wood for various purposes, primarily fuelwood, revealing the importance of it to their livelihoods. It also assisted with explaining the higher percentage that perceived themselves as the reason for most of the changes in the mangrove forests.

The results of my socio-economic study revealed that local people not only recognised their dependence on mangrove forests but also thought that they were one of the primary reasons for the spatial and temporal changes in relation to them. It is important to clarify that locals do not generally refer to the salt flats as a part of the mangrove forests; the *playones* are generally considered to be distinct from the mangrove forests. Therefore, local coastal people do not necessarily link the conversion of the salt flats for shrimp production as the cause of mangrove deforestation and, therefore, are not as likely to attribute blame to the commercial shrimp industry. The last point might explain the lower percentage of respondents who thought that mangrove deforestation was due to shrimp farming. However, through observation it was apparent that people in communities where CODDEFFAGOLF had been more active over the years had a tendency to attribute more blame to industry practices.[101]

---

[101] The socio-economic analysis conducted was a stratified random sample of southern Honduras. I did not do any cross-comparisons of communities, as it was not the purpose of the analysis. The statement is based on observations and interviews while conducting local fieldwork. For example, Guapinol was a centre of CODDEFFAGOLF campaigns over the years and, therefore, people in this community were exposed more frequently to the discursive positions related to those campaigns.

In question 82, when asked if their quality of life was threatened due to the deterioration of mangrove forests 75 per cent of the respondents (269/357) stated yes. If the respondents stated yes to question 82, they were then asked an open-ended question as to how the deterioration of mangrove forests affected their quality of life. Gladis Amador stated, "We wouldn't be able to fish or get fuelwood".[102] Horacio González indicated that "deforestation isn't good, we wouldn't have places where we could collect crabs, shrimp, and wood".[103] Juana Cabrera argued that "the environmental impacts, if the mangrove forests are destroyed, would diminish the production of fish and shellfish, bringing bad consequences to the economies of the families that depend on these products for work, to earn an income in order to subsist".[104] Each of these statements reveals that local people have a very clear understanding of the links between their livelihoods and the importance of intact mangrove forests. These data could help to explain why 33 per cent of the respondents (122/365) stated that they have participated in some type of mangrove reforestation since 2004. Of those who have participated in reforestation efforts, 63 per cent stated that their reason for participating was for species conservation (Wilburn, 2005). In sum, the results indicate that local perspectives are counter to the dominant discourse articulated by CODDEFFAGOLF and shared by others at the governmental, inter-governmental, and NGO levels in regard to mangrove deforestation in the GOF.[105]

The previous point leads to the conclusion that CODDEFFAGOLF's emphasis on mangrove deforestation proved to be more effective at garnering international support to protect the wetlands, than issues concerning competition over access to and control over fisheries resources. CODDEFFAGOLF was more successful in gaining support when

---

[102] Gladis Amador (age 29), Respondent #2 from Valle Nuevo, Municipality of Alianza, Department of Valle, Socioeconomic Analysis, July 2005. Original quotation was: "No podríamos sacar comida (pescar) o leña".

[103] Haracio González (age 51), Respondent #4 from Las Playitas, Municipality of Alianza, Department of Valle, Socioeconomic Analysis, July 2005, Original quotation was "El deforestar no es bueno, debido que no habría lugares de donde se pueda sacar cangrejo, camarones y madera".

[104] Juana Cabrera (age 45), Respondent #190 from La Pintadillera, Municipality of Amapala, Deparment of Valle, Socioeconomic Analysis, July 2005, Original quotation was "El ambiente se perjudica, si se destruye el mangle disminuiría la producción de peces y mariscos trayendo malas consecuencias a las economías de las familias que dependen de estos productos para trabajar y generar ingresos para subsistir".

[105] This statement is based on the results of over 100 open-ended interviews throughout the Central American region. I generally asked what the interviewee thought was the reason for the conflicts in the GOF. More often than not, the respondents mentioned the conflict between CODDEFFAGOLF and ANDAH. When probed further, they generally stated that the mangrove forests were being cleared for shrimp farming and that artisan fishermen were opposed to the destruction of the wetlands. The perceptions at the institutional level throughout Central America were fairly consistent on this point.

emphasising the discourse of deforestation as a result of industry practices rather than the effects of the aquaculture industry on local livelihoods. Furthermore, these results assist with explaining why CODDEFFAGOLF rarely emphasised the impact of local use on the wetlands to generate support for its cause, although locals perceived themselves to be a greater threat than the industry. Finally, it was apparent that CODDEFFAGOLF successfully politicised livelihood issues situated around fisheries resources by linking the industry's practices to the discourses of violence and human rights violations.

Consequently, the incongruence between local perceptions and those of actors at the national, regional, and international levels reveal the importance of understanding how and why explanations of mangrove deforestation have become more dominant as well as the narrative of violence. The implications of these are serious, given that programmes initiated by external actors may fail to consider local perceptions of mangrove deforestation, leading to management approaches inconsistent with the local-level reality. My last point is particularly important given that the state has more recently established protected areas in the region and, thus should create management plans aligned with the local-level reality. The more recent emergence of new forms of environmental governance in the region emphasise the need to conduct studies of this nature so that management plans are directed to address conservation issues important to local resource dependent populations.

CODDEFFAGOLF's Tactics

Beginning in 1999, CODDEFFAGOLF not only introduced new discourses on the industry's role in violence and mangrove degradation but also began to focus on ensuring that the government was enforcing its laws and forcing shrimp companies to comply. This was combined with direct action at the local level. The more direct actions can probably be attributed to training and support from Greenpeace, which advocates direct intervention to stop environmental degradation. Greenpeace, in conjunction with Red Manglar Latinoamericana, a newly created organisation for the protection of mangroves in Latin America, initiated the Campaign against the Shrimp Farming Industry and for the Defense of Tropical Coasts. Most of the published accounts in the media of CODDEFFAGOLF's actions from this point forward were noticeably more direct and frequent, illustrating the influence of external actors on local movements and *vice versa* (CODDEFFAGOLF Boletin Informativo No. 51 September/October 1999 *CODDEFFAGOLF en Congreso Internacional*: 11).

In addition to working with Greenpeace, CODDEFFAGOLF continued to work with other international networks of NGOs opposed to private industrial shrimp farming, such as the Mangrove Action Project (MAP) and Mangroves 2000. A number of organisations began inviting Varela to attend and speak at numerous workshops and international conferences in order to localise the issues for constituent members in the developed world. These efforts were designed to encourage external actors to apply more pressure on the Honduran government to increase enforcement and the industry to change its practices (CODDEFFAGOLF Boletin Informativo No. 51 September/October 1999 *CODDEFFAGOLF en Congreso Internacional*: 11).

Attention focused on the largest commercial shrimp farms in southern Honduras. ANDAH continued to be a primary target, but CODDEFFAGOLF and international ENGOs shifted their attention to protesting against the expansion of GGMSB, Hondufarms and El Faro's operations and linked them to mangrove deforestation and violent practices. GGMSB and Hondufarms became the most heavily targeted; an effective tactic on the part of CODDEFFAGOLF as President Maduro was an investor in both companies, targeting their practices applied political pressure where it counted: financially and through the electorate.

Throughout the period, CODDEFFAGOLF argued that SERNA should be setting aside any space available in the region for the fishermen, not the larger industry, since they had been affected most by the continuing expansion of it. CODDEFFAGOLF also decried activities contrary to the terms of the RAMSAR Convention. Signatories to RAMSAR are obligated to develop national policies for the conservation and wise use of mangroves, whilst incorporating considerations concerning environmental conservation into land use policies (Hernandez, 1999: 10).[106] The government had not yet defined land use nor created management plans in relation to these newly protected spaces.

CODDEFFAGOLF's actions forced the government to respond and take measures against commercial businesses violating national laws while it also worked on the development of management plans (*La Tribuna*, 28 October, 2002). By 2006, they had also succeeded in pushing the government to pass a new aquaculture law. CODDEFFAGOLF successfully used reconstruction as a political opportunity to bring attention to the south and utilised the images from reconstruction efforts to illustrate that the industry was destroying the wetlands.

---

[106] Hernandez, Gabriela (Editor) IUCN, Meso-American Wetlands RAMSAR Sites in Central America and Mexico, 1999: 10.

On 19 March 2004 in San Jose, Costa Rica, the Central American Water Tribunal released its non-binding ruling on the case: *Destruction and Contamination of Coastal Wetlands of the Gulf of Fonseca, Departments of Choluteca and Valle, Republic of Honduras*. The tribunal found in favor of CODDEFFAGOLF, validating their accusations against the IFC, World Bank, GGMSB and El Faro (the second largest company after GGMSB) on the grounds that they were in violation of a number of Honduran laws as well as several international treaties and conventions to which the government was signatory.[107] The Tribunal verified that the industry's production of shrimp, particularly that of GGMSB and El Faro, had seriously impacted the wetlands and the ecology of the GOF and that the Government of Honduras had failed to fulfill it's agreed obligations under the RAMSAR convention. Although the ruling was non-binding, the Tribunal made several recommendations that mirrored those repeatedly offered in the past, such as continuing a moratorium, delimiting areas where shrimp could be produced, developing a management plan for the GOF, and developing a new aquaculture law, enforcing existing laws, among others (Mangrove Action Project Newsletter, 12 April 2004).

The Emergence of New Forms of Environmental Governance

In the aftermath of Hurricane Mitch the combination of continuing pressure from CODDEFFAGOLF and its international partners and the increase in political will to address the environment as part of the rebuilding efforts (although it was not the priority) led to a surge in government actions on the environmental front.

The same trend was visible beyond the level of the state as environmental management issues were increasingly incorporated at the regional level and addressed by the IFIs. A prime example was the establishment of the MBC, which was led jointly by USAID, The Nature Conservancy (TNC), WWF and the Rainforest Alliance in order to create a contiguous habitat corridor throughout the CA region, building on the *Paseo Pantera*, through the *Programa Regional Ambiental para Centro América* (Regional Environmental Programme for Central America, PROARCA). The USAID-funded programme emphasised environmental management through market-based initiatives in accord with neoliberal objectives in the region. The programme was created to support efforts to create

---

[107] The political Constitution of Honduras, the International Treaties such as the Universal Declaration of Human Rights, the American Convention on Human Rights, the International Pact about economical, social and cultural rights, the World Letter of the Nature, the Rio Declaration on Environment and Development, the Stockholm Declaration, the Agreement on Biologic Biodiversity, among others, establish that men and women have the right to the common patrimony of all human beings, the natural resources.

protected areas within the newly created MBC as *ad hoc* regional environmental governance institutions emerged.[108]

Prior to Hurricane Mitch (1997) the IUCN, through its PROGOLFO[109] programme, funded by DANIDA, had completed studies identifying potential protected areas and multiple use areas in the GOF region. CODDEFFAGOLF lobbied SERNA, AFE-COHDEFOR and the National Congress to approve the plan that emerged from the PROGOLFO study once it had been agreed upon (*La Tribuna*, 15 October 1998). Less than a year after the hurricane, the Flores administration declared the region of the GOF as RAMSAR Site 1,000, a part of the MBC, and the wetlands as protected areas.

Within the same time period the Pacific coast was declared a part of the MBC (*El Heraldo*, 23 June 1999). The designation included a proposal to create seven new areas for the management of habitats and species, a new national marine park, and two areas for multiple use located in the Bahia de Chismuyo, San Lorenzo, El Jicarito, Las Iguanas, Punta Condega, San Bernardo, La Berberia, Los Delgaditos, the Archipelago of the Gulf of Fonseca, *la Isla del Tigre* (Tiger Island) and Cerro Guanacaure, covering a total territory of 76,208 ha. SERNA stated that the declaration was the result of consultations with government institutions, local government, private industry, conservation groups and the coastal communities (*El Heraldo*, 23 June 1999). At this point, the CVC was reinvigorated as a mechanism for collaboration between CODDEFFAGOLF and ANDAH. Each of these organisations agreed to work together to improve protected areas, specifically those that lay near fishing communities close to the shrimp farms in San Bernardo that continued to be contested (*La Tribuna*, 15 October 1998).

The Flores government also made great strides in improving the enforcement of environmental protection laws and by 2001 his administration began to prosecute violations more frequently (*Economist*, Intelligence Unit 'Riskwire', 29 October 2003). Beginning in June 1999, Clarissa Vega, Special Environmental Prosecutor for the Honduran government released a public statement indicating that there would be increased enforcement of the rule of law (*El Heraldo*, 27 June 1999). The 'rule of law' was increasingly emphasised as a fundamental principle of neoliberal reform as it was believed to create a more secure environment for foreign investment and strengthen

---

[108] The information pertaining to PROARCA was obtained through several interviews between 2003 and 2005 with Edas Munos, Technical Director for PROARCA's programme in the GOF which concluded in 2005.

[109] Proyecto Conservación de los Ecosistemas Costeros en el Golfo de Fonseca (PROGOLFO).

democratic rule as long as there was a strong judiciary to prosecute violations. Furthermore, the government indicated that it would begin creating management plans for the protected areas, national parks and reserves to protect the country's natural resources.

A few years later in 2003, Vega was appointed Special Prosecutor for Organised Crime, placing her in a position to pursue those in violation of the law and giving her the power to begin imposing fines (*La Prensa*, 28 July 2003). In her new role, Vega worked with the new Environmental Prosecutor, Elmer Lizardo, and immediately imposed a fine of $57,700 on the shrimp company Tony's Mar for degrading mangrove forests in the south. It was stated that the fine was for violating the protected areas law and for not complying with the terms of their environmental licence (*La Tribuna*, 26 November 2003). The government appeared to be taking actions to enforce compliance and to respond to CODDEFFAGOLF's demands articulated in formal legal complaints and through protests against the company. By 2004, CODDEFFAGOLF's position was justified when the Central American Water Tribunal released its ruling in relation to its case against the World Bank and the industry.

As the government sought to strengthen environmental institutions in order to address the negative impacts of economic reform efforts, laws were passed to increase state agencies' power to act. In September 1999, the government passed an amendment to the General Environmental Law, to clarify the roles and responsibilities of SINAPH. SINAPH was given the power and authority to apply elements of the General Environmental Law and propose areas of cultural and natural resources for conservation by creating protected areas (Chapter I, Section I, Article 1). Specific roles and responsibilities regarding the management of protected areas were also clarified (*La Gaceta*, 25 September 1999). The Vice Minister of the Secretariat of Agricultural and Ranching (SAG) also encouraged SERNA to complete environmental audits of the businesses in the south to determine whether they had appropriate operating licences, finding that a number of them did not (*La Tribuna*, 9 February 2001). He also stated that they would introduce a new version of the Fisheries Law that took into consideration current issues surrounding aquaculture and fisheries resources (*El Heraldo*, 14 January 2001). Negotiations between the actors commenced to revise the 1959 Fisheries Law and to prepare new aquaculture legislation that would address both socio-economic and environmental concerns.

Negotiations of the Aquaculture Law began in 2001 and lasted until the autumn of 2006 (see Appendix VI). The law linked aquaculture with environmental conservation and sustainable development. To ensure that the industry's practices were socially and environmentally responsible, the National Congress created the Consejo Nacional de Acuicultura de Honduras (National Council for Honduran Aquaculture, CONAACUIH) with representatives from SAG, DIGEPESCA and other relevant agencies (Article 13). The law came into effect in tandem with the election of a new President, Manuel Zelaya of the PLH. The first task of the administration was to build consensus around the management plans, which had been under negotiation since the creation of the protected areas. Restructuring DIGEPESCA was also a priority to begin establishing management regimes. COHDEFOR, with lead responsibility for the protected areas, has continued to play an important role in the management of the mangrove ecosystem.

Despite the rapid neoliberal reform efforts (discussed in Chapter 12), these examples demonstrate that there existed the will, and certainly an increased level of environmental conscientiousness, on the part of some government officials during this period. The strengthening of environmental law and increased enforcement is aligned with larger trends in the Central American region. An example of this point is the Central American Free Trade Agreement's (CAFTA) environmental component which has coincided with the rise and strengthening of institutions such as the Central American Commission on Environment and Development (CCAD) within the System for Regional Integration (SICA). The last is the overarching governing body in the region and focuses on both political and economic integration in Central America. Although the state is playing an active role in these processes, civil society continues to expand the scope of its efforts to reshape the future of the neoliberal landscape.

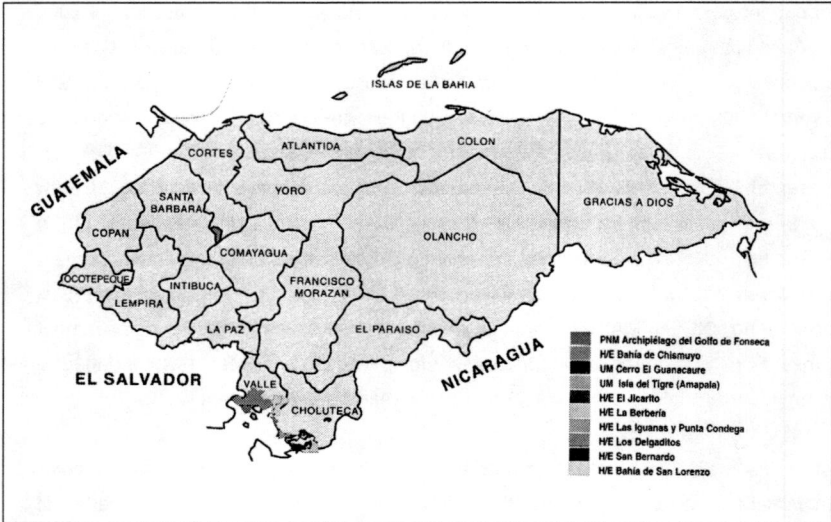

ISLAS DE LA BAHIA

GUATEMALA

CORTES
ATLANTIDA
COLON

SANTA BARBARA
YORO

COPAN
COMAYAGUA
OLANCHO
GRACIAS A DIOS

OCOTEPEQUE
INTIBUCA
FRANCISCO MORAZAN

LEMPIRA
LA PAZ
EL PARAISO

EL SALVADOR
VALLE
CHOLUTECA

NICARAGUA

PNM Archipiélago del Golfo de Fonseca
H/E Bahía de Chismuyo
UM Cerro El Guanacaure
UM Isla del Tigre (Amapala)
H/E El Jicarito
H/E La Berbería
H/E Las Iguanas y Punta Condega
H/E Los Delgaditos
H/E San Bernardo
H/E Bahía de San Lorenzo

MAP 13.1: PROTECTED AREAS IN THE SOUTH
Source: Areas Protegidas de la Zona Sur de Honduras, 2001: 26

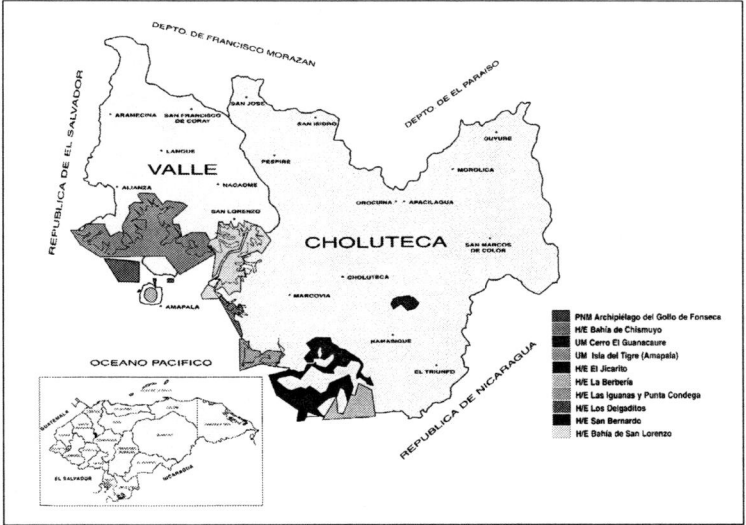

MAP 13.2: PROTECTED AREAS IN THE DEPARTMENTS OF VALLE AND CHOLUTECA
Source: AFE-COHDEFOR, 2002

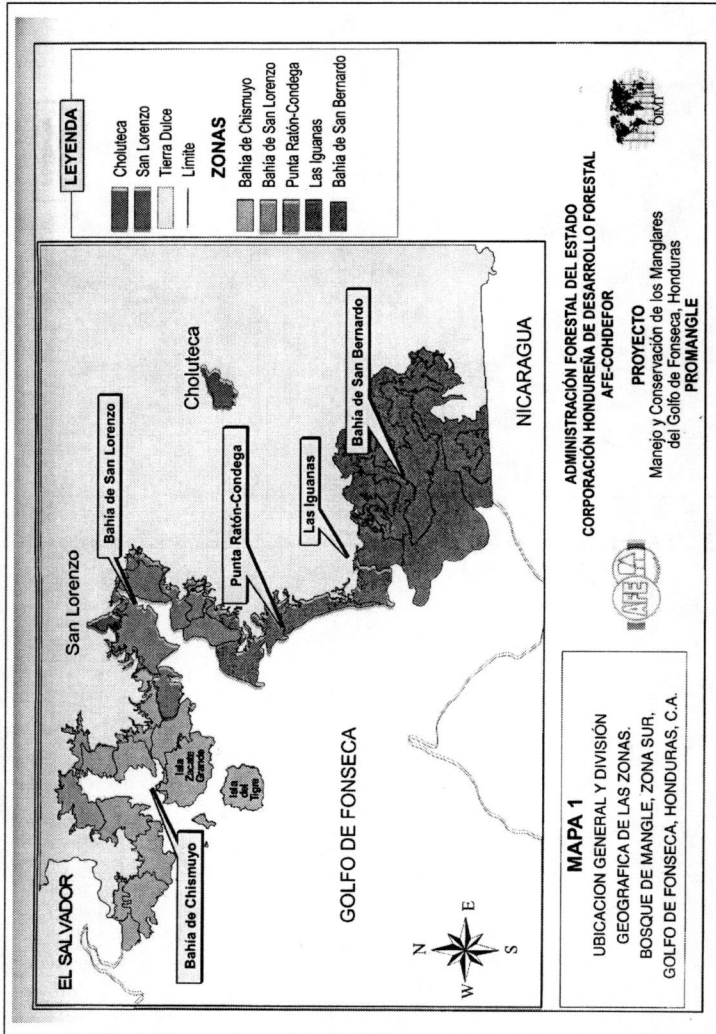

MAP 13.3: GEOGRAPHIC DIVISIONS OF THE MANGROVE FOREST ZONES IN SOUTHERN HONDURAS

Source: AFE-COHDEFOR, 2002

LEYENDA

Bahía de Chismuyo
1 Delta del Río Goascorán
2 Monñueca
3 Llano Largo
4 Caguin
5 Jiote Grande
6 El Aceituno
7 Boca de La Brea
8 Río Viejo
9 Isla Chocolate
10 Puerto Soto (El Relleno)
11 El Morey
12 Islas del Pacífico

Bahía de San Lorenzo
13 Rincón de Judas
14 San Lorenzo
15 Zona Litoral Laure
16 Paso Crucero
17 Zona Litoral Las Arenas
18 Isla Barracones
19 Zona Litoral Isla de Piedra

Punta Ratón-Condega
20 Punta Ratón
21 Cedeño
22 Guapinol

Las Iguanas
23 Playa Negra

Bahía de San Bernardo
24 El Pedregal
25 Puna Glabiéis
26 La Berbería
27 Áreas Protegidas
28 Industria Acuícola

☐ Tierra Dulce
— Límite

EL SALVADOR

San Lorenzo

Choluteca

GOLFO DE FONSECA

NICARAGUA

N
W — E
S

MAPA 2
UBICACION GENERAL Y DIVISIÓN DE
LAS SUB-ZONAS

BOSQUE DE MANGLE, ZONA SUR,
GOLFO DE FONSECA, HONDURAS, C.A.

AFE

ADMINISTRACION FORESTAL DEL ESTADO
CORPORACIÓN HONDUREÑA DE DESARROLLO FORESTAL
AFE-COHDEFOR

PROYECTO
Manejo y Conservación de los Manglares
del Golfo de Fonseca, Honduras
PROMANGLE

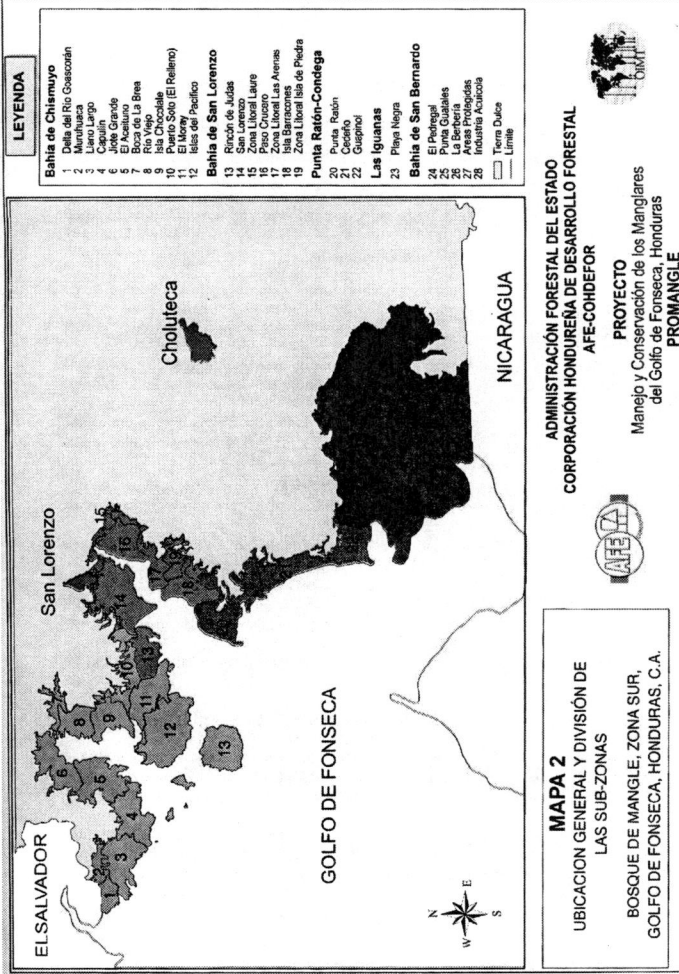

MAP 13.4: PROTECTED AREAS AND GENERAL DIVISION OF FORESTRY ADMINISTRATION

Source: AFE-COHDEFOR, 2002

There is no doubt that neoliberal futures will continue to be negotiated and contested in Honduras and regionally as they have been for over two decades. Social and environmental issues will remain central to these debates as cultures and societies attempt to adjust to the effects of neoliberalisation. Interestingly, the Wildlife Conservation Society (WCS) has initiated a new project, the Ecological Corridor of the Americas project (EcoAméricas) which is an intentional counter to the FTAA. The initiative seeks to create a hemispheric system of protected areas and other wild areas. The goal is to integrate efforts such as the MBC into EcoAméricas from the northernmost point in Alaska to Tierra del Fuego, just as the FTAA seeks to integrate the region's economy in the same space through trade.

FIGURE 13.2: FISHERMAN/PROTESTOR BEING ARRESTED
Source La Tribuna, 7 February 2001

The goal of EcoAméricas is the conservation of biodiversity, maintenance of ecological viability and evolutionary processes and the provision of environmental services, through the creation of a conceptual framework for cooperative action that would assist and connect local, regional and national conservationist efforts along the corridor route.

It also aims to contribute to social, economic, cultural and scientific development of the Western Hemisphere, through conservation and restoration of biological diversity.[110] However, impediments lie ahead as regional infrastructure development projects such as Plan Pueblo Panama are implemented to facilitate trade and development. As infrastructure development is pursued to support trade and tourism, it is likely that forests and tropical coastal ecosystems will continue to be encroached upon.

Dénouement

The shrimp industry has undergone restructuring as it faces increased pressure due to global market forces that have affected prices. It also continues to face adversity from international ENGOs worldwide that continue to link its practices with violence and mangrove degradation. As more consumers in the US, EU, and Japan become increasingly conscious of where their products originate and what practices are involved in their production, the industry will have to continue adapting to their demands, global market forces, national politics, and contestation against its practices, which they continue to claim affect the environment and society.

The US will continue to play a pivotal role in the CA region. Bilateral relations have strengthened as a result of the large amount of assistance provided by the US to reconstruct. These relations continue to be particularly important when taking into consideration that Honduras' largest trading partner is the US, 70 per cent of Honduran exports are bound for the US (CIA World Factbook, 2007). US exports to Central America in 2000 reached $8.8 billion; this is more than the US exported to Russia, Indonesia and India combined (USAID Mexico and Central America Regional Strategy, FY 2004-2009). Furthermore, the US accounted for 40 per cent of Honduras' *non-maquila* exports and over 95 per cent of *maquila* exports, making access to the US market extraordinarily important (*Economist*, 2002).[111] The recent passage of the CAFTA will assist but will have little bearing on the shrimp industry, as import tariffs, duties, and taxes on aquaculture were all but eliminated with the passage of the Caribbean Basin Initiative Act in 1983. However, it is clearly indicative of US intentions to continue the process of neoliberal reform which will have implications for the country and the region.

---

[110] http://www.ecoamericas.com/en/issue.aspx
[111] http://www.eiu.com/

One need only look towards the language used in the statement made by US Trade Representative Robert Zoellick[112], on 8 January 2003 when CAFTA negotiations were launched to identify neoliberal intentions on the part of the US. Zoellick stated that:

> CAFTA will give Americans better access to affordable goods and services and **promote US exports** and **jobs**, even as it advances Central America's prospects for development. This **free trade** agreement will reinforce **free-market reforms** in the region. The **growth stimulated by trade** and the **openness** of an agreement will **deepen democracy**, the **rule of law**, and **sustainable development**. This agreement will further the regional integration that the Central Americans have begun, and complement our vital work on the **Free Trade Agreement of the Americas** (FTAA)" (USAID Mexico and Central America Regional Strategy, FY 2004-2009).

The strategy is underpinned by the "overarching goal of a *more stable and prosperous Mexico and Central America, sharing the benefits of trade-led growth broadly among their citizens*" in order to reduce poverty (USAID Mexico and Central America Regional Strategy, FY 2004-2009). The strategy is linked with US goals under the first Bush administration's Enterprise for the Americas Initiative, which established the FTAA. The passage and ratification of CAFTA by the US Congress in 2005 was one more step to achieving US foreign policy interests in the region. However, the US Congress has agreed to provide on average $40 million annually through the Environmental Cooperation Agreement (ECA) for labour and environment-related activities as CAFTA is implemented to address concerns raised by numerous civil society actors concerned about both. Questions about who will benefit and how the money will be spent are undetermined, as is the historical trajectory of the future of the region.

The environmental movement's efforts to resist the effects of neoliberal reform have more recently coincided with the resurgence of populist sentiment in the region as new social actors continue to resist US interests, the foundation of which is neoliberal economic and political reform. On 4 May 2007 the *Financial Times* reported that this contemporary revolution "comes at a time when Latin America is less dependent on the multilateral institutions than at any time since the debt crisis of the early-1980s", which may be one of the reasons the space has opened up for change in the region (*Financial Times*, 4 May 2007).

---

[112] On 30 May 2007, Robert Zoellick was nominated by President George W. Bush to become President of the World Bank, with Paul Wolfowitz formally stepping down on 30 June.

As these issues play out, it will likely remain apparent that the forms of resistance to neoliberal ideology are expressed in heterogeneous forms at a variety of levels. As the forces contesting and reshaping neoliberal futures come together, US hegemony in the region may be diminished as everyday people and political leaders' commitment to market-based reforms wanes. One thing is certain: civil society will continue to play an active role at the local, national and international levels in terms of putting forth new social imaginaries within the public arena. Democratisation of the region is an on-going process that takes place in numerous fora as social actors at various levels seek to carve out new spaces within the public arena, while continuing to make claims about the physical spaces in which they reside.

It seems likely that CODDEFFAGOLF, along with numerous other social actors, will continue to play a critical role in reshaping the future of the country and the region through initiatives such as EcoAméricas. Civil society continues to demand that the government, IFIs and supermarket chains like Marks and Spencer consider social and environmental issues. Whom shall the English public believe? Should they believe the industry that supplies the shrimp to the supermarket or the environmental organisation that claims that the English public is eating the blood of Honduran children, as Varela stated in the 2004 BBC documentary *Price of Prawns*? One point is certain, each, along with the unassuming customer buying Honduran King Prawn's in Cambridge, will increasingly be required to take into consideration social and environmental factors when making decisions that affect people and the environment, often thousands of miles away from where products originate.

Chapter 14
Conclusion

Introduction

In this thesis I have aimed to deepen understanding of conflicts around marine and coastal resources by exploring the discourses and associated actions of the actors involved, while assessing their relationship to the physical geographic space contested. A number of questions were explored through a case study of a conflict over the expansion of shrimp aquaculture in the coastal zone of the GOF, southern Honduras between 1973 and 2006. Special attention was given to how this conflict played out within discursive space and its relation to the more observable aspects of conflicts in relation to physical space.

The role of the actors involved in the conflict, and how they produced and reproduced it through processes of politicisation, were focal points. In a departure from traditional conflict analysis, I approached the research from an historical perspective, based on the premise that conflicts are naturally manifested within the context of socio-economic and political change. In this regard, this conflict was historically situated, socially process-oriented, and an inherent aspect of social change that has occurred as actors discursively negotiated shared meanings and understandings in relation to the physical spaces for which they have been competing.

Through careful analysis of the discourses, actions, and interests of the various actors involved in the conflict, I have demonstrated that the conflict in the GOF has partially been over the effects related to the implementation of neoliberal economic policies. Particular attention was focused on the claims related to the consequences of capitalist transformation on local livelihoods in the south as aquaculture was promoted, which became a case of neoliberal futures being contested.

The analysis illustrated how the conflict in southern Honduras has been directly associated with two distinct but inter-related social imaginaries – neoliberalism and environmentalism – competing to influence how the wetlands of southern Honduras *ought* to be used, managed and conserved over the last three decades. These two ideologies eventually converged in a biased compromise due to power disparities that are still being negotiated within the public arena. State, civil society and private-sector actors have increasingly worked together to identify ways to integrate social and environmental concerns into NDPs alongside economic growth.

A central issue on commencement of the research was the lack of an adequate approach to assess the discursive aspects of an environmental conflict over physical spaces, which was a primary goal of the research. To resolve this analytical dilemma I combined a number of theoretical and methodological approaches, mainly linking an actor-network approach to discourse analysis, and events history and ecology, within a poststructuralist political ecology framework (discussed in Chapters 1 and 2). The outcome was an explication of how and why local people in southern Honduras have opposed commercial shrimp farming and their role in the global aquaculture and environment debate, one of the stated goals of the thesis. It also elucidated the discourses that have underpinned both neoliberal and environmental ideologies and their performance in the social construction of environmental conflict.

Case Study Summary

In Chapter 4, I argued that the government focused on market expansion as a means of increasing foreign exchange revenue. I argued that the conflicts that emerged around the coastal wetlands of the GOF in the 1980s were directly attributed to the Honduran government's efforts to promote NTAX as it sought to pursue export-led growth in accord with neoliberal economic reform. The promotion of NTAX was an expression of the government's intention to transform the physical-spatial reality of southern Honduras. New collective social imaginaries in relation to the wetlands emerged within the public arena as those affected by processes of capitalist transformation sought to redefine the physical-spatial reality around the GOF.

Two state discourses – economic development and food security – were identified, both of which were used to establish state legitimacy to intervene in the coastal wetlands between 1973 and 1998. These discourses sought to create an image of rationality behind state actions and were thus a method of justifying what *ought* to be done within the physical space in which aquaculture development commenced.

In Chapters 5 and 6, I illustrated how the state translated neoliberal ideas into law, policy and action through various techniques of power which had a transformative effect on the coastal landscape of southern Honduras. The effects of the techniques of power used by the state illustrated its role in the arrangement and regularisation of spaces as it sought to establish new meanings in relation to the use, management and conservation of the coastal wetlands (Radcliffe, 2001a: 236). I demonstrated that the government's interpretation of the legal language in various laws was aligned with its objectives to pursue economic development and food security, two discourses that remained dominant until Hurricane Mitch devastated the region.

The actions taken by the state in conjunction with other international actors to promote aquaculture led to the formation of CODDEFFAGOLF. Davis has argued that collective action results through movements such as CODDEFFAGOLF due to local actors' lack of access to the state, which is often bureaucratically and territorially distant from the citizens it seeks to serve (Davis, 1999: 608 and 609). Once established, CODDEFFAGOLF contested the expansion of the shrimp industry on the premise that it was infringing on the rights of local people, who were being politically and economically marginalised from state decisions as the government promoted shrimp farming on the coastal wetlands of southern Honduras.

CODDEFFAGOLF consistently made two claims. First, access to the seasonal lagoons where locals fished was diminished due to the introduction of aquaculture. Second, the capture of wild postlarvae for the purpose of shrimp production was affecting the fisheries resources upon which a faction of the artisan fishing community depended for subsistence. Contrary to some propositions that NSMs' struggles lie outside the sphere production, it was clear that CODDEFFAGOLF's goals, strategies and tactics were directly associated with the problem of access to or control of the means of production in relation to the fisheries sector (Hellman, 1995: 169). The latter point is consistent with findings that contemporary conflicts are the expression of social groups claiming access to representation within the public arena and redistribution of access to resources rather than specific antagonism towards the state (Melucci, 1995: 118).

The dominant narrative articulated by CODDEFFAGOLF, as it politicised the issues of enclosure of common pool resources and the impact of by-catch on fisheries resources, was mangrove degradation. Mangrove degradation was consistently attributed to industry practices rather than local use, which was actually the more important concern of the Honduran environmental movement in the early 1980s. Mangrove degradation became a relevant issue as industry expansion increased throughout the 1990s. However, I illustrated that the initial discourse on the degradation of mangrove forests in the early 1980s was not attributed to industry practices but arose as a result of the AHE's concerns regarding local use of mangrove resources for fuelwood and construction materials. Consequently, the conflict was not initially associated with mangrove degradation but rather was directly associated with the changing dynamics of the fisheries sector as aquaculture development influenced the socio-economic conditions and labour market dynamics in the south. State intervention in the local fisheries sector was the direct cause of the divisions in the labour market that led to conflict.

Counter to the notion that the enclosure of material reality coincides with discursive enclosure, my results indicate that it does in some ways and does not in others. The enclosure of material space within the context of democratic transformation in Honduras actually led to the opening of new discursive fora that transcended the physical and political boundaries in which the conflict has taken place. However, those living in rural coastal communities often remained marginalised from these spaces, although actors such as CODDEFFAGOLF purportedly represented their interests. My points were exemplified in Chapters 7 and 9.

In Chapter 7, I discussed the emergence of the Honduran environmental movement, local actors and alternative discourses on the mangrove wetlands that began to appear in the media as actors sought access within the public arena. I discussed how the democratic transition in Honduras made it possible for the emergence and strengthening of the Honduran environmental movement. In Chapter 9, I illustrated how the formation of transnational environmental political networks opened the discursive terrain further as CODDEFFAGOLF forged links to ENGOs external to the country. I also argued that the rise of contestation to aquaculture in Honduras was congruent with the emergence and strengthening of the global environmental movement.

My findings are partially consistent with those of Salman and Assies, who have argued that globalisation and the capitalist transformations taking place have opened up new opportunities and facilitated a non-territorial democratisation of social and environmental issues (Salman and Assies, 2000: 291). Although this may be the case for CODDEFFAGOLF, local people living in rural coastal communities in southern Honduras remain on the periphery, marginalised from the discursive fora in which decisions are made and interests articulated. Davis has argued that rural communities and their citizens can be marginalised from or connected to state institutions, policies, practices and even discourses geographically, culturally, institutionally and by social class (Davis, 1999: 603). Consequently, processes of decentralisation will need to reach into the coastal communities of southern Honduras while directing adequate resources to strengthen their participation within newly emerging fora within civil society to rectify the effects of geographies of distance. Until this occurs, the rural poor will remain discursively and politically marginalised from the public arena, where the decisions regarding their livelihoods are made, often without their input.

The exclusion of local people from the opening of discursive fora and political spaces was illustrated through several examples. I demonstrated that representations of the socio-economic and environmental reality put forth by CODDEFFAGOLF and ENGOs through the media, and various other initiatives were inconsistent with local perceptions of the issues, illustrated through the results of the socio-economic analysis. For example, local perceptions of the conflict and the degree to which it was considered to be violent were inconsistent with how it was depicted by CODDEFFAGOLF, the media or within recent academic literature. The socio-economic data also revealed that local actors perceived themselves to be the most significant threat to mangrove resources rather than the shrimp industry, which was also counter to the dominant narratives used by CODDEFFAGOLF to explain the socio-economic and environmental reality in the south.

For example, CODDEFFAGOLF claimed to represent artisan fishermen and fostered the 'received wisdom' that the origins of this conflict were due to the shrimp industry's deforestation of the mangrove wetlands for shrimp pond construction. However, as I demonstrated, this claim does not appropriately represent the local level reality. The conflict was initially a conflict between artisan fishermen due to changes in the labour market, that were the result of the introduction of shrimp aquaculture as a commercial activity as the Honduran government gradually embraced neoliberal economic reform. Later, some fishermen began to oppose aquaculture expansion due to enclosure of traditional fishing grounds, such as the seasonal lagoons and estuaries, and due to concerns about deforestation of the mangrove wetlands as the commodification of the area for the production of shrimp supervened. These claims were analysed in detail in Parts II, III, and IV. Here, Foweraker's claim that NSMs do not always represent the perceptions or interests of local people was substantiated, further justifying my choice of an actor-network approach that focused on a broader conception of social actors and agency.

In Chapter 8, I addressed the rise and strengthening of the neoliberal agenda after President Callejas gained power in 1989, which led CODDEFFAGOLF and ENGOs to continue contesting state practices and the neoliberal imaginaries that underpinned them. The state, IFIs and USAID continued to use the discourses of economic development and food security to justify expansion of the industry. In Chapter 9, I illustrated how the Callejas administration's strong support for neoliberal economic reform increased the contestation to shrimp farming in the south. However, the resistance to the administration's practices revealed that neoliberal futures were not preordained, their political effects were undetermined, and the trajectories of neoliberal state formation were influenced by both local actors and transnational environmental political networks. Both were actively involved in the production, reproduction and re-imagining of the state and human interventions related to the coastal wetlands of southern Honduras (Radcliffe, 2001b: 20).

In Chapter 10, I illustrated how subsequent changes in Honduran environmental law were partially a response to the demands articulated from within civil society but were made in accord with neoliberal objectives. The state's attempts to respond to the demands expressed from within civil society eventually led to the formation of the CVC, one of the first marine and coastal management regimes established in southern Honduras that incorporated state, civil society and private-sector actors. I argued that although efforts were made to address some of the issues, the continued expansion of the industry exacerbated the conflict between 1988 and 1998. However, the creation of the CVC and changes in environmental policy demonstrated that the government was taking a more active role in relation to the environment as the issues became more pronounced.

Although the state had taken steps to address shared resource issues, it was clear that uncertain territorial boundaries and overlapping legal jurisdictions continued to make it difficult for ministries to regulate the harvest and trade in marine and coastal resources, such as shrimp postlarvae, fish, molluscs and mangrove wood. As illustrated, most of these resources have fallen under one or more systems of state *de jure* control, but in practice rules governing these extensive harvests have been difficult to enforce. In the end, in light of the lack of state capacity to regulate the industry, various organisations, the industry, and other actors within civil society have collectively prepared the way for changes to the laws as well as the shrimp industry's behaviour. To what extent contestation of the state's actions has provided the legitimacy for government bureaucrats to change social and environmental policies in Honduras is left for future research. However, more recent research in political science and international relations has indicated that advocacy groups have been able to exercise power effectively, represent the interests of marginalised groups and enable political changes at the level of the state (Radcliffe, 2001b: 26).

In Chapter 11, I demonstrated how the debates in southern Honduras were increasingly linked to similar protests and resistance taking place worldwide against the aquaculture industry, which eventually led to the formation of ISA-net. The increased threats to the commercial aquaculture industry, due to a global boycott and a moratorium on shrimp farming, led to the formation of the GAA. The formation of these groups, or discursive coalitions, marked the elevation of disparate local conflicts to the international arena, unifying each under a common cause, and resulting in the global aquaculture and environment debate. Subsequently, new discursive formations have arisen through the process and in the context of politicisation and democratic transformation.

In Chapters 12 and 13, I discussed the convergence of neoliberal and environmental social imaginaries as the state and international actors responded to the impacts of Hurricane Mitch. I argued that although the hurricane raised important questions in relation to past economic transformative processes, the implementation of neoliberal reform sped up rather than slowed down after the hurricane. A number of actors allied with CODDEFFAGOLF continued to pressure the Honduran government to implement previously articulated goals in relation to the management of the marine and coastal environment of the GOF. After decades of protests, the area was finally designated as a part of the MBC, a RAMSAR Site, and a protected area. The more recent attention to the socio-economic and environmental issues exemplifies the state's efforts to redefine the coastal wetlands in accordance with a wider-set of political and economic interests. Impediments still exist but the blending of environmental discourse into the neoliberal agenda is apparent. Currently, the state has made efforts to work collaboratively with the private sector and civil society to develop management plans for the region. Until these plans are completed and implemented, the state's creation of multiple use areas and protected areas in the south will remain rhetorical.

Discussion

Environmental conflict is a natural process that occurs as various socio-economic and environmental imaginaries are negotiated and contested in the public arena. In this case study, I illustrated that the conflict was associated with various conceptions of access to and control over the mangrove wetlands of the GOF. As social actors seek to construct their social and natural realities in alignment with their interests they came into conflict with the perspectives and interests of other actors. Actors' political and economic interests were affected through their interpretations of the discourses and actions of others and by the 'natural agency' of the environment (see Chapter 3). As actors' interests changed these interests generally corresponded with how the issues were framed and reframed in relation to the social situation within which the actors continued to act. Consequently, this conflict was revealed as an iterative process, as meaning around the issues was socially constructed.

The means by which actors perceived the socio-economic and environmental issues and participated in the conflict was the result of shared meanings, ideas, rules (formal and informal) and norms that governed and guided their interactions. Since meaning was contrived incongruently by the social actors, and ideas were not always understood equally, conflict resulted. However, the conflicts that emerged as a result of inconsistencies in the way that actors framed their worlds were understood as an inherent process of social change that produced varied effects. As actors were confronted by the effects produced, their knowledge and perceptions of the physical space and the issues central to the conflict changed accordingly.

The politicisation of the environmental and socio-economic issues were the means by which the conflict was manifested, produced, and reproduced by a variety of social actors. The results revealed that actors utilised specific sets of discourses in order to achieve their political and economic interests in relation to the coastal wetlands. Politicisation of the issues was conducted in a manner consistent with each actor's interests in relation to the space disputed. The concepts used by the actors ultimately became political forces in their own right as they co-constructed the socio-historical and political terrain in which the conflict took place.

The poststructuralist emphasis on discourse made it possible to illustrate the ways in which the co-construction of narratives was both a socio-economic and political process in relation to the wetlands. Particular attention was focused on the discourses that influenced shared meanings, differing interpretations of specific events, and the actions pursued by the actors in relation to those events. However, I found that the discursive positions of various actors were not always consistent with their actions in relation to the physical geographic space in dispute and, as a result, this was a prominent factor that contributed to the emergence and evolution of environmental conflict.

Although discourses are often indicative of the intentions, interests, and strategies of the actors in relation to the physical geographic space in dispute, conflict at the discursive level does not always translate into specific actions that affect physical geographic space. For this reason the research took care to link claims related to social and ecological outcomes to the interests and actions of the social actors associated with the conflict. By so doing, I provided the basis for identifying the substantive power of discourses.

The inconsistencies between discourses and actions affected the interpretative process between social actors. When what was stated discursively was not consistent with what occurred through action, trust between the actors was diminished and, therefore, affected each actor's capacity to influence the other throughout the course of the conflict. When consistency between actors' discourses and actions were more congruent, social and political capital was strengthened, permitting the actors to collaborate more effectively in relation to their common interests.

In sum, the discourses and actions of the social actors' involved in this conflict were important analytical variables. However, the social history of the region; patterns of human environmental interaction; human-induced and natural changes in the environment; and macro-economic and political changes were also important variables that conditioned the possibilities for conflict between various actors in relation to the coastal wetlands. Social actors' interests were continuously influenced by these factors and, therefore, influenced the ways in which the interpretative process evolved throughout the socio-historical trajectory of the conflict.

Conclusion

The contestation performed by environmentalists in the context of economic change in southern Honduras has resulted in a convergence between neoliberalism and environmentalism, articulated as neoliberal environmental governance in current academic literature. However, I also argue that it a biased convergence due to the hegemonic disposition of neoliberal ideology that has assimilated environmental discourses in the pursuit of the commodification of nature. Global institutions and the state remain behind these practices and although there has been an emphasis on global environmental management discourses they continue to be neoliberal in terms of espousing market-oriented solutions (Adger, 2001: 701). Forced to pursue conservation initiatives in the language of the market (e.g. biodiversity offsets, green markets, carbon trading, economic valuation of wetlands, etc.) environmentalism is increasingly commoditised and capitalised.

Those that live outside of the market, through subsistence, are increasingly confronted by the *invisible hand of the market*' as it seeks to enroll them in the capitalist project, as if it did not have people behind it. An obvious material manifestation of the same theme is the corporation, permitted to occupy the world of the private sphere rather than the public. Usually that sphere is also discussed as if there were not people that embody it. Consequently, I argue that neoliberalism not only disembodies the natural world, diminishing its agency as 'risk management', but also places the human subject outside of the forces behind the capitalist project of the disembodiment of nature, relieving humans of the responsibility for the natural world and turning it over to the market.

Appendix I
Research Design and Methodology

In this appendix I discuss general considerations associated with the fieldwork as well as how it was completed in various phases. It then turns towards a description of how the data were collected and organised. The primary methods for data collection were: archival research; interviews and a socio-economic assessment of the GOF region. Additional sources of information such as digital maps and photographs and grey literature were collected through other government, educational, non-profit and international institutions in the region and in Washington DC. The archival research is discussed within the context of the development of an historical timeline of the key events associated with the conflict. The appendix proceeds with a section describing the socioeconomic assessment that was conducted in 21 coastal communities utilising 368 household questionnaires. The use of semi-structured and open-ended interviews is then discussed in conjunction with how the key social actors were identified. Lastly, the role of participant observation is discussed.

A number of comparative qualitative analytic methods were chosen due to their compatibility with the theoretical structure of the research. The goal was to use a combination of theories and methods in order to consider a number of "interrelated ideas about various patterns, concepts, processes, relationships, or events" associated with conflicts having to do with access and control over the physical geographic space of the Gulf of Fonseca (GOF) (Berg, 2001: 15).

The methodological intent of this research was to develop a new set of propositions that explains the social phenomenon of environmental conflict whilst accepting that "the criteria set up to judge whether or not a particular explanation is reasonable and satisfying is highly subjective" (Harvey, 1969: 15). The goal was to select variables that would assist with establishing a coherent explanation of environmental conflict, while recognising that "poststructuralist research [is inherently subjective since] the researcher interprets, activates, and transmutes meanings and their contexts" through a subjective lens (Duncan et al., 2004: 89-90). The primary variables associated with this analysis are the discourses, characteristics, interests, and the actions of the various actors who have interests in the mangrove ecosystem of the GOF.

Research Design

PHASING OF THE FIELDWORK/GENERAL CONSIDERATIONS:

I was first exposed to the GOF while employed as the Senior Program Manager for Latin American and Caribbean with the National Oceanic and Atmospheric Administration. In this position, I was responsible for Sea Grant program development in the region, however, the conflicts surrounding marine and coastal resource use made this task extraordinarily complex. Subsequently, I made the decision to research these issues in more depth. Due to the willingness on the part of a number of individuals and institutions within and outside Honduras, I was able to accomplish the goals of the research.

The selection of the GOF as a research site was based on the literature review and fact-finding visits begun as early as 2002. Its geographic isolation and size naturally limited the number of variables while a relatively small and identifiable group of key actors associated with the conflict made it easier to consider the struggles between actors over the physical space associated with the marine and coastal environment, specifically the mangrove ecosystem. Another consideration was accessibility to the research site as well as prior knowledge of the area. The GOF is a relatively small area with easily identifiable communities and access to sample populations large enough to solicit the necessary information. In turn, there was sufficient access to local, national, and international documents and actors to complete the goals of the research.

DATA COLLECTION AND ORGANISATION

The research methods for this study were selected with the aim of reconstructing a timeline of events associated with the emergence and evolution of environmental conflict in the GOF. Furthermore, it was important to identify key actors at the local, national and international levels who have been involved in events that define the historical trajectory of conflict surrounding mangroves in southern Honduras. Lastly, appropriate local case study sites in the GOF had to be selected.

ARCHIVAL RESEARCH:

The point of the archival research was to begin reconstructing an accurate timeline of events during the conflict, acquire information on the changes in the socio-economic changes in the region, identify major issues that have affected southern Honduras and locate relevant information that could assist with explaining social and ecological changes from the discursive positions of various social actors. Information was collected from newspapers, legal documents and a number of other sources for these reasons. The archival information has made a significant contribution towards the discursive analysis of the conflict, given that newspapers and other sources often include quotations from various social actors. Analysing these documents allowed the use of these sources to trace changes in the actors' discourses and their role in the conflict, at least as conveyed through the discourses drawn from these various media.

When assessing the discourses of actors it was important to keep in mind that quotations are often used out of their original context and efforts were made to substantiate information where possible, usually through interviews or through the reading of additional materials.   Furthermore, I recognise that politicians and government agencies often receive more press attention than the thoughts and programmes of organizations within civil society.   For this reason, I sought out alternative archival sources including pamphlets, brochures, newsletters, and other forms of documentation to obtain greater access to the perspectives from within civil society organisations.

The majority of the archival research took place at the Hermeroteque (archives) at the Universidad Nacional Autonomous de Honduras (UNAH) which has the best archival collection of newspapers in the nation including *El Cronista*, *La Tribuna* and *El Heraldo*. Newspaper research focused on different periods depending on the circulation rates and prominence of the newspapers at different times. From 1970 to 1980 I focused on *El Cronista* as the most prominent daily newspaper of the time; between 1980 and 1995 I focused on *La Tribuna*, the most widely circulated paper at the time, and after 1995 the focus was on *El Heraldo*, which has the highest current circulation and is one of the most respected. Although it would have proven beneficial to review each newspaper circulated in Honduras since 1970 for information related to the southern part of the country and topics relevant to this research, it was decided that for the sake of time these three newspapers would be targeted for information during each period mentioned. Finally, the archives of the Dirección General del Medio Ambiente (DGA), the Inter-American Institute for Cooperation on Agriculture (IICA) and the archives of other institutions and organisations throughout the region were consulted and documents acquired.

Archival research was also conducted to acquire the complete text of relevant laws through their original publication in *La Gaceta*, the official national record of the law. Once a law is made public and printed in *La Gaceta* it goes into effect. Laws are important because they outline the key characteristics of state agencies and their responsibilities associated with the mangrove wetlands. Laws also establish the rules and/or regulations associated with state interventions in a specific physical geographic space and, therefore, are important not only as discursive interventions, but also as being the basis of state 'legitimacy' for intervening in a particular physical geographic space where the law permits.

Linking Key Historical Events with Environmental Conflict:

An analysis of environmental conflict in the GOF required reconstructing an historical timeline of key events associated with it. The key events were identified through the archival research and the open-ended interviews. Key events were also identified by analysing the various documents and materials collected during the fact-finding visits and through subsequent research which was designed to "make use of historical and time-series data sets; which pays serious attention to inhabitants' own experiences and opinions" since 1973 (Leach and Mearns, 1996: 30).

To complete this task, the research drew upon the archival sources, government and project documents. A number of methodological approaches borrowed from 'event history' and 'events ecology' practitioners proved beneficial to develop the timeline. These methods work towards the development of a causal historical analysis. Event history practitioners "compile databases from published accounts in the contemporary press on those events that they consider worthy of notice" and proceed by creating a "temporal map of incidents through which the... activities and interactions [of various actors] can be traced" and further considered (Tarrow, 1996: 875). This approach was used to isolate key events associated with conflict in the GOF. Events ecology has been described as "the analytical methodology for evaluating causal links between historical events and integrating relevant biophysical and socio-economic information" and is guided by "open-ended questions about why specific environmental changes of interest (events) have occurred" (Walters, 2003: 295). Forsyth has stated that in this regard events ecology "adopts a partly phenomenological attitude by seeking to understand "events" as local changes of significance, rather than as "facts" that can be incorporated into preexisting theories, or "factors" that imply events have causal significance" (Forsyth, 2003: 223).

SOCIOECONOMIC ASSESSMENT: ASSESSING THE INTERESTS AND ACTIONS OF LOCAL COASTAL RESOURCE USERS

A socio-economic analysis of the area was conducted in order to obtain a snapshot of human activities in coastal communities throughout southern Honduras as well as perspectives related to the issues surrounding the conflict. It was also used to gather some baseline data on the coastal communities of southern Honduras. The information collected through the assessment was coupled with that obtained through semi-structured interviews, which were conducted separately from the questionnaires. Both were used to identify the discourses and interests of local people in relation to the mangrove ecosystem and, as a result, served as the basis for determining the extent to which they were congruent or incongruent with those of more dominant actors associated with the conflict. The socio-economic assessment was conducted in 21 coastal communities around the Gulf of Fonseca through the use of household questionnaires (368 in the end).

The questionnaire used in the socio-economic assessment was designed employing methodological tools presented in a manual published by the University of Rhode Island Coastal Resources Center (URI/CRC) entitled "Assessing the Behavioral Aspects of Coastal Resource Use" and another manual published by the Global Coral Reef Monitoring Network (GCRMN) entitled "Socioeconomic Manual for Coral Reef Management". These two manuals were developed expressly for practitioners working in coastal communities and take into consideration the methodological issues associated with communities that lie on or near the marine and coastal environments, and both include research techniques relevant to this specific type of geographic setting. The variables adapted from both manuals and incorporated into the questionnaire are listed below (The questionnaire is located in Appendix II):

1) socio-demographic information;
2) household characteristics;
3) characteristics associated with material style of life;
4) migration;
5) social/civil society participation;
6) use of coastal resources including the production of shrimp and salt, collection of crustaceans and mollusks, fishing, and mangrove use and replanting; and
7) perceptions and attitudes pertaining to mangrove degradation and conflicts surrounding marine and coastal resource use

The 21 coastal communities (See Table 2.1) used in the socio-economic assessment were selected carefully to obtain a sample throughout the southern portion of the country that represented all of the variations in social and economic activities that occur in the region. Information gathered during interviews with key informants assisted in the selection of communities. During the initial phase of lengthier fieldwork the mangrove ecosystem of the GOF was investigated by vehicle. Each of the major estuaries, rivers, bays and islands in the GOF were explored to determine the location of local settlements, communities or towns that are directly dependent upon mangroves or where people use the space in or around mangroves for other social and economic activities. A map assisted with identifying the location of the places that were particularly relevant to the overall goals of the research. This was often completed with the assistance of local individuals and key informants based in the region.

The goal in the selection of these 21 coastal communities was to ensure that a diversity of perspectives in relation to the use, management and conservation associated with the mangrove wetlands was as diverse as possible. For example, the integrity of the research could have been compromised if all of the selected sites had been located in close proximity to large-scale private industrial shrimp farming vs. places that are not.

INTERVIEWS: OPEN ENDED AND SEMI-STRUCTURED

My research accepted the inherent subjectivity that comes with 'explanation' as described by Harvey, while at the same time accepting that "the essence of explanation... lies in providing a network of connections between events" (Harvey, 1969: 14). Making the connections between events was accomplished by conducting open-ended and semi-structured interviews with specific actors associated with the dominant events that have defined the conflict, as well as those events that local people articulated as being relevant to explaining their perception of the local level 'reality'.

A number of key informants working in municipal, regional, national, and international organisations were interviewed using both semi-structured and open-ended interviews. Through the open-ended and semi-structured interviews the research drew out the discursive positions of 'experts' affiliated with various organisations. The framework allowed me to work toward penetrating the major ideas behind specific decisions related to national and international interventions. In turn this allowed me to identify the discursive formations that influenced these larger interests and decision-making processes. It also allowed me to begin unraveling how various actors were connected and the influence exercised over each other throughout decision-making processes.

The secondary sources described in detail in the section on 'data collection and organisation' were particularly important to consider comparatively the statements of various actors against data compiled on the spatial and temporal impacts on the mangrove ecosystem of the GOF. Another goal was to consider the ways in which the legal and policy framework has changed in relation to the mangrove ecosystem of the GOF. Relevant officials were interviewed to determine how various policies have influenced or affected the ways in which the physical geographic space of the mangrove ecosystem in southern Honduras has been exploited, degraded, conserved or managed. The goal was to understand the 'why' behind specific legal and policy decisions. The goal was also to determine the bases upon which specific policies were developed and by whom, as well as their relationship to other actors.

Interviews were held at convenient locations for the participants, which in most cases were their places of employment or homes or in restaurants or hotels that were centrally located to minimise the travel distance for participants. Transcripts were developed using information recorded with a digital audio recorder. The transcriptions for recorded interviews included information on the date and location of the interview, the names of the participants involved, and their respective organisations, if appropriate. The transcripts of the interviews have also been hyperlinked to any field notes that were written as a result of observations made during the interviews (e.g. body language). All of the information resulting from the interviews and observations was thoroughly read, and annotated into coded topics, themes and issues to facilitate analysis of the data.

Identification of the Key Informants/Social Actors:

The research began by targeting key informants within major organisations associated with activities in the GOF and in local communities. These informants, or opinion leaders, can be viewed as people who played a significant role in decision-making processes, influenced the opinions of other individuals within a community or organisation, and were often seen by others as being very knowledgeable about topics relevant to the research. Interviews with informants are a rich source of information for those reasons. Selection of informants for interviews was accomplished through conversations with local inhabitants or individuals within larger organisations operating in the region.

The initial group of key informants was identified through conversations held during fact-finding trips with individuals at the University of Zamorano, US Embassy Honduras, USAID and a number of other institutions and organisations. This approach also provided an easy way to identify additional and appropriate respondent groups. Participant observation also provided the means to consider the actual visible behavior and actions of local people. This allowed me to contrast their observed actions and behaviours with their discourse during the semi-structured interview process. Time was spent interacting with each resource-user group, participating in their daily practices when possible, and asking open-ended questions in relation to their productive activities that directly or indirectly affected the mangrove ecosystem in southern Honduras. Attending meetings of various organisations representing various sectors in the south, as well as municipal, NGO or any other meetings relevant to the goals of this research provided the possibility to consider how groups of people interacted, and their relationship with other social actors.

Appendix 2: Socioeconomic Questionnaire and List of Respondents

NÚMERO DE ENCUESTA _____

ZAMORANO

ANÁLISIS SOCIOECONÓMICO EN EL SUR DE HONDURAS

ENCUESTA DE HOGARES

Buen día: Soy estudiante de la Escuela Agrícola Panamericana Zamorano, estamos levantando una encuesta sobre el uso de los recursos naturales, especialmente el manglar. La encuesta se esa llevando a cabo en varias comunidades de los departamentos de Valle y Choluteca. La información recogida durante esta entrevista es solo para el uso del estudio. Se espera que la entrevista no tarde más de 25 minutos. El resultado de todas las entrevistas será compartido con organizaciones locales, nacionales, e internacionales que trabajan en el sur de Honduras. Por adelantado les damos las gracias por su tiempo.

Código de encuestador y encuesta: _____ Fecha: _____

Departamento: _____ Municipio: _____ Comunidad: _____

SECCIÓN I: Información sociodemográfica

1. Nombre completo de la persona entrevistada:

1/_/_/_/_/_/_/_/_/_/_/_/_/_/_/_/

2/_/_/_/_/_/_/_/_/_/_/_/_/_/_/_/

2. Género de la persona entrevistada   (1) Masculino   (2) Femenino

3. Edad_____

SECCIÓN II: Características de la familia

4. **Cuadro 1**: Información de la familia.

A	B	C		D	E	F
		Género		Edad	Escolaridad	
Relación familiar	Estado civil jefe	M	F	(años cumplidos)	(último año aprobado)	Ocupación principal
4.1Jefe de familia		(1)	(2)			
4.2		(1)	(2)			
4.3		(1)	(2)			
4.4		(1)	(2)			
4.5		(1)	(2)			
4.6		(1)	(2)			
4.7		(1)	(2)			
4.8		(1)	(2)			
4.9		(1)	(2)			
4.10		(1)	(2)			
4.11		(1)	(2)			

Codigos

A. Relación Familiar	B Estado civil del Jefe de Familia:	E. Escolaridad:	F. Ocupación principal:
1. Esposa, esposo o compañero	1. Soltero, viudo, separado	1.No asistió	1. Profesionales, técnicos y
2. Hija o hijo	2. Casado, con el cónyuge	2. Primaria incompleta	personas en ocupaciones afines
3. Padre o madre	siempre presente en el	3. Primaria completa	2. Empleados de oficinas del
4. Nieto o nieta	hogar / Unión libre	4. Secundaria incompleta	estado, y empresa privada
5. Abuelo o	3. Casado, con el cónyuge	5. Secundaria completa	3. Comerciantes y vendedores
abuela	emigrante (es decir, el	6. Técnico	4. Agricultores, ganaderos y
6. Otro pariente	cónyuge tiene todavía	7. Universitaria incompleta	trabajadores agropecuarios
7. Otra persona	vínculos económicos con	8. Universitaria completa	5. Trabajadores de la industria
que no es	el jefe de familia)		textil, albanileria, mecánica,
pariente			electricidad, etc.
			6. Industria de Acuacultura
			7. Industria de Producción de Sal
			8. Pescador
			9. Desempleado
			10. Retirado / Discapacitado/ No puede trabajar

SECCIÓN III: Características de la vivienda, servicios e ingresos

5. ¿Su vivienda es?
   (1) Propia   (2) Arrendada   (3) De un familiar (prestada)   (4)
   Otro _____

6. ¿De cuántas habitaciones está compuesta su vivienda, tomando en cuenta: sala, cocina, comedor, cuando son separados o independientes? _____

7. ¿Cuál es el material predominante del techo (de la vivienda principal)?
   (1) Lámina de Zinc/ asbesto   (2) Tejas de barro   (3) Madera de mangle
       (4) Otro tipo de madera   (4) Cemento   (5) Otro_____

8. ¿Cuál es el material predominante de las paredes exteriores (de la vivienda principal)?
   (1) Barro/ adobe (2) Bloque de cemento   (3) Ladrillo   (4) Madera de Manglar   (4)
   Otro_____

9. ¿Cuál es el material predominante del piso?
   (1) Tierra/ barro   (2) Sólo cemento   (3) Ladrillo/Gravito   (4) Cerámica   (5)   Madera de
   Manglar

   (5) Otro_____

10. ¿Tiene acceso a agua potable?   (1) Sí   (0) No

11. El agua potable : (1) Llega a su hogar (tiene llave en la casa)   (2) La recoge de un lugar público (chorro, cantarera, etc)

12. ¿Con qué tipo de instalaciones sanitarias cuenta su hogar?
   (1) Servicio sanitario   (2) Letrina   (3) Fosa séptica   (4) No tiene   (5)
   Otro_____

13. ¿Con qué tipo de suministro de electricidad cuenta su casa?
    (1) Sin suministro   (2) Con conexión compartida con otros   (3)

A	B	C	D
Usar el mismo numero **del cuadro 1 , columna A)**	Actividad para ganar dinero *(Escriba la respuesta)*	**Código**	Cantidad recibida en el mes pasado (Lps)
4.1Jefe de familia			
4.2			
4.3			
4.4			
4.5			
4.6			
4.7			
4.8			

14. A continuación le preguntaremos sobre cuántos lempiras contribuyó al hogar cada miembro de la familia?

Cuadro 2

A. Relación Familiar	C. Ocupación principal:
1. Esposa, esposo o compañero	1. Profesionales, técnicos y personas en ocupaciones afines
2. Hija o hijo	
3. Padre o madre	2. Empleados de oficina del estado, y empresa privada
4. Nieto o nieta	
5. Abuelo o abuela	3. Comerciantes y vendedores
6. Otro pariente	4. Agricultores, ganaderos y trabajadores agropecuarios
7. Otra persona que no es pariente	
	5. Trabajadores de la industria textil, albanileria, mecánica, electricidad, etc.
	6. Industria de Acuacultura
	7. Industria de Producción de Sal
	8. Pescador
	9. Desempleado
	10. Retirado / Discapacitado/ No puede trabajar

SECCIÓN IV: Migración

15. ¿Cuánto tiempo lleva viviendo en la comunidad? _____ años

16. ¿Usted recibe algún tipo de ayuda, en efectivo o en especies (ropa, medicina etc.), de algún familiar que reside fuera del hogar?

(1) Si **[Siga]**   (0) No **[Pase a 20]**

17. ¿Dónde reside su familiar?

(1) En otro lugar de Honduras   (2) Otro país Latinoamericano   (3)   EE.UU.

(4) Otros

18. ¿Con qué frecuencia rd18feecibe la ayuda (en efectivo o en especies)?

(1) Semanalmente   (2) Mensual   (3) Cada 2 o 3 meses   (4)

Dos veces al año

(5) Una vez al año   (6) Otro _____

19. ¿Si usted recibe ayuda de algún familiar en efectivo, cuanto le envían normalmente? _____ (1) Dólares   (2) Lempiras

SECCIÓN V: Participación social.

20. ¿Usted, o algún miembro del hogar, miembro de alguna cooperativa, organización, grupo social o político?   (1) Sí   (0) No

Por favor indique el tipo de organizacion:	Sí	No	Cual?
21. Una organización social, política y/ o gremial			
22. Una organización (cooperativa o asociación)			
23. Organización de pescadores			
24. Organización de acuicultura			
25. Grupo Ambiental			
26. ANDAH			
27. CODDEFFAGOLF			
28. Otro			

SECCIÓN VI: Patrones de uso de los recursos.

29. ¿Del terreno de su propiedad, cuanto utilizan para agricultura? *[Incluir terrenos en otros lugares]*

Manzanas totales _____   o   Hectáreas totales _____

30. ¿Cuánto de ese terreno tiene título?

Manzanas totales _____   o   Hectáreas totales _____

SECCIÓN VII: Producción de camarones y sal.

31. ¿Usted, o algún miembro del hogar, **cultiva camarones**?

(1) Si *[Siga]*   (0) No *[Pase a 34]*

32. ¿Si responde "si", cuantas hectáreas dedican a **cultivar camarones**? _____

33. Cuanto tiempo llevan **cultivando camarones** en esta propiedad?_____años

34. ¿Usted, o algún miembro del hogar, utiliza alguna porción de su propiedad para producción **de sal**?

    (1) Si *[Siga]*   (0) No *[Pase a 37]*

35. ¿Si responde "si", cuantas hectáreas dedican a **producción de sal**? _____

36. Cuanto tiempo llevan **produciendo sal** en esta propiedad?_____años

SECCIÓN VIII: Recolección de moluscos/crustáceos.

37. ¿Usted o sus familiares recolectan crustáceos o moluscos en los ríos o estuarios?
    (1) Si *[Siga]*   (0) No *[Pase a 44]*

38. ¿Adonde lo recolectan?
    (1) Río (2) Estero   (3) Manglar   (4) Mar

39. ¿Cómo llegan a este lugar?
    (1) Caminando   (2) Bote con motor   (3) Vehículo   (4)   Canoa
        (5) Otro

40. ¿Cuánto tiempo se demora en llegar a este lugar?
    (1) Menos de 30 minutos   (2) de 30 minutos a 1 hora   (3) 1-2 horas
        (4) mas de 2 horas

41. ¿Cuánto tiempo se demora en llegar a este lugar en los años 80 o alrededor del tiempo cuando el José Azcona estaba en la presidencia?
    (1) Menos de 30 minutos   (2) de 30 minutos a 1 hora   (3) 1-2 horas
        (4) mas de 2 horas

42. ¿Cada cuanto recolectan los crustáceos o moluscos?
    (1) Una vez a la semana   (2) de 2 a 3 veces por semana   (3) diariamente
    (4) cada 2 o 3 semanas   (5) Mensualmente.

43. ¿Hace cuanto recolectan moluscos o crustáceos?
    (1) Menos de 1 año   (2) de 1 a 5 años   (3) 6 a 10 años   (4) 11 a 15 años   (5) 16 a 20 años
    (6) Mas de 20 años.

SECCIÓN IX Pesca.

44. ¿Usted o alguien de su familia se dedica a la pesca en los estuarios o en el Golfo de Fonseca como forma de ingreso?
    (1) Si *[Siga]*   (0)   No *[Pase a 54]*

45. ¿Adonde pescan?
    (1) Río (2) Estero   (3) Manglar   (4) Mar

46. ¿Cómo llegan a este lugar?
    (1) Caminando   (2) Bote con motor   (3) Vehículo   (4)   Canoa
        (5) Otro

47. ¿Cuánto tiempo se demora en llegar a este lugar?

(1) Menos de 30 minutos   (2) de 30 minutos a 1 hora   (3) 1-2 horas

     (4) mas de 2 horas

48. ¿Cuánto tiempo se demora en llegar a este lugar en los años 80 o alrededor del tiempo cuando el José Azcona estaba en la presidencia?
(1) Menos de 30 minutos   (2) de 30 minutos a 1 hora   (3) 1-2 horas

     (4) mas de 2 horas

49. ¿Cada cuanto se dedican a la pesca?
(1) Una vez a la semana   (2) de 2 a 3 veces por semana   (3) diariamente

(4) cada 2 o 3 semanas   (5) Mensualmente.

50. ¿Hace cuanto se dedican a la pesca?
(1) Menos de 1 año   (2) de 1 a 5 años   (3) 6 a 10 años   (4) 11 a 15 años   (5) 16 a 20 años

(6) Mas de 20 años.

51. ¿Ha existido alguna razón por la cual usted no ha podido llegar al lugar donde normalmente pesca o pescaba?
(1) Si *[Siga]*   (0) No *[Pase a 54]*

52. ¿Porqué no ha podido tener acceso al lugar?
_____
_____
_____
_____

53. ¿Hace cuanto tiempo no pueden llegar a ese lugar?_____años

SECCIÓN X: El Manglar.

54. Responda todos los enunciados verticales (55 – 61) para cada opcion

Especies de Manglar utilizadas, compradas, vendidas o cortados para diferentes

A	B		C	D		D		E	
Usted o su familia usa el manglar para:			Que tipos de especies usan ustedes? Escriba los nombres de la especies que ellos usan? (si no saben escriba "no saben")	¿Usted lo compra?		¿Usted lo vende?		¿Usted lo corta?	
	Si	No		Si	No	Si	No	Si	No
1. Leña	(1)	(0)		(1)	(0)	(1)	(0)	(1)	(0)
2. Construcción	(1)	(0)		(1)	(0)	(1)	(0)	(1)	(0)
3. Postes de cercas o Vigas	(1)	(0)		(1)	(0)	(1)	(0)	(1)	(0)
4. Teñido para cuero	(1)	(0)		(1)	(0)	(1)	(0)	(1)	(0)
5. Medicina	(1)	(0)		(1)	(0)	(1)	(0)	(1)	(0)
6. Tintes o colorantes	(1)	(0)		(1)	(0)	(1)	(0)	(1)	(0)
7. Otro	(1)	(0)		(1)	(0)	(1)	(0)	(1)	(0)

propósitos

55. ¿ Hay algún otro uso o beneficio de los manglares que no hemos discutido?
_____
_____
_____
_____
_____

56. Hace cuanto cortan usted o su familia manglares por alguna de las razones que usted menciono?
(1) Menos de 1 año   (2) de 1 a 5 años   (3) 6 a 10 años   (4) 11 a 15 años   (5) 16 a 20 años
(6) Mas de 20 años.

57. ¿Que clase de herramienta utiliza para cortar el manglar?
(1) Machete   (2) Hacha   (3) Motosierra   (4) Otro

58. ¿Cada cuanto cortan mangle?
(1) Una vez a la semana   (2) de 2 a 3 veces por semana   (3) diariamente
4) cada 2 o 3 semanas   (5) Mensualmente.

59. ¿En temporada del año corta mas mangle?
(1) Verano (época seca)   (2) Invierno (época lluviosa)

60. ¿Cómo llegan a este lugar?
    (1) Caminando   (2) Bote con motor   (3) Vehículo   (4)   Canoa
    (5) Otro

61. ¿Cuánto tiempo se demora en llegar a este lugar?
    (1) Menos de 30 minutos   (2) de 30 minutos a 1 hora   (3) 1-2 horas
    (4) mas de 2 horas

62. ¿Cuánto tiempo se demora en llegar a este lugar en los años 80 o alrededor del tiempo cuando el José Azcona estaba en la presidencia?
    (1) Menos de 30 minutos   (2) de 30 minutos a 1 hora   (3) 1-2 horas
    (4) mas de 2 horas

63. ¿Qué tan difícil es para usted cortar mangle en este lugar?
    (1) Muy difícil   (2) Difícil   (3) Mas o menos difícil   (4) Fácil
    (5) Muy fácil

64. ¿Porqué usted corta mangle en ese lugar?
    _____
    _____

65. De quien es el terreno donde corta mangle?
    (1)Propio   (2) Privado (otra persona)   (3) Del estado   (4)   Área protegida
    (5) De la comunidad   (6) Otro_____.

SECCIÓN XI: Resiembra de Manglar.

66. ¿Usted ha participado alguna vez en una reforestación de mangle?
    (1) Si *[Siga]*   (0) No *[Pase a 79]*

67. ¿Con que frecuencia resiembra mangle? _____

68. ¿Qué cantidad de mangle siembra? _____

69. ¿Porque reforesta mangle?
    _____
    _____

70. ¿Qué tipo de mangle reforesta? _____

71. ¿Desde hace cuanto tiempo lo hace? _____años

SECCIÓN XII: Percepciones y actitudes.

72. ¿En los últimos años, ¿ha cambiado los manglares? Desde cuándo?
    (1) Si *[Escriba]*   (0) No *[Pase a 81]*
    _____
    _____
    _____

73. ¿Cómo y por qué ha cambiado los manglares?

_____

_____

74. Cuales piensa usted que son las 5 amenazas mas grandes de los manglares?(en orden de importancia)

      1.)_____

      2.)_____

      3.)_____

      4.)_____

      5.)_____

75. ¿Usted siente que su calidad de vida se ve amenazada por el deterioro del manglar?
(1) Si *[Siga]*   (0) No *[Pase a 84]*

76. Puede explicare como el deterioro del manglar amenaza su calidad de vida?

_____
_____
_____

77. ¿Hay áreas protegidas en su municipio?

    (1) Si   (0) No   *(2) No Sabe*

78. Cuales piensa usted que son los 5 problemas mas grandes de su comunidad (en orden de importancia)

      1.)_____

      2.)_____

      3.)_____

      4.)_____

  5.)_____¿Usted piensa que los

                manglares son:

(1) Para nada importantes   (2) No muy importantes   (3) Importantes
      (4) Algo importantes

(5) Muy importantes

79. Sabe usted de algún conflicto (por ejemplo: peleas, argumentos, disputas, demandas legales) en relación al uso, acceso, administración, conservación, o control, de los manglares o la tierra alrededor de estos?
      (1) Si *[Continue]* (0) No *[Encuesta Completa]*

80. Puede explicar la naturaleza del conflicto o cómo empezó?

_____
_____
_____

81. ¿Desde hace cuanto tiempo existen los conflictos? _____años

82. Quien o qué grupos han estado asociados a los conflictos?

Por favor indique el tipo de organizacion:	Sí	No	Cual?
83. Alcalde, UMAS			
84. Una organización (cooperativa o asociación)			
85. Organización de pescadores			
86. Organización de acuicultura			
87. Grupo Ambiental			
88. ANDAH			
89. CODDEFFAGOLF			
90. Otro			

91. ¿Cómo describe los conflictos en la zona por el manglar:
    (1) Muy violento   (2) Algo violento   (3) En ocaciones violento   (4) No muy violento

    (5) Para nada violento

92. Si respondió 1, 2, 3, . Qué grupos están relacionados con esa violencia?

favor indique el tipo de organizacion:	Sí	No	Cual?
Alcalde, UMAS			
Una organización (cooperativa o asociación)			
Organización de pescadores			
Organización de acuicultura			
Grupo Ambiental			
ANDAH			
CODDEFFAGOLF			
. Otro			

Age range of the respondents to the household questionnaire:

e Range	Number of respondents	%
– 25	57	15.7
– 35	61	16.8x
– 45	100	27.5
– 55	68	18.6
– 65	44	12.1%
– and older	34	9.3
tal	364	100x

List of the Respondents Interviewed using the Questionnaire: Individuals were randomly selected for lengthier open-ended interviews and each is cited in the 'interviews' appendix.

N	Fecha	Departamento	Municipio	Aldea	Nombre	Género	Edad
1	02.07.05	Valle	Alianza	Valle Nuevo	Maria Granados	Femenino	40
2	02.07.05	Valle	Alianza	Valle Nuevo	Gladis Amador	Femenino	29
3	02.07.05	Valle	Alianza	Las Playitas	Delmis Sanchez	Femenino	42
4	02.07.05	Valle	Alianza	Las Playitas	Horacio González	Masculino	51
5	02.07.05	Valle	Nacaome	Campamento	Alexis García	Masculino	27
6	02.07.05	Valle	Nacaome	Campamento	Daysi Inostrosa	Femenino	35
7	02.07.05	Choluteca	Nacaome	Campamento	Milagro Viera	Femenino	49
8	02.07.05	Valle	Nacaome	Valle Nuevo	Maribel Acosta	Femenino	41
9	02.07.05	Valle	Alianza	Valle Nuevo	José Bustillos	Masculino	66
10	02.07.05	Valle	Alianza	Los Guatales	Jefley Josui	Masculino	16
11	02.07.05	Valle	Nacaome	Campamento	Gidmina Burgos Isaguirre	Femenino	65
12	02.07.05	Valle	Alianza	Valle Nuevo	Santos Antonio Viera Vanegas	Masculino	44
13	02.07.05	Valle	Alianza	Valle Nuevo	Teófilo Amador Ferrufino	Masculino	64
14	02.07.05	Valle	Alianza	Los Guatales	Alcelmo Mejía Vanegas	Masculino	67
15	02.07.05	Valle	Nacaome	Campamento	Doris Esperanza Cruz	Femenino	32
16	02.07.05	Valle	Alianza	Los Guatales	Antonio Maldonado	Masculino	20
17	02.07.05	Valle	Nacaome	El Relleno/Puerto Soto	Omar Alexander Hernández Oliva	Masculino	20
18	02.07.05	Valle	Alianza	Los Guatales	Luis Alonso Reyes	Masculino	66
19	02.07.05	Valle	Nacaome	Campamento	Gledis Guerrero	Femenino	28

20	02.07.05	Valle	Alianza	Los Guatales	Armando Torres	Masculino	65
21	02.07.05	Valle	Alianza	Los Guatales	Reina Isabel Reyes	Femenino	44
22	02.07.05	Valle	Alianza	Valle Nuevo	Delmi Danilia Velásquez	Femenino	42
23	02.07.05	Valle	Alianza	Valle Nuevo	Arwen Reyes	Femenino	49
24	02.07.05	Valle	Nacaome	Campamento	Hipocita Une	Femenino	51
25	02.07.05	Valle	Alianza	Valle Nuevo	Erika Ferrufino	Femenino	48
26	02.07.05	Valle	Alianza	Los Guatales	Jorge Luis Torres Torres	Masculino	21
27	02.07.05	Valle	Nacaome	Campamento	Rafael Rivera Acosta	Masculino	39
28	02.07.05	Valle	Nacaome	Campamento	Hilda Ramos	Femenino	28
29	02.07.05	Valle	Alianza	Valle Nuevo	Concepción Aguilar Frutillo	Femenino	51
30	02.07.05	Valle	Nacaome	Campamento	Heydi Carolina García Rivera	Femenino	18
31	02.07.05	Valle	Alianza	Valle Nuevo	Marcelino Bustamante	Masculino	60
32	02.07.05	Valle	Alianza	Valle Nuevo	Alejandro Bustillos	Masculino	66
33	02.07.05	Valle	Alianza	Las Playitas	María Cárcamo	Femenino	68
34	02.07.05	Valle	Alianza	Las Playitas	Elia Burgos	Femenino	53
35	02.07.05	Valle	Nacaome	Campamento	Juan Maldonado	Masculino	27
36	02.07.05	Valle	Nacaome	Campamento	Asencio García	Masculino	54
37	02.07.05	Valle	Nacaome	La Brea	Irma Rosa Hernán	Femenino	53
38	02.07.05	Valle	Nacaome	La Brea	Rita Fernández	Femenino	54
39	02.07.05	Valle	Nacaome	Playa Negra	Maximina Flores	Femenino	67
40	02.07.05	Valle	Nacaome	Playa Grande	Olga Lidia Alemán	Femenino	45
41	02.07.05	Valle	Nacaome	Playa Grande	Rosa Osorio	Femenino	37
42	02.07.05	Valle	Nacaome	La Brea	Martha Lidia Gómez	Femenino	26
43	02.07.05	Valle	Nacaome	La Brea	Ramón Rivas	Masculino	62
44	02.05.07	Valle	Nacaome	La Brea	Arnold Mejía	Masculino	15
45	02.07.05	Valle	Nacaome	Playa Grande	María D. Martínez	Femenino	50

#	Fecha	Departamento	Municipio	Lugar	Nombre	Sexo	Edad
46	02.07.05	Valle	Nacaome	Playa Grande	Glenda Saraí Flores	Femenino	23
47	02.07.05	Valle	Nacaome	Playa Grande	Felicito Hernández	Masculino	38
48	02.07.05	Valle	Nacaome	Playa Grande	Rita Rodriguez	Femenino	50
49	02.07.05	Valle	Nacaome	La Brea	Joel Aranda	Masculino	17
50	02.07.05	Valle	Nacaome	La Brea	Elías Cárcamo	Masculino	50
51	02.07.05	Valle	Nacaome	Playa Grande	Yajaira Bonilla	Femenino	24
52	02.07.05	Valle	Nacaome	Playa Grande	Teresa Contrera	Femenino	47
53	02.07.05	Valle	Nacaome	Playa Grande	Florida López	Femenino	25
54	02.07.05	Valle	Nacaome	La Brea	María Mercedes	Femenino	65
55	02.07.05	Valle	Nacaome	La Brea	Santos Martínez	Femenino	29
56	02.07.05	Valle	Nacaome	Playa Grande	Maria Antonia Vargas	Femenino	66
57	02.07.05	Valle	Nacaome	Playa Grande	Julian Santos	Masculino	85
58	02.07.05	Valle	Nacaome	La Brea	Hipolita Rubio	Femenino	78
59	02.07.05	Valle	Nacaome	La Brea	Gloria Vargas	Femenino	50
60	02.07.05	Valle	Nacaome	Playa Grande	Alfredo Martínez	Masculino	sin edad
61	02.07.05	Valle	Nacaome	Playa Grande	Lucia Vanegas	Femenino	28
62	02.07.05	Valle	Nacaome	Playa Grande	Maria Mendoza	Femenino	60
63	02.07.05	Valle	Nacaome	Playa Grande	Christina Matamoros	Femenino	39
64	02.07.05	Valle	Nacaome	Playa Grande	Leoncia Granados López	Masculino	65
65	02.07.05	Valle	Nacaome	La Brea	Sandra Cárdenas	Femenino	29
66	02.07.05	Valle	Nacaome	La Brea	Noel Vaquedano	Masculino	36
67	02.07.05	Valle	Nacaome	La Brea	Leonor Pineda	Femenino	29
68	02.07.05	Valle	Nacaome	La Brea	Ernesto Hernández	Masculino	60
69	02.07.05	Valle	Nacaome	La Brea	María de la Cruz	Femenino	51
70	02.07.05	Valle	Nacaome	Playa Grande	Anselma Moreno	Femenino	46
71	04.07.05	sin departamento	sin municipio	Playa Negra	Roger Carranza	Masculino	24
72	04.07.05	Choluteca	Namasigue	San Bernardo	Elbisa Ochoa Estrada	Femenino	49

#	Fecha	Depto	Municipio	Comunidad	Nombre	Sexo	Edad
73	02.07.05	Valle	Nacaome	Playa Grande	Cristina Flores	Femenino	65
74	02.07.05	Valle	Nacaome	Playa Grande	Cecilia Vargas	Femenino	37
75	02.07.05	Valle	Nacaome	Playa Grande	Martin Vargas	Masculino	42
76	02.07.05	Valle	Nacaome	La Brea	Héctor Hernández	Masculino	64
77	02.07.05	Valle	Nacaome	Playa Grande	Maria Damaris Rivas Mendoza	Femenino	23
78	02.07.05	Valle	Nacaome	La Brea	Candida Rosa Pereira	Femenino	59
79	02.07.05	Valle	Nacaome	La Brea	Procerpina Cárcamo	Femenino	36
80	02.07.05	Valle	Nacaome	La Brea	Grabriela Mesa Ortiz	Femenino	68
81	04.07.05	Choluteca	Marcovia	Cedeño	Jesse Roble	Femenino	23
82	04.07.05	Choluteca	Namasigue	Guamerú	Victoria Ochoa	Femenino	40
83	04.07.05	Choluteca	Nacaome	Playa Negra	Trancito Gonzalez	Femenino	53
84	04.07.05	Choluteca	Nacaome	Playa Negra	Florentina Flores	Femenino	72
85	03.07.05	Choluteca	Marcovia	Guapinol	Maria Lada	Femenino	44
86	03.07.05	Choluteca	Marcovia	Guapinol	Herman Serrato	Masculino	40
87	03.07.05	Choluteca	Marcovia	Guapinol	Milagros Cruz	Femenino	40
88	03.07.05	Valle	Alianza	Las Playitas	Hilda Serrato	Femenino	57
89	03.07.05	Valle	Alianza	Las Playitas	Isa Francisca	Femenino	27
90	03.07.05	Choluteca	Marcovia	Guapinol	Maria Grande	Femenino	38
91	04.07.05	Choluteca	Namasigue	Playa Negra	Trinidad Castillo	Femenino	58
92	04.07.05	Choluteca	Namasigue	Playa Negra	Salvadora González	Femenino	39
93	04.07.05	Choluteca	Namasigue	Playa Negra	Juan de la Cruz González	Masculino	50
94	04.07.05	Choluteca	Marcovia	Guapinol	María Zamora	Femenino	78
95	04.07.05	Valle	Alianza	Las Playitas	Enma Matute	Femenino	50
96	04.07.05	Choluteca	Namasigue	San Bernardo	Rafael Rios	Masculino	33
97	04.07.05	Choluteca	Namasigue	San Bernardo	Sixto Perez	Masculino	50
98	04.07.05	Choluteca	Namasigue	San Bernardo	Jose de la Cadena Martinez	Masculino	59
99	04.07.05	Choluteca	Namasigue	San Bernardo	José Carranza	Masculino	5

101	03.07.05	Choluteca	Marcovia	Guapinol	Wilson Noel Ávila Guzmán	Masculino	19
102	03.07.05	Choluteca	Marcovia	Guapinol	Martha Guzmán	Femenino	50
103	03.07.05	Choluteca	Marcovia	Guapinol	Meselemia Moudragon	Femenino	sin edad
104	03.07.05	Choluteca	Marcovia	Guapinol	Candida Ramos	Femenino	54
105	03.07.05	Choluteca	Marcovia	Guapinol	Gloribel Zerón	Femenino	24
106	03.07.05	Choluteca	Marcovia	Guapinol	Tomasa Rivas	Femenino	41
107	03.07.05	Choluteca	Marcovia	Guapinol	Martha Jiménez	Femenino	47
108	03.07.05	Choluteca	Marcovia	Guapinol	Santos Soliz	Masculino	39
109	03.07.05	Choluteca	Marcovia	Guapinol	Reina Blanco	Femenino	35
110	03.07.05	Choluteca	Marcovia	Guapinol	Alba Amador	Femenino	23
111	03.07.05	Choluteca	Marcovia	Guapinol	Jose Santos	Masculino	50
112	03.07.05	Choluteca	Marcovia	Guapinol	Juan Elchino	Femenino	27
113	03.07.05	Choluteca	Marcovia	Guapinol	Suyapa Méndez	Femenino	35
114	03.07.05	Choluteca	Marcovia	Guapinol	Darina Rarilo	Femenino	18
115	03.07.05	Choluteca	Marcovia	Guapinol	Sandra Avile	Femenino	36
116	03.07.05	Choluteca	Marcovia	Guapinol	Olga Hernández	Femenino	34
117	03.07.05	Choluteca	Marcovia	Guapinol	Blanca Santos	Femenino	67
118	03.07.05	Choluteca	Marcovia	Guapinol	Marcos Pérez	Masculino	47
119	03.07.05	Choluteca	Marcovia	Guapinol	Ana Morán	Femenino	43
120	03.07.05	Choluteca	Marcovia	Guapinol	Gloria Flores	Femenino	29
121	03.07.05	Valle	Marcovia	Guapinol	Fermin Flores	Masculino	70
122	03.07.05	Choluteca	Marcovia	Guapinol	Luz Maria Soriano	Femenino	31
123	04.07.05	Choluteca	Marcovia	Guapinol	Amado Hernandez	Masculino	37
124	03.07.05	Choluteca	Marcovia	Guapinol	Karen Geneda	Femenino	25
125	03.07.05	Choluteca	Marcovia	Guapinol	Francisca Diaz	Femenino	43
126	03.07.05	Choluteca	Marcovia	Guapinol	Maria Paula Gonzalez	Femenino	42
127	04.07.05	Choluteca	Namasigue	San Jerónimo	Julio César Río	Masculino	45

#	Fecha	Departamento	Municipio	Aldea	Nombre	Sexo	Edad
131	03.07.05	Choluteca	Marcovia	Guapinol	Bessy Alvarez	Femenino	22
132	02.07.05	Choluteca	Marcovia	Guapinol	María Morales	Femenino	37
133	03.07.05	Choluteca	Marcovia	Guapinol	Doricia Gerón Pérez	Femenino	21
134	03.07.05	Choluteca	Marcovia	Guapinol	María Pérez	Femenino	32
135	03.07.05	Choluteca	Marcovia	Guapinol	Marcelina Mendoza	Femenino	61
136	03.07.05	Choluteca	Marcovia	Guapinol	Diana Cerrato	Femenino	43
137	03.07.05	Choluteca	Marcovia	Guapinol	María Palma	Femenino	37
138	03.07.05	Choluteca	Marcovia	Guapinol	Enta Ortiz	Femenino	23
139	03.07.05	Choluteca	Marcovia	Guapinol	Satiel Chavarría	Masculino	53
140	03.07.05	Choluteca	Marcovia	Guapinol	María de la Cruz Ordoñez	Femenino	36
141	03.07.05	Choluteca	Marcovia	Guapinol	Segundina Martínez	Femenino	45
142	03.07.05	Choluteca	Marcovia	Guapinol	Sonia Margarita Mendoza	Femenino	41
143	03.07.05	Choluteca	Marcovia	Guapinol	Vilma Sánchez	Femenino	35
144	03.07.05	Choluteca	Marcovia	Guapinol	José Vicente Galeas	Masculino	40
145	04.07.05	Choluteca	Namasigue	Costa Azul	María Luisa Linares	Femenino	50
146	04.07.05	Choluteca	Namasigue	Costa Azul	Doris Aguilar	Femenino	25
147	04.07.05	Choluteca	Namasigue	Guamerú	Esther Hernández	Femenino	21
148	04.07.05	Choluteca	Namasigue	Guamerú	Dicsia Liliana Lindo Vallejo	Femenino	21
149	04.07.05	Choluteca	Namasigue	San Bernardo	Bernardo Anael Ordóñez	Masculino	36
150	04.07.05	Choluteca	Namasigue	Guamerú	César Mendoza	Masculino	50
151	04.07.05	Choluteca	Namasigue	Costa Azul	Erasmo Pastrana	Masculino	53
152	04.07.05	Choluteca	Namasigue	Playa Negra	Eva Dalila Velásquez	Femenino	21
153	04.07.05	Choluteca	Namasigue	Guamerú	Luis Manuel Ayala Armas	Masculino	63
154	04.07.05	Choluteca	Namasigue	Guamerú	Nora Ochoa	Femenino	42
155	04.07.05	sin departamento	sin municipio	Costa Azul	Madonio Velazquez	Masculino	53

ID	Fecha	Departamento	Municipio	Comunidad	Nombre	Sexo	Edad
156	04.07.05	Choluteca	Marcovia	Guapinol	Gina Rivera Nuñez	Femenino	44
157	04.07.05	Choluteca	Marcovia	Guapinol	José Fr.Armendariz	Masculino	23
158	04.07.05	Choluteca	Marcovia	Guapinol	Hernán Amador	Masculino	52
159	04.07.05	Choluteca	Marcovia	Guapinol	Antonio Barahona Ordoñez	Masculino	36
160	04.07.05	Choluteca	Marcovia	Guapinol	José Luis Gereda	Masculino	37
161	03.07.05	Choluteca	Marcovia	Guapinol	José Edas Ortiz	Masculino	31
162	03.07.05	Choluteca	Marcovia	Guapinol	Cristobal Rodríguez	Masculino	21
163	04.07.05	Choluteca	Namasigue	Guamerú	María González	Femenino	59
164	04.07.05	Choluteca	Namasigue	Costa Azul	María Teresa Cruz	Femenino	32
165	04.07.05	Choluteca	Namasigue	Guamerú	Rigoberto Ordoñez	Masculino	49
166	02.07.05	Valle	Nacaome	El Relleno/Puerto Soto	Olga Marina Pozada García	Femenino	56
167	02.07.05	Valle	Amapala	La Pintadillera	Juan Pablo Velásquez Cardenas	Masculino	44
168	02.07.05	Valle	Amapala	La Pintadillera	Juana Rosa García Oseguera	Femenino	68
169	02.07.05	Valle	Amapala	Coyolito	Marlene Hernández López	Femenino	24
170	02.07.05	Valle	Amapala	Puerto Grande	Joselito Zavala	Masculino	40
171	02.07.05	Valle	Amapala	La Pintadillera	Rosendo Laínez	Masculino	54
172	02.07.05	Valle	Amapala	La Pintadillera	Martha Laínez	Femenino	24
173	02.07.05	Valle	Amapala	El Coyolito	Melvin Hernández	Masculino	20
174	02.07.05	Valle	Amapala	Puerto Grande	Angelica Berrillo	Femenino	54
175	02.07.05	Valle	Amapala	Puerto Grande	Danilo Corrales	Masculino	43
176	02.07.05	Valle	Amapala	Coyolito	José Luis Hernández	Masculino	42
177	02.07.05	Valle	Amapala	La Pintadillera	María Ramírez	Femenino	72
178	02.07.05	Valle	Amapala	La Pintadillera	Santos Núñez	Masculino	39
179	02.07.05	Valle	Nacaome	El Relleno/Puerto Soto	Omar Alexander Hernández Oliva	Masculino	20
180	02.07.05	Valle	Nacaome	El Relleno/Puerto Soto	José Francisco	Masculino	25

No.	Fecha	Departamento	Municipio	Comunidad	Nombre	Sexo	Edad
181	02.07.05	Valle	Nacaome	El Relleno/Puerto Soto	Mariano Sierra	Masculino	40
182	02.07.05	Valle	Amapala	La Pintadillera	Mariana Velazquez	Femenino	40
183	02.07.05	Valle	Amapala	La Pintadillera	Dolores Llanos	Femenino	65
184	03.07.05	Valle	Amapala	Coyolito	Paulina Rodriguez	Femenino	37
185	02.07.05	Valle	Amapala	Playa Grande	Merlinda Yani	Femenino	16
186	02.07.05	Valle	Nacaome	El Relleno/Puerto Soto	Eva Matamoros	Femenino	52
187	02.07.05	Valle	Nacaome	El Relleno/Puerto Soto	Ama Hernandez	Femenino	20
188	03.07.05	Valle	Amapala	La Pintadillera	Amalia Castillo	Femenino	39
189	02.07.05	Valle	Amapala	La Pintadillera	Maria Oliva	Femenino	39
190	02.07.05	Valle	Amapala	La Pintadillera	Juana Cabrera	Femenino	45
191	03.07.05	Choluteca	Marcovia	Punta Ratón	Oscar Mejía	Masculino	65
192	03.07.05	Valle	amapala	Coyolito	Henry Palavichini	Masculino	38
193	02.07.05	Valle	Amapala	Puerto Grande	Nicolasa Avila	Femenino	36
194	02.07.05	Valle	Amapala	Puerto Grande	Ingrid Isidro	Femenino	19
195	02.07.05	Valle	Amapala	Coyolito	Florinda Flores	Femenino	52
196	02.07.05	Valle	Amapala	Puerto Grande	Estela Rodas	Femenino	67
197	02.07.05	Valle	Nacaome	El Relleno/Puerto Soto	Odilia Concepción Umuña	Femenino	29
198	02.07.05	Valle	Nacaome	El Relleno/Puerto Soto	Dorotea Argentina Castellóon	Femenino	32
199	02.07.05	Valle	Amapala	La Pintadillera	Mercedes Estrada	Femenino	33
200	02.07.05	Valle	Amapala	La Pintadillera	Martha Gladis Cerrato	Femenino	31
201	02.07.05	Valle	Amapala	Puerto Grande	Imelda Corre Meléndez	Femenino	54
202	02.07.05	Valle	Amapala	Coyolito	Juan Carlos Zepeda Valladares	Masculino	20
203	02.07.05	Valle	Amapala	Coyolito	Claudia de la Cruz García	Femenino	44
204	02.07.05	Valle	Amapala	La Pintadillera	José Santos Estrada Manzanero	Masculino	39

#	Fecha	Depto	Municipio	Comunidad	Nombre	Sexo	Edad
205	02.07.05	Valle	Amapala	La Pintadillera	Ana Lisbeth Escobar Alvarado	Femenino	22
206	02.07.05	Valle	Amapala	La Pintadillera	José Lino Estrada Manzanares	Masculino	32
207	02.07.05	Valle	Nacaome	El Relleno/Puerto Soto	Ana Rosa Velasco	Femenino	68
208	02.07.05	Valle	Nacaome	El Relleno/Puerto Soto	Magdaleña Armendaris Rvias	Femenino	27
209	02.07.05	Valle	Amapala	Coyolito	Cristina Isabel Santos	Femenino	52
210	02.07.05	Valle	Amapala	El Relleno/Puerto Soto	Manuel Medina	Femenino	28
211	03.07.05	Choluteca	Marcovia	Punta Ratón	Marina Montúfar	Femenino	52
212	03.07.05	Choluteca	Marcovia	Punta Ratón	Ingrid Ayala	Femenino	24
213	03.07.05	Choluteca	Marcovia	Punta Ratón	Orlando Vargas	Masculino	36
214	03.07.05	Choluteca	Marcovia	Punta Ratón	María Cárcamo	Femenino	44
215	03.07.05	Choluteca	Marcovia	Punta Ratón	Elisa Avariba	Femenino	52
216	03.07.05	Choluteca	Marcovia	Pueblo nuevo	Felida Moreno	Femenino	56
217	03.07.05	Choluteca	Marcovia	Pueblo nuevo	Abelardo Serrano	Masculino	36
218	03.07.05	Choluteca	Marcovia	Pueblo nuevo	María Aguilera	Femenino	66
219	03.07.05	Choluteca	Marcovia	Pueblo nuevo	Maria Isabel Maradiaga	Femenino	21
220	03.07.05	Choluteca	Marcovia	Pueblo nuevo	Paula Maradiaga	Femenino	25
221	03.07.05	Choluteca	Marcovia	Pueblo nuevo	Gladys Margarita Arriola Gallo	Femenino	.
222	03.07.05	Choluteca	Marcovia	Punta Ratón	Constantino Cruz	Masculino	85
223	03.07.05	Choluteca	Marcovia	Pueblo nuevo	Santos Gallo	Masculino	70
224	03.07.05	Choluteca	Marcovia	Punta Ratón	Brenda Marissa	Femenino	30
225	03.07.05	Choluteca	Marcovia	Pueblo nuevo	Santos Hernández Manzanarez	Masculino	67

No.	Fecha	Departamento	Municipio	Lugar	Nombre	Sexo	Edad
226	03.07.05	Choluteca	Marcovia	Pueblo nuevo	Patrocinia Martínez	Femenino	73
227	03.07.05	Choluteca	Marcovia	Pueblo nuevo	Martha Gonzalez	Femenino	60
228	03.07.05	Choluteca	Marcovia	Punta Ratón	Ana Julia Cabrera	Femenino	64
229	03.07.05	Choluteca	Marcovia	Punta Ratón	Suyapa Ramírez	Femenino	47
230	03.07.05	Choluteca	Marcovia	Punta Ratón	Rosadilia Cabrera	Femenino	42
231	03.07.05	Choluteca	Marcovia	Punta Ratón	Susana Castillo	Femenino	39
232	03.07.05	Choluteca	Marcovia	Pueblo nuevo	Ana Castro	Femenino	37
233	03.07.05	Choluteca	Marcovia	Pueblo nuevo	Rolando Flores	Masculino	50
234	03.07.0	Valle	Marcovia	Pueblo nuevo	José Herrera	Masculino	24
235	03.05.07	Choluteca	Marcovia	Pueblo nuevo	Francisca Velásquez	Femenino	58
236	03.07.05	Choluteca	Marcovia	Pueblo nuevo	Pablo Maradiaga Castillo	Masculino	63
237	03.07.05	Choluteca	Marcovia	Pueblo nuevo	Julio Aguirre Sosa	Masculino	50
238	03.07.05	Choluteca	Marcovia	Pueblo nuevo	Sandra Salinas Velásquez	Femenino	37
239	03.07.05	Choluteca	Marcovia	Pueblo nuevo	Alma Janeth Alvarez	Femenino	35
240	03.07.05	Choluteca	Marcovia	Punta Ratón	Gustavo Adolfo Escoto Amador	Masculino	41
241	03.07.05	Choluteca	Marcovia	Punta Ratón	Leonardo Quiroz Maradiaga	Masculino	76
242	03.07.05	Choluteca	Marcovia	Punta Ratón	Luzmida Montúfar	Femenino	39
243	03.07.05	Choluteca	Marcovia	Punta Ratón	Ada Quiroz	Femenino	28
244	sin fecha	Choluteca	Marcovia	Pueblo nuevo	Felicita Flores	Femenino	52
245	03.07.05	Choluteca	Marcovia	Pueblo nuevo	María Muñoz	Femenino	35
246	03.07.05	Choluteca	Marcovia	Pueblo nuevo	Antonia Flores	Femenino	15
247	03.07.05	Choluteca	Marcovia	Pueblo nuevo	Rosalma Rivera	Femenino	61
248	03.07.05	Choluteca	Marcovia	Cedeño	Telma Flores	Femenino	23
249	03.07.05	Choluteca	Marcovia	Cedeño	Sara Majano	Femenino	27
250	03.07.05	Choluteca	Marcovia	Cedeño	Cristino Flores	Masculino	30
251	03.07.05	Choluteca	Marcovia	Cedeño	María Rivas	Femenino	57
252	03.07.05	Choluteca	Marcovia	Cedeño	Silvano Ramirez	Masculino	5

ID	Código	Departamento	Municipio	Apellido	Nombre	Sexo	Edad
254	03.07.05	Choluteca	Marcovia	Cedeño	Santos Julio Valladares	Masculino	58
255	03.07.05	Choluteca	Marcovia	Cedeño	Francis Azucena Cruz	Femenino	38
256	03.07.05	Choluteca	Marcovia	Cedeño	Dora Alicia Hernández	Femenino	66
257	03.07.05	Choluteca	Marcovia	Cedeño	Yamilet Espinal	Femenino	29
258	03.07.05	Choluteca	Marcovia	Cedeño	José Alvarez Galindo	Masculino	36
259	03.07.05	Choluteca	Marcovia	Cedeño	Azucena Alemán	Femenino	23
260	03.07.05	Choluteca	Marcovia	Cedeño	José Anacléto Rubí	Masculino	87
261	03.07.05	Choluteca	Marcovia	Cedeño	María Zambrano	Femenino	38
262	03.07.05	Choluteca	Marcovia	Cedeño	Cristino Laínez	Masculino	76
263	03.07.05	Choluteca	Marcovia	Cedeño	Agustín Sauceda	Masculino	75
264	03.07.05	Choluteca	Marcovia	Cedeño	Gregorio Rivera	Masculino	56
265	03.07.05	Choluteca	Marcovia	Cedeño	Roxana Cerna	Femenino	18
266	03.07.05	Choluteca	Marcovia	Cedeño	Oliva Canles	Femenino	35
267	03.07.05	Choluteca	Marcovia	Cedeño	Alberto Rivas	Masculino	23
268	03.07.05	Choluteca	Marcovia	Cedeño	Guillermina Pineda	Femenino	45
269	03.07.05	Choluteca	Marcovia	Cedeño	Irma Merina	Femenino	36
270	03.07.05	Choluteca	Marcovia	Cedeño	María Catalina Amador	Femenino	42
271	03.07.05	Choluteca	Marcovia	Cedeño	Moisés Vivar	Masculino	45
272	03.07.05	Choluteca	Marcovia	Cedeño	Rudy Zamora	Femenino	18
273	03.07.05	Choluteca	Marcovia	Cedeño	Santos Colinares	Femenino	48
274	03.07.05	Choluteca	Marcovia	Cedeño	José Aguilar	Masculino	43
275	03.07.05	Choluteca	Marcovia	Cedeño	Santos Majano	Masculino	48
276	03.07.05	Choluteca	Marcovia	Cedeño	Víctor Espinal	Masculino	37
277	03.07.05	Choluteca	Marcovia	Cedeño	Irene Quevedo	Femenino	40
278	03.07.05	Choluteca	Marcovia	Cedeño	Mabi Espinoza	Femenino	35
279	03.07.05	Choluteca	Marcovia	Cedeño	Francisca Calderón	Femenino	30
280	03.07.05	Choluteca	Marcovia	Cedeño	Wendy Cruz	Femenino	2

283	03.07.05	Choluteca	Marcovia	Cedeño	Elva Virginia Maradiaga	Femenino	26
284	03.07.05	Choluteca	Marcovia	Cedeño	Yolanda Salmerón	Femenino	32
285	03.07.05	Choluteca	Marcovia	Cedeño	Manuel de Jesús Alvarez	Masculino	71
286	03.07.05	Choluteca	Marcovia	Cedeño	Maria Emilia Blandis	Femenino	49
287	03.07.05	Choluteca	Marcovia	Cedeño	Rosa Amelia Fuentes	Femenino	64
288	03.07.05	Choluteca	Marcovia	Cedeño	María Auxiliadora Maldonado	Femenino	40
289	03.07.05	Choluteca	Marcovia	Cedeño	Elsa Marina Cardenas	Femenino	44
290	03.07.05	Choluteca	Marcovia	Cedeño	Angie Campos	Femenino	18
291	03.07.05	Choluteca	Marcovia	Cedeño	Petronilo Ramírez	Masculino	33
292	03.07.05	Choluteca	Marcovia	Cedeño	Leonardo Flores	Masculino	57
293	03.07.05	Choluteca	Marcovia	Cedeño	María Solan	Femenino	21
294	03.07.05	Choluteca	Marcovia	Cedeño	Marciano Rivera	Masculino	36
295	03.07.05	Choluteca	Marcovia	Cedeño	Bradys Canales	Masculino	45
296	03.07.05	Choluteca	Marcovia	Cedeño	Reye Ramírez	Masculino	28
297	03.07.05	Choluteca	Marcovia	Cedeño	Oscar Carcamo	Masculino	30
298	03.07.05	Choluteca	Marcovia	Cedeño	Delmi Xillagra	Femenino	20
299	03.07.05	Choluteca	Marcovia	Cedeño	Reina Mendoza	Femenino	47
300	03.07.05	Choluteca	Marcovia	Cedeño	sin nombre	sin género	sin edad
301	03.07.05	Choluteca	Marcovia	Cedeño	Claudio Manzanarez	Masculino	60
302	03.07.05	Choluteca	Marcovia	Cedeño	Jerónimo Cerrato	Masculino	48
303	03.07.05	Choluteca	Marcovia	Cedeño	Martín Colindres	Masculino	38
304	03.07.05	Choluteca	Marcovia	Cedeño	María González	Femenino	40
305	03.07.05	Choluteca	Marcovia	Cedeño	José Hernández	Masculino	40
306	03.07.05	Choluteca	Marcovia	Cedeño	Carlos Gutiérrez	Masculino	52
307	03.07.05	Choluteca	Marcovia	Cedeño	Josefino Bautista	Masculino	63
308	03.07.05	Choluteca	Marcovia	Cedeño	Angélica Ramírez	Femenino	55

309	03.07.05	Choluteca	Marcovia	Cedeño	Rolando Sierra	Masculino	35
310	03.07.05	Choluteca	Marcovia	Cedeño	Noe Rivas	Masculino	44
311	03.07.05	Choluteca	Marcovia	Cedeño	René Aguilera	Masculino	40
312	04.07.05	Choluteca	Namasigue	San Jerónimo	Leopoldo Corrales	Masculino	48
313	03.07.05	Choluteca	Marcovia	Cedeño	Petrona Joya	Femenino	43
314	03.07.05	Choluteca	Marcovia	Cedeño	Suyapa Martínez	Femenino	42
315	03.07.05	Choluteca	Marcovia	Cedeño	Innés Funez	Masculino	59
316	03.07.05	Choluteca	Marcovia	Cedeño	Martina Rivas	Femenino	82
317	03.07.05	Choluteca	Marcovia	Cedeño	Ismenda Osorio.	Femenino	35
318	03.07.05	Choluteca	Marcovia	Cedeño	Juan López.	Masculino	55
319	03.07.05	Choluteca	Marcovia	Cedeño	Senubia Amparo	Femenino	56
320	03.07.05	Choluteca	Marcovia	Cedeño	Nelson Ruiz	Masculino	65
321	04.07.05	Choluteca	Choluteca	El Tulito	Alfredo Salomón Solano	Masculino	52
322	04.07.05	Choluteca	Choluteca	El Carrizo	Ana Patricia Aguilera	Femenino	26
323	04.07.05	Choluteca	Choluteca	El Carrizo	Katy Marcela Serrano	Femenino	27
324	04.07.05	Choluteca	Marcovia	El Tulito	José Fernando Castillo	Masculino	18
325	04.07.05	Choluteca	Marcovia	El Tulito	Marlene Sánchez Lula	Femenino	34
326	04.07.05	Choluteca	Marcovia	El Tulito	Marcel Amador	Masculino	42
327	04.07.05	Choluteca	Marcovia	El Tulito	Paulina Martínez	Femenino	43
328	04.07.05	Choluteca	Marcovia	El Tulito	Pedro Chávez	Masculino	42
329	04.07.05	Choluteca	Marcovia	El Tulito	María Monte	Femenino	54
330	04.07.05	Choluteca	Marcovia	El Tulito	Vicenta Bertilia Méndez Pérez	Femenino	62
331	04.07.05	Choluteca	Namasigue	San Jerónimo	Ana Ariata	Femenino	23
332	03.07.05	Choluteca	Namasigue	San Jerónimo	Silvio Estrada	Masculino	27
333	03.07.05	Choluteca	Marcovia	Pueblo nuevo	Sonia Marina Jimenez	Femenino	20
334	03.07.05	Choluteca	Marcovia	Pueblo nuevo	Florentino Gutiérrez	Masculino	43
335	03.07.05	Choluteca	Marcovia	choluteca	Santos Montoya	Masculino	5

336	03.07.05	Choluteca	Marcovia	El Carrizo	Santos Clorral	Masculino	29
337	04.07.05	Choluteca	Choluteca	El Carrizo	Jorge Antonio Muñoz	Masculino	17
338	04.07.05	Choluteca	Marcovia	El Tulito	Hermilio Núñez	Masculino	60
339	04.07.05	Choluteca	Marcovia	El Tulito	Rosa Perez	Femenino	43
340	04.07.05	Choluteca	Choluteca	El Carrizo	Pablo Valle	Masculino	65
341	04.07.05	Choluteca	Namasigue	san Jerónimo	Oscar Martínez	Masculino	48
342	04.07.05	Choluteca	Namasigue	san Jerónimo	José Santos García	Masculino	52
343	04.07.05	Choluteca	Namasigue	san Jerónimo	Fredyvilma Blandis	Femenino	40
344	04.07.05	Choluteca	Namasigue	san Jerónimo	Valeska Zepeda	Femenino	26
345	04.07.05	Choluteca	Namasigue	San Jerónimo	Delia Muñoz	Femenino	33
346	04.07.05	Choluteca	Namasigue	San Jerónimo	Ronalberto Paladino.	Masculino	48
347	04.07.05	Choluteca	Namasigue	San Jerónimo	Andrés Abelino Blandin	Masculino	50
348	04.07.05	Choluteca	Namasigue	san Jerónimo	Marco Valladares.	Masculino	77
349	04.07.05	Choluteca	Namasigue	san Jerónimo	Blanca Muñoz.	Femenino	48
350	04.07.05	Choluteca	Namasigue	san Jerónimo	Nolbenia Morena.	Femenino	43
351	04.07.05	Choluteca	Marcovia	El Tulito	Germán Núñez	Masculino	60
352	04.07.05	Choluteca	Marcovia	El Tulito	Edis Pérez	Femenino	35
353	04.07.05	Choluteca	Choluteca	El Carrizo	Patricia Flores	Femenino	23
354	04.07.05	Choluteca	Choluteca	El Carrizo	Alejandrina Pérez	Femenino	38
355	04.07.05	Choluteca	Choluteca	El Carrizo	Jackeline del Carmen Reyes	Femenino	35
356	04.07.05	Choluteca	Nacaome	El Carrizo	Surize de Estrada	Femenino	20
357	04.07.05	Choluteca	Choluteca	El Carrizo	Rhina Isabel Martínez	Femenino	35
358	04.07.05	Choluteca	Choluteca	El Carrizo	Rolando Reyes Luna	Masculino	40
359	04.07.05	Choluteca	Choluteca	El Carrizo	Carolina Mondrago	Femenino	29
360	04.07.05	Choluteca	Namasigue	San Jerónimo (2)	Ana Yolanda Espinal	Femenino	47
361	04.07.05	Choluteca	Marcovia	San Jerónimo	Marleny Landar	Femenino	33
362	04.07.05	Choluteca	Namasigue	San Jerónimo	Juana María Aguilera	Femenino	37

364	04.07.05	Choluteca	Namasigue	San Jerónimo	Martha Flores	Femenino	49
365	04.07.05	Choluteca	Namasigue	San Jerónimo	Olivia Flores	Femenino	50
366	04.07.05	Choluteca	Namasigue	San Jerónimo	Ana Reyes	Femenino	62
367	04.07.05	Choluteca	Namasigue	San Jerónimo	Ermelinda Medina	Femenino	74
368	04.07.05	Choluteca	Namasigue	San Jerónimo	Modesto Aguirre	Masculino	62

Appendix 3

Estimates of Mangrove Cover in the Gulf of Fonseca

The following table provides a list of estimates of mangrove cover in the Gulf of Fonseca back to 1958. The left hand column is the data source, followed by year, author, where their data originated, and the amount of mangrove cover cited in the document as well as any relevant information as to how the study was conducted.

Data Source:	Where data originated:	Amount of Mangrove Cover Presented:
CODDEFFAGOLF (2001). Areas Protegidas de la Zona Sur de Honduras (Pg. 6)	Information originated from CODDEFFAGOLF with no explanation as to how the estimate was arrived at.	78,000 hectares of mangrove of high quality in all three countries of the GOF.
PROMANGLE, A.-C. -. O. P. (1997). Manejo y Conservacion de los Manglares en el Golfo de Fonseca, Honduras: Tercera Revision del Documento de Proyecto PD 44/95 (F) Para La Implementacion del Proyecto por Fases. Santa Cruz de la Sierra, Bolivia, International Tropical Timber Organization: Comite de Repoblacion y Ordenacion Forestales: 97.	Assume it is an original study since there is no citation along with the data presented.	1. The study estimated that in 1997 that there were 5,757 km² in southern Honduras, GOF or 4,211 km² in the Department of Choluteca and 1,546 km² in the Department of Valle
Vasquez Cristobal, J. and F. Wainwright W. (2002). Zonificacion de los Bosques de Mangle del Golfo de Fonseca, Honduras, C.A. Choluteca, Honduras, Administracion Forestal del Estado Corporacion Hondurena de Desarrollo Forestal (AFE-COHDEFOR)Organizacion Internacional delas Maderas Tropicales (OIMT) Manejo y Conservacion de los Manglares del Golfo de Fonseca, Honduras (PROMANGLE)	This is an original study conducted by COHDEFOR.	Total area of wetlands including the salt flats is estimated to be 90,150 hectares in southern Honduras.

Data Source:	Where data originated:	Amount of Mangrove Cover Presented:
Vindel Calix Felipie, L. (1997). Sobreexplotacion de la Vegetacion Manglar y Sus Efectos al Ecosistema de La Bahia de Chismuyo, Honduras: Comparacion Estructural del Bosque Manglar Segun el Tipo de Explotacion Economica. Guatemala City, Guatemala, The Nature Conservancy, World Wildlife Fund, and Univ. of Rhode Island Coastal Resources Center (PROARCA/Costas)	This is an original study.	1. Estimated 71,409 hectares of mangrove wetlands including salt flats, seasonal lagoons, and mangrove forests. 2. The document cites the COHDEFOR study completed in 1987 which estimated that out of the total wetlands there were 46,000 hectares of mangrove forests in the same year.
Sanchez, Alexis (1999). Evaluacion de los Manglares del Golfo de Fonseca Mediante un Analisis Multitemporal de Imagenes del Satelite LANDSAT-TM Entre los Anos (1989-1995), (1995-1998), y (1989-1998). Tegucigalpa, Honduras, Administracion Forestal del Estado Corporacion Hondurena de Desarrollo Forestal (AFE-COHDEFOR): 45.	1. Segun J. Prats Llaurado (1958) Page 18 in Sanchez "Informe sobre los Manglares Hondureños del Golfo de Fonseca, Honduras" 2. Silviagro, 1996 quoted a 1965 FAO Study. His study was called: "Analisis del Sub-Sector Forestal de Honduras. Cooperacion Hondurena Alemana, Program Social Forestal. AFE-COHDEFOR. Honduras. 3. Economic Intelligence Unit, Ltd. 1977 information was cited by B. Rollet in 1986 4. Inventario Forestal del Mangle-Zona Sur, 1987 study completed by COHDEFOR cited by Omar Oyuela in 1997. 5. Forest Map by GAF of Germany for COHDEFOR using satellite images from LANDSAT-TM 1993-1995 verified in the field 1991 – 1996.	1. 28,000 hectares of mangrove in 1958 according to this study. Scale was 1:62:000 and 1:64:000. 2. 91,800 hectares of mangrove forests in the southern zone was estimated by FAO in 1962 and 1965. Scale 1:500,000 3. 32,000 hectares of mangrove forests were estimated to cover the GOF region in the data presented. 4. The 1997 information cited by Oyuela stated that there were an estimated 41,320 hectares of mangrove forests in southern Honduras in 1995. This was contrasted with the 1987 COHDEFOR study which estimated 46,710 hectares. 5. 47,200 hectares of mangrove forests in southern Honduras according to this study which used a scale of 1:500,000.

Data Source:	Where data originated:	Amount of Mangrove Cover Presented:

Consultores en Gestion Ambiental (CONGESA) (2001) Valoracion Economica de los Manglares del Golfo de Fonseca, Honduras. Choluteca, Honduras, Administracion Forestal del Estado Corporacion Hondurena de Desarrollo Forestal (AFE-COHDEFOR) Organizacion Internacional de las Maderas Tropicales (OIMT) Manejo y Conservacion de los Manglares del Golfo de Fonseca, Honduras (PROMANGLE) Consultores en Gestion Ambiental (CONGESA)	**The original source is identified as AFE-COHDEFOR**	1. Mangrove Forest estimates for southern Honduras in this document are on page 30 -32.
		   a. 1987: 46,710 hectares
		   b. 1992: 43,678 hectares
		   c. 1995: 41,320 hectares
		2. This document cites the Sanchez study and states that the total area of mangrove forests in 1998 was 30,350 hectares.
		3. The document also cites Vergne *et al.*, 1993 study which indicated:
		   a. 1973: 30,697 hectares
		   b. 1982: 28,776 hectares
		   c. 1992: 23,937 hectares
		4. The document also provides estimates of the total amount lost.
		5. It quotes the 1965 FAO study as having identified a total of 91,800 hectares of mangrove forests of which 57,300 were located in Choluteca and 34,500 in Valle. In 1999, the document states that there were
		6. 21,800 hectares of mangrove forests in Choluteca and 25,400 hectares in Valle totaling 47,200 hectares of mangrove forests.
		7. The document indicates that 35,500 hectares of mangrove forests have been lost in Choluteca since 1965 and that 9,100 hectares have been lost in Valle totaling 44,600 hectares of mangrove forests lost.

Appendix 4

Latin American Agribusiness Development Corporation S.A.

Agribusiness Journal, October 1981, Volume 2 Number 10, Robert H. Holden, "Central America Is Growing More Beef and Eating Less – As the Hamburger Connection Widens http://multinationalmonitor.org/hyper/issues/1981/10/holden.html

With the exception of Nicaragua, the grazing land throughout Central America is largely owned by an elite of Latin and US business that prospers by producing beef for the export market. Although it is impossible to penetrate fully the secrecy that cloaks private US investment abroad, one important investor is the Latin American Agribusiness Development (LAAD) Corp., S.A., a Panamanian-registered corporation with offices in Coral Gables, Fla.; Santo Domingo; Guatemala City, and Santiago, Chile. Its shareholders consist of 14 US-based banks and agribusiness multinationals (see sidebar), plus one Dutch and one Latin American concern: LAAD describes itself as an investor "in private enterprises located in countries in Central America and the Caribbean." Its objective is "To improve the production, distribution and marketing of agriculture based products", with high priority to investment in businesses that produce for export.

The second-largest target industry for LAAD investments (after food processing) is beef cattle. In 1980, the corporation was investing 15 percent of its capital, or about $7.5 million, in 20 different cattle raising ventures. Although the company does not identify the location of the cattle herds, it does point out that Guatemala, Costa Rica, Honduras and Nicaragua alone account for 62 percent of its investments.

Shareholders in the Latin America Agribusiness Development Corporation S.A.:

Adela Investment Co. S.A., Luxembourg

Bank America International Financial Corporation, San Francisco Ca.

Borden Inc., New York

Cargill Inc., Minneapolis Mn.

Castle & Cooke Inc. San Francisco Ca.

Caterpillar Tractor Company, Peoria Il

Centrale Robobank, Curacao NV Willemstad, Curacao, Netherlands Antilles

Chase Manhattan Overseas Banking Corporation, Newark Del.

CPC International Inc., Englewood Cliffs, NJ

Deere and Company, Moline Il.

Gerber Products Company, Freemont Mi.

Girard International Bank, New York

Goodyear Tire and Rubber Co., Akron, Oh.

Monsanto Co., St. Louis Mo.

Ralston-Purina Co., St. Louis Mo.

Southeast First National Bank of Miami Fl.

Appendix 5

Choluteca Declaration 16 October 1996

CHOLUTECA DECLARATION (Choluteca, Honduras - 16 October, 1996)

We, the delegates of twenty-one non-governmental and community organizations from Latin America, North America, Europe and Asia, the participants in the Choluteca Forum Aquaculture and its Impacts, express the following declarations and demands to the international community:

WE   **DECLARE**

Our concern over the increasing environmental destruction evident world wide, and in particular the destruction of mangrove forests, estuaries, and lagoons; and, in general, we declare our deep concern over the conversion of coastal wetlands and areas to shrimp farms, an unsustainable activity that is growing in an uncontrolled manner throughout the tropical and subtropical regions of the world.

WE   **DECLARE**

Our concern over the deprivation, displacement and marginalization of native communities that depend on coastal wetlands, due to the establishment of shrimp farms in these areas.

WE   **DECLARE**

That the lack of planning for an integrated marine and coastal area development is an assault against biological diversity, by allowing for the destruction or contamination of habitats and, amongst other acts, allowing for the wholesale elimination of marine life that is captured incidentally in the harvest of wild shrimp post-larvae used to supply the majority of shrimp farms in the world.

WE   **DECLARE**

Our desire for the imposition of a global moratorium to halt the further establishment or expansion of shrimp farming and that responsible parties undertake a study to determine the best means to transform shrimp aquaculture into an activity which is equitable and ecologically compatible with the principles of sustainable development. During the moratorium, a period of transition or conversion from the use of destructive technologies or practices to more responsible practices should be initiated to ensure the long term survival and health of ecosystems and the viability of a variety of sustainable human activities dependent on those systems.

WE   **DECLARE**

Our support for the criteria set forth in the NGO Declaration on Unsustainable aquaculture to the United Nations Commission for Sustainable Development in May 1996. Therefore:

WE   **DEMAND**

The implementation of the aforementioned criteria and those which are detailed as follows:

DEMAND   **#1**

Guarantee that the development and operation of various types of aquaculture, in particular shrimp aquaculture, does not, in the short, medium or long term, adversely affect biotic and abiotic environments and artisanal fisheries and the communities which depend on them.

DEMAND   **#2**

Ensure that shrimp aquaculture and other forms of coastal development are conducted within an integrated plan of management that includes the real and effective participation of all groups that benefit from coastal resources, in particular local communities.

DEMAND   **#3**

Ensure that the development of shrimp aquaculture is compatible with the structure and function of natural ecosystems and with the socio-cultural and economic interests of coastal communities and their areas of concern.

DEMAND   **#4**

Ensure that multilateral development banks, bilateral aid agencies, agencies of international cooperation, the FAO and other relevant national and international organizations and institutions do not finance or promote in any way the development of shrimp aquaculture practices which are not consistent with these demands.

DEMAND   **#5**

Prior to the development of any shrimp aquaculture activities, require an evaluation of the environmental, social and cultural impacts involving the participation, in a decision-making capacity, of NGOs and peoples potentially impacted; in addition, require environmental audits of existing operations on an ongoing basis with the same mechanisms for participation.

DEMAND   **#6**

Guarantee respect for the Human Rights of all people affected by shrimp aquaculture. Where human rights violations occur, they must be investigated by the competent authorities and processed in accordance with the civil, administrative and judicial responsibilities of the country concerned in compliance with the laws, treaties and international agreements to which such countries are parties.

DEMAND   **#7**

Regulate the use of fresh water in shrimp aquaculture operations with due regard to the ecological costs or impacts on the watersheds and basins where they are located, the need for freshwater supplies for human consumption, and the supply of water for other activities (e.g. agriculture, industry, tourism, urban development); and prohibit the salinization of fresh waters supplies including groundwater and reservoirs.

DEMAND   **#8**

Ensure the protection of wetland areas, in particular mangrove forests, rivers, lagoons, inlets, bays, estuaries, swamps, marshes and tidelands.

DEMAND   **#9**

Prohibit the use of substances harmful to the biodiversity of the area impacted by shrimp aquaculture operations.

DEMAND   **#10**

Apply the precautionary principle to every step in the development of shrimp aquaculture.

DEMAND   **#11**

Prohibit the contamination of surrounding areas as a result of the excessive discharge of organic and inorganic wastes.

DEMAND   **#12**

Prohibit the introduction and the use of organisms modified by genetic engineering and support the establishment of strict international measures relating to biosafety.

DEMAND   **#13**

Prohibit the use of exotic (non-native) species and stimulate research into methods for the in-vitro reproduction of native species.

DEMAND   **#14**

Prohibit the conversion of land used for agriculture or livestock production to use for shrimp aquaculture production, with particular attention placed on land-use management planning and national food security.

DEMAND   **#15**

Prohibit the use of feeds for farmed shrimp consisting of fish that could be used for human consumption.

DEMAND   **#16**

Ensure that the capture of shrimp larvae and any other type of activity does not adversely affect the survival and diversity of other species.

DEMAND   **#17**

Ensure that the shrimp aquaculture industry assumes all civil, legal (and corporate) responsibilities for the socio-environmental damages caused by the industry's installations, operations and production. The burden of proof that no damage has been done rests with the industry.

DEMAND   **#18**

Finally, we demand a global moratorium on any further expansion of shrimp aquaculture in coastal areas until the criteria for sustainable shrimp aquaculture are put into practice. We also demand the formation of an independent body of national, regional and international organizations, including non-governmental organizations, to monitor the implementation of this process at the global level.

Based on these demands, the organizations that endorse this declaration agree to unite their efforts and to work together to pressure governments, producers, and financial institutions as well as consumers to put these demands into practice as soon as possible.

The Non-Governmental Organizations present have approved a plan of action to begin a process of investigation, denunciation, education and public pressure to confront the activities of unsustainable shrimp aquaculture.

The following organizations support the present declaration:

Acción Ecológica, Ecuador, Amigos en defensa de la Gran Sabana (AMIGRANSA), Venezuela Asociación Ecológica Santo Tomas, México, Asociación Unionense para el Medio Ambiente (ASUMA), El Salvador, Centro de Estudios de Tecnología

Aplicada (CESTA), El Salvador, Centro de Estudios Integrales del Ambiente (CENAMB), Universidad Central de Venezuela, Comité para la Defensa y Desarrollo de la Flora y Fauna del Golfo de Fonseca (CODDEFFAGOLF), Honduras, Desarrollo Ambiente y Sociedad (DAS), México, Fundación Ecológica de Muisne (FUNDECOL), Ecuador, Greenpeace Internacional, Instituto Derramar, Brasil, Jóvenes Ambientalistas (JA), Nicaragua, Mangrove Action Project (MAP), USA, Movimiento Ambientalista de Nicaragua (MAN), Movimiento de Pescadores Ribereños y Aguas, Interiores, México, Natural Resourses Defense Council (NRDC), USA, Organización de Comunidades Negras, Colombia, PREPARE, India, Swedish Society for Nature Conservation, Sweden, Tumbes Silvestre, Perú, Unión Nacional de Organizaciones Ecologistas (UNES), El Salvador.

Source: http://darwin.bio.uci.edu/~sustain/shrimpecos/declare1.html

Appendix 6

2006 Aquaculture Law

The 2006 aquaculture law links aquaculture with environmental conservation and sustainable development. To ensure that the industry's practices are socially and environmentally responsible, the National Congress created the Consejo Nacional de Acuicultura de Honduras (National Council for Honduran Aquaculture, (CONAACUIH) with representatives from SAG, DIGEPESCA and other relevant agencies (Article 13).

The new law recognises the relationship between artisan fishing and aquaculture and states that norms should be established that promote inter-institutional and inter-sector cooperation for the purpose of preventing problems associated with overlapping legal jurisdictions (Article 15). It also requires that environmental evaluation and impact statements be completed for all aspects of the business, including the laboratory production of the postlarvae (Article 21). It also requires the industry to comply with environmental, forestry and labour legislation (Article 22). The law also makes mandatory certification of aquaculture products by applying the norms of the International Standardisation Organisation (ISO) to reduce the negative impacts on the environment and improve prices on the international market (Article 23).

In Title IV, Chapter I the law specifies the regulations and ordinances that will govern aquaculture activities and specifically pertains to Concessions and Permissions. A number of CODDEFFAGOLF's demands made it into the law. SAG through DIGEPESCA is granted the authority to oversee national lands for the purpose of aquaculture development. The agency is given responsibility for providing authorisations, permissions, licences and concessions to any businesses, artisanal cooperatives, and micro-enterprises; and can suspend, expire, modify, renew and cancel the concessions and permits (Article 29).
Aquaculture production that takes place on private land is subject to the same laws as those on public land. Concessions can be given for 20 years and renewed for an additional 10, and require that appropriate environmental licenes be obtained at a cost to be determined (Article 24). The government, in an effort to eliminate the black market in land speculation, made it illegal to transfer concessions partially or totally (Article 25). All

concessions will be publicly obtainable and any concessions to coastal areas will require the approval of the General Prosecutor of the Republic (Article 26). Article 27 (4) makes it illegal for any operations to impede or restrict the free movement of people who have utilised the land legally or by custom to access the beaches for the purpose of fishing or for the collection of crustaceans and mollusks. Local people in the rural communities are provided with priority use of the hydrological resources (referring to the wetlands) (Article 32).

Title IV, Chapter III states that any revenue generated through the issuance of concessions and the provision of permits shall be used to implement programmes and projects for the recuperation and restoration of zones that have been altered as a result of aquaculture-related activities in favour of restoring the evolution and continuation of natural processes (Article 41). In order to promote aquaculture practiced by artisan groups or for subsistence, DIGEPESCA is required to work with them to elaborate and execute an Aquaculture Training Plan that is participatory in nature (Article 47).

Title VI, Chapter I addresses control, surveillance, infractions and sanctions. DIGEPESCA and other relevant agencies are given the authority to inspect the industry to ensure compliance, including in relation to conservation, health and the protection of the environment (Article 50). Chapter II, Articles 53-56, gives the agency authority to issue sanctions for infractions. The law also requires DIGEPESCA to restructure the agency to implement the new authorities outlined in the law (Article 59). Finally, the law establishes a policy that does not permit the government to lease national lands for the expansion of aquaculture in coastal areas adjacent to the GOF. The government will be responsible for initiating a process to restore the zone by utilising technical and scientific studies that have been completed and implement recommendations related to the use of the existing resources (Article 60).

Appendix 7
World Wildlife Fund for Nature's
Position Statement on Aquaculture

![WWF logo]

December 1998

WWF Position Statement on shrimp aquaculture

The last two decades have witnessed a rapid expansion of commercial shrimp aquaculture throughout the world, particularly in Asia: this has been driven primarily by demand from consumers in Japan, North America, and Europe However, in many locations commercial shrimp farming has adversely affected local livelihoods, and has devastated fragile coastal ecosystems, causing mangrove destruction, coastal erosion, pollution of surface and ground waters including salinisation of vital coastal freshwater aquifers, and in some cases introduction of exotic species.

Many shrimp farms have to be abandoned completely after 5-6 years, often rendering the land unfit for any other productive use. Substantial areas of ecologically and economically productive coastal lands have met this fate and, while some proportion of this may be restored in the future, there is urgent need for research on and, where appropriate, investment in reclamation activities.

Shrimp aquaculture, as currently practised in many areas, provides a striking example of unsustainable use of natural resources for export markets, undermining food security at the local level and reducing prospects for development and poverty alleviation. In some countries, farmed shrimp are fed on fishmeal, processed from fish that are consumed by local people. The industry has triggered serious social conflicts in some locations by marginalizing and sometimes displacing village communities and the poor. In many cases, while a few individuals benefit from this industry, many more see their livelihoods and local environment damaged or destroyed

The full range and magnitude of the impacts of shrimp farming on the costal environment, biodiversity and in socio-economic terms have yet to be fully documented and understood. However, the few cost-benefit analyses performed to date have indicated that the cost of natural resource depletion and environmental damage far outweighs the direct economic returns.

WWF's mission is to conserve nature ecological processes, whilst ensuring the sustainable sound shrimp aquaculture in terms of providing food security, revenue, and as alternative-food source to wild caught fish now threatened by the global fishers crisis. However, shrimp farming is often ecologically, socially and economically unsuitable, both in the short-term and in the long-term, and the consequences may prove to be disastrous.

WWF considers that:
- Shrimp aquaculture should harm neither coastal nor marine ecosystems, and should be environmentally, economically and socially sustainable.
- Where the industry is inherently unsustainable, it should be closed down and appropriate measures taken for regeneration of damaged coastal zones;
- Best management practices, including those based on traditional methods or technological innovations, should be promoted;
- There should be no net mangrove loss as a result of shrimp aquaculture, is terms both of area coverage and of environmental services provided by this habitat type.

WWF seeks:
- A review by all countries producing farmed shrimp of the social and environmental impacts of this industry, in fulfilment of the requirements of Article 9 of FAO's Code of Conduct for Responsible Fisheries;
- The closure, or conversion to sustainable production methods, of those farms causing damage, in line with recommendations in the Jakarta Mandate of the Convention on Biological Diversity;

- The implementation by all governments of national legislation relevant to the environmental effects of shrimp aquaculture, and the consideration of this activity, in integrated coastal area planning and management;
- The refinement and application of best management practices including strict controls on effluents, careful siting of ponds and prohibition on the use of mangroves, use of indigenous (rather than exotic) species, prohibition of routine use of chemicals and antibiotics, better feeding practices including reduction in use of fishmeal, development of hatcheries whenever possible to reduce the use of wild populations as broodstock, consideration of all social, impacts, and introduction of measures to ensure that industry and consumers bear the full cost;
- The review, by international financial institutions such as the World Bank, other multilateral development banks, and bi-lateral aid agencies in collaboration with the relevant environmental and sustainable development bodies, of she social and environmental effects of promoting and subsidising this industry;
- Increased financing by private investors, both domestic and international, of more eco-friendly forms of shrimp production, through for example comprehensive environmental assessment of, and multi-stakeholder consultation prior to, investment, the agencies responsible for funding or otherwise supporting shrimp aquaculture operations should commit further finance and support to remedy the negative impacts and to develop models for sustainable shrimp production through research and development.

WWF will help to:

- Identify and demonstrate specific methods, techniques and policies that contribute to sustainable shrimp aquaculture.
- Established mechanism to ensure that guidelines and best management practices are effectively implemented.
- Develop globally acceptable guidelines for sustainable aquaculture through a multi-stakeholder process; shrimp consumers, who are ultimately responsible for the dramatic growth of this industry, can play an important role and WWF will support efforts to develop an independent third party certification system to provide an economic incentive, for sustainable shrimp aquaculture;
- Raise awareness of key issues in shrimp aquaculture among consumers, the general public, policy makers, investors, the industry itself, and other key target groups.

Appendix 8
Honduran Laws Consulted

Name of Law or regulation	Date Published in *La Gaceta*
Fisheries Law	9 June 1959
Forestry Law	4 March 1972
Honduran Corporation for Forestry Development Law	10 January 1974
Agrarian Reform Law	18 August 1975
Convention on Dumping Solid Waste in the Sea	6 February 1980
Law on the Use of Marine Natural Resources	13 June 1980
Law on the Declaration, Planning and Development of Tourism Zones	22 July 1980
Reform to Decree Law Number 135	24 July 1980
Law on Seeds	15 October 1980
Regulation on the Law on the Declaration, Planning and Development of Tourism Zones	18 November 1981
Declaration of National Tourism Zoning	13 January 1983
Law on Hydrocarbons	28 February 1985
Declaraion of Protected Forest Areas	5 August 1987
Law on municipalities, including all regulations	19 November 1990
Reform of the Municipalities Law	23 May 1991
Law Creating the General Direction for Fisheries and Aquaculture	18 July 1991
Law for the Modernization and Development of Agricultural Development	6 April 1992
Regulation for the Law on Municipalities	18 February 1993
Creation of the Council on Protected Areas	25 June 1993
General Environmental Law	30 June 1993
Forest Related Aspects of Decree 31-92 from 5 March 1992	20 July 1993
Regulation on the General Environment Law	5 February 1994
Law on Incentives for Aforestation and Protection of the Forest	29 March 1994
Regulation on the Rights of Populations Living in National Forested Territories	25 March 1997
Law for the Protection of the Cultural Heritage of the Nation	21 February 1998
General Mining Law	6 February 1999
Law on Tourism Incentives	23 April 1999
Regulation on the National System of Protected Areas of Honduras	25 September 1999
Property Law	29 June 2004
Regulation for the Law on Tourism Incentives	4 October 2004
Aquaculture Law of 2006	November 2006

Appendix 9

Open-ended and Semi-Structured Interviews, Personal Correspondence, and Conversations: 2003 – 2007

Interviews were primarily conducted in English. The * denotes those interviews that occurred in Spanish and English. In these cases, interviews were generally done in Spanish but English was spoken when needed.

1. Alarid, David (2002 – 2004). Diplomat, US Embassy Costa Rica, Central American and Caribbean Regional Environmental Hub Officer, US Department of State (personal correspondence, interviews, and dialogues in San Jose, Costa Rica; Washington D.C.; Belize City, Belize; Managua, Nicaragua; Tegucigalpa, Honduras; and Miami, Florida).

2. *Allamirano, Jose Orlando (2003). Ministry of Environment and Natural Resources. San Salvador, El Salvador. Personal interview.

3. Alm, Anders (2004). World Bank, Miami, Florida. Personal interview.

4. Alvarez, Ramon (2005). USAID Tegucigalpa, Honduras. Personal interview.

5. Amador, Gladis (2005). Respondent #2 from Valle Nuevo, Municipality of Alianza, Department of Valle, Honduras. Personal interview during the Socioeconomic Analysis, July.

6. Anderson, Stephen (2002). Economic Officer, US Diplomat, American Embassy, Guatemala City, Guatemala. Personal interview.

7. *Aguilar, Ricardo (2002). Consultant, USAID's PROARCA-SIGMA. Personal interview.

8. *Aguero, Joaquin (2005). Senior Advisor to the Minister of Environment and Natural Resources (SERNA), Government of the Republic of Honduras. Personal interview at the White Water to Blue Water Conference, Miami, Florida 25 March 2004.

9 . *Altamirano, Orlando (2003). Advisor to the Ministry of Environment and Natural Resources (MARN). San Salvador, El Salvador. Personal interview.

10 . *Alvarez, Maria (2003). ANDAH, Choluteca, Honduras. Personal interview.

11 . *Ammour, Tania (2006). Tropical Agronomic Center for Higher Education and Research. San Jose, Costa Rica. Personal interview.

12 . *Arenas, Antonio (2003). Director of the National Territorial Studies Service (SNET), Government of El Salvador, San Salvador. Personal interview.

13 . *Arias, Celeo Emilio (2003). United Nations Organization for Industrial Development (ONUDI). Choluteca, Honduras. Personal interview.

14 . Astralaga, Margarita (2004). Senior Advisor for the Americas, Latin American Representative, RAMSAR Convention. Miami, Florida. Personal interview and correspondence.

15 . *Avila, Luz (2003). Executive Director, Association for the Sustainable Development of the Gulf of Fonseca (ADESGOLFO). San Salvador, El Salvador. Personal interview.

16. Baird, Ronald (2001 – 2005). Director, Nacional Sea Grant Collage Program. Personal conversations and correspondence.

17. *Barraza, Jose Enrique (2003). Ministry of Environment and Natural Resources, San Salvador, El Salvador. Personal interview.

18 . *Barraza, Sandra de (2003). Plan de Nación, San Salvador, El Salvador. Personal interview.

19 . *Barria, Eduardo Valdes (2003). President, University of Central America, Managua, Nicaragua. Personal interview.

20 . *Barrientos, Eduardo (2003). Advisor to the Vice Minister of Maritime Transportion Issues. San Salvador, El Salvador. Personal interview.

21. Bender, Steven (2003). Principal, Unit for Environment and Sustainable Development, Organization of American States, Washington D.C. Personal interview.

22. Best, Barbara (2002 – 2004). Marine and Coastal Specialist, USAID. Washington D.C. Personal interviews and correspondence.

23. Blake, Byron (2002). The Caribbean Community Secretariat (CARICOM). Belize City, Belize. Dialogue.

24. Borel, Andrea (2002 – 2007). Foreign Service Nacional, American Embassy, San Jose, Costa Rica. Personal correspondence and interviews between 2002 – 2007.

25. Boyd, Claude E. (2004). Aquaculture Expert, Auburn University. Tegucigalpa, Honduras. Personal interview, 26 August.

26. Brandão, Miriam (2004). Central American Representative, Inter-American Foundation (IAF). Personal interview. August.

27. Brands, William (2004). Agricultural Development Officer, USAID's Broad-Based Growth Team for Latin America and the Caribbean. Personal interview.

28. Bransky, Joel (2003). Deputy Representative for Plan Pueblo Panama. Inter-American Development Bank, San Salvador, El Salvador. Personal interview, February.

29. Brennan, William (2004 – 2007). Deputy Assistant Secretary for International Affairs, National Oceanic and Atmospheric Administration. Washington D.C. Personal conversations.

30. Bristow, Jeremy (2004). British Broadcasting Corporation (2004). *Price of Prawns*. Senior Producer, BBC Natural History Unit. Cambridge, England. Personal interviews, correspondence, and conversations.

31. Bunce, Leah (2004 – 2005). National Ocean Service, Washington D.C. Personal correspondence between 2004 and 2005 regarding socio-economic monitoring in coastal communities.

32. Burke, Lauretta (2004). Senior Associate, World Resources Institute. Washington D.C. Personal interview.

33. Caballero, Luis (2002 – 2005). University of Zamorano, Department of Socioeconomics, Environment, and Development. Honduras. Personal interview and conversations.

34. Cabrera, Juana (2005). La Pintadillera, Municipality of Amapala, Deparment of Valle, Honduras. Personal interview during the Socioeconomic Analysis, July.

35. Canfield, Eloise (2003 – 2007). Director for International Programs, National Fish and Wildlife Foundation (NFWF) and later with the Millenium Challenge Corporation (MCC). Washington D.C. Personal conversations.

36. *Cardenal, Jorge (2003). Nicaraguan Minister of MARENA. Managua, Nicaragua. Personal interview, January.

37. Carner, George (2003). USAID's Guatemalan and Central American Project (GCAP). San Salvador, El Salvador. Personal interview. February.

38. Carr, Brad (2003). US Diplomat, Economic Officer, American Embassy. San Salvador, El Salvador. Personal interview.

39. Carr III, Archie (2004). Wildlife Conservation Society. Personal interview, White Water to Blue Water Conference. Miami, Florida. March.

40. *Chanlatte, Marino (2003). Inter-American Institute for Cooperation on Agriculture (IICA). Managua, Nicaragua. Personal interview.

41. Chamberlain, G. (2004). President, Global Aquaculture Alliance (GAA), Tegucigalpa, Honduras. Personal interview. 25 August.

42. Chamberlain, S. (2004). Office Manager, GAA, Tegucigalpa, Honduras. Personal interview. 25 August.

43. Chaparro, Ruperto (2002 – 2005). University of Puerto Rico Sea Grant, Coordinated Hurricane Mitch efforts on behalf of Sea Grant in Central America. Personal conversations, travel, correspondence, and interviews occurred throughout the course of the research.

44. *Chavez, Araceli (2003). SINIA-Ministry of Environment and Natural Resources (MARENA). Managua, Nicaragua. Personal interview.

45. Contreras, Mario (2002 – 2005). Director of External Affairs, University of Zamorano. Honduras. Personal interviews.

46. *Contreras, Fabrisio (2003). Extension Agent, Center for Aquatic Ecosystems Research (CIDEA), University of Central America (UCA), Puerto Morazon, Nicaragua. Personal interview and field exploration.

47. *'Consuela' (2005). Local Coastal Resident and Salt Producer, North of Cedeño, Honduras. Personal interview and tour of salt production. April.

48. *Cordoba, Rocio (2003). World Conservation Union (IUCN). San Jose, Costa Rica. Personal interview.

49. *Corea, Leana (2005). Organizer, Committee for the Defense and Development of the Flora and Fauna of the Gulf of Fonseca (CODDEFFAGOLF). San Lorenzo, Honduras. Personal interviews and correspondence.

50. Corrales, Hector (2004). Grupo Granjas Marinas San Bernardo (GGMSB), Personal interview. August.

51. Costa-Pierce, Barry (2002 – 2007). Aquaculture Expert, Director of the University of Rhode Island Sea Grant College Program, Personal interviews and correspondence.

52. *Coto, Sandra Leon (2003). National University of Heredia. San Jose, Costa Rica. Personal interview.

53. *Coze, Agnes Saborio (2002 – 2007) Personal communication occurred between Ms. Agnes Saborio and I throughout the course of the research.

54. Crawford, Brian (2003 – 2005) Senior Program Manager, University of Rhode Island Coastal Resources Center. Personal communication.

55. *De Rodriguez, Mary Latino (2004). Foreign Service National, USAID's Office of Environment. San Salvador, El Salvador. Personal interviews.

56. Del Castillo, Christina (2004). USAID Latin America and the Caribbean. Personal conversations.

57. Dix, Anne (2006). USAID Guatemala, Guatemala City, Guatemala. Personal interview.

58. *Dominguez, Jorge (2003). Manager of Education, Ministry of Environment and Natural Resources (MARN). San Salvador, El Salvador. Personal interview.

59. Donald, Michael (2003). USAID Guatemalan Central American Project (GCAP). Guatemala City, Guatemala. Personal interview.

60. Donoso, Maria (2002). Director, CATHALAC. Miami, Florida. Personal interview.

61. Drazba, Monica (2004). USAID/Hurricane Mitch Coordinator for Nicaragua. Personal interview.

62. *Espinoza, Eloisa (2003). Director of Aquaculture, General Directorate for Fisheries and Aquaculture (DIGEPESCA), Secretariat of Agriculture and Ranching (SAG), Government of Honduras. Personal interview.

63. *Esquivel, Juan Calvia (2003). Inter-American Institute for Cooperation on Agriculture (IICA). San Jose, Costa Rica. Personal interview.

64. Flagg, Melissa (2004). US Department of State, Office of Science and Technology. Washington D.C. Personal interview.

65. *Funes, César (2003). Global Environment Facility/United Nations Environment Program Representative for the Government of El Salvador, Manager of Environmental Systems, Ministry of Environment and Natural Resources. San Salvador, El Salvador. Personal interview.

66. Galo, Patricia Panting (2004). Ministry of Environment and Natural Resources (SERNA), Tegucigalpa, Honduras. Personal interview.

67. Gammage, Sarah (2004 – 2006). Senior Trade Economist, Development and Training Services, Inc. Washington D.C. Personal interview, correspondence, and conversations.

68. Genizzotti, Ana Gomez (2003 and 2004). US Embassy Honduras, US Department of Agriculture's Foreign Agricultural Service (FAS). Tegucigalpa, Honduras. Personal interview.

69. *González, Conrado (2003). Protected Areas, AFE-COHDEFOR, Honduran Government. Tegucigalpa, Honduras. Personal interview and conversations.

70. *González, Haracio (2005). Las Playitas, Municipality of Alianza, Department of Valle, Honduras. Personal interveiw during the Socioeconomic Análisis. July.

71. *González, Lisa (2002 and 2004). Nicaraguan Ministry of Environment and Natural Resources (MARENA). Personal interviews in Belize City, Belize (2002) and in Managua, Nicaragua (2004).

72. González, Carmen Aida (2003 – 2006). USAID's Guatemalan Central American Project (GCAP). Guatemala City, Guatemala. Personal interviews and correspondence.

73. *González, Jose Rigoberto (2003). Government of Honduras. Tegucigalpa, Honduras. Personal interview.

74. *González, Marco (2006). Central American Commission on Environment and Development (CCAD). San Salvador, El Salvador. Personal interview.

75. *González, Mario (2003 and 2006). Director of SINDPESCA, Regional Coordinator for OSPESCA, Central American Commission on Environment and Development (CCAD). San Salvador, El Salvador. Personal interview.

76. *Granados, Mauro A. (2003). Grupo Gestor La Union, Plan de Nacion. San Salvador, El Salvador. Personal interview.

77. *Granados, Antonieta de (2003). Grupo Gestor La Union, Plan de Nacion. San Salvador, El Salvador. Personal interview.

78. Green, Bart. (2004). Aquaculturist, Auburn University. Personal interview at the Central American Aquaculture Conference, Tegucigalpa, Honduras. 26 August.

79. Guerrero, Aldo. (2004). Project Coordinator, Dos Mares. Personal interview. August.

80. *Guillen, Leonel. (2003 and 2005). Regional Director, AFE/COHDEFOR PROMANGLE. San Lorenzo, Honduras. Personal interview February 2003 and July 2005.

81. *Gutierrez, Alejandro B. (2003). International Oceans Institute, National Autonomous University. San Jose, Costa Rica. Personal interview.

82. Hatziolos, Marea (2002 – 2004). Coastal and Marine Specialist, World Bank. Washington D.C. Personal interview.

83. Haws, Maria (2005). Director of the Pearl Research and Training Program. University of Hawaii, Hilo (Aquaculture Consultant GOF). Personal interview in Rhode Island.

84. *Hayen, Ernesto (2003). Ministry of Environment and Natural Resources (MARN). San Salvador, El Salvador. Personal interview.

85. Healy, Kevin (2003). Inter-American Foundation (IAF). Washington D.C. Personal interview.

86. Hearne, Peter (2002 – 2005). USAID Environment Officer, Tegucigalpa, Honduras. Personal interviews and correspondence.

87. Heartney, Edward (2004). US Diplomat, American Embassy. San Salvador, El Salvador. Personal interview.

88. Heidelberg, Karla (2003 - 2004). Marine Scientist Policy Analyst, US Department of State, Ocean Affairs. Washington D.C. Personal interview.

89. Heerin, Jim (2004). Chairman of Sea Farms International (SFI). Telephone conversation and personal correspondence. 17 March.

90. Herrera, Luis E. Torres (2004). ANAM-CATHALAC. Miami, Florida. Communication.

91. Hoadley, Kenneth L. (2004 – 2005). President of the University of Zamorano. Honduras. Personal interviews.

92. Hogan, Nadine (2003). Inter-American Institute for Cooperation on Agriculture (IICA). Washington D.C. Personal interview.

93. Hohn, Janet (2004). World Conservation Union (IUCN). Telephone interview.

94. Hooten, Anthony (Andy) (2003). World Bank Marine Consultant (AJH Environmental Services). Washington D.C. Personal interview.

95. Ibrekk, Hans (2003). World Bank. Washington D.C. Personal interview.

96. *Izquierdo, Aura (2003). Coordinator for the Meso-American Biological Corridor, Ministry of Environment and Natural Resources (MARENA). Managua, Nicaragua. Personal interview.

97. *Jurado, Alba Margarita Salazar de (2003). Central American System for Regional Integration (SICA). San Salvador, El Salvador. Personal interview.

98. Kelley, Jill (2006). USAID Environment Officer, Guatemala City, Guatemala. Personal interview.

99. Kosmal, Danelle (2003). The Nature Conservancy. San Jose, Costa Rica. Personal interview.

100. Laughlin, Tom (2003 and 2004). Director of International Affairs, US Department of Commerce, National Oceanic and Atmospheric Administration. Washington D.C. Personal conversations.

101. *Lahman, Enrique (2003). World Conservation Union (IUCN) Meso-American Regional Office. San Jose, Costa Rica. Personal interview.

102. *Lancho, Frederico (2003). Inter-American Institute for Cooperation on Agriculture (IICA) Regional Office, San Jose, Costa Rica. Personal interview.

103. Leaman, Kevin (2002 and 2004). University of Miami, Florida. Personal interview.

104. LeMay, Michelle (2005). Latin America and the Caribbean, Inter-American Development Bank. Tegucigalpa, Honduras. Personal interview.

105. *Lexias, Mario A. (2003). Director, Inter-American Institute for Cooperation on Agriculture (IICA). San Jose, Costa Rica. Personal interview.

106. *Maltez, Norma (2002). Nicaraguan Ministry of Environment and Natural Resources (MARENA). Managua, Nicaragua. Personal interview in Belize City, Belize.

107. *Martinez, Franklin (2002 – 2005). University of Zamorano Aquaculture Program. Personal interviews and on-going conversations in relation to aquaculture and the GOF between 2002 and 2005.

108. *Martinez, Laura (2003). Extension Agent, Center for Aquatic Ecosystems Research (CIDEA), University of Central America (UCA). Puerto Morazon, Nicaragua. Personal interview and field exploration.

109. *Martinez, Sergio (2003). Director of Fisheries Research Center, World Wildlife Fund (WWF). Managua, Nicaragua. Personal interview.

110. Mateo-Vega, Javier (2003 and 2004). Organization for Tropical Studies. San Jose, Costa Rica. Personal interviews.

111. McGirr, Mike (2004 – 2006). US Department of Agriculture (International). Washington D.C. Personal interviews and correspondence.

112. McGuire, Harriet (2003). US Department of State, Western Hemisphere Affairs. Washington D.C. Personal interview.

113. Meassick, Mark (2003). Strategic Planning Specialist, Inter-American Institute for Cooperation on Agriculture (IICA). Washington D.C. personal interview.

114. Meganck, Richard (2002). Organization of American States (OAS). Washington D.C. Personal interview.

115. Membreno, Tomas (2005). USAID. Tegucigalpa, Honduras. Personal interview.

116. *Mendez, Jacobo Sanchez (2003). Coordinator for PROGOLFO Nicaraguan Ministry of Environment and Natural Resources (MARENA). Managua, Nicaragua. Personal interview.

117. *Mendoza, Hector (2003). Honduran Association for Community Service. Tegucigalpa, Honduras. Personal interview.

118. *Mendoza, Sandra (2003). Representative, The Nature Conservancy. Tegucigalpa, Honduras. Personal interview.

119. Mecurio, Cristina (2002). Environmental Protection Agency, International Affairs Office. Personal interview in Belize City, Belize.

120. Meyer, Daniel E. (2002 - Present). Director of Aquaculture, University of Zamorano. Personal interviews, correspondence, and on-going conversations regarding the GOF between 2002 and 2007.

121. *Meyer, Suyapa Triminio de (2002 – Present). University of Zamorano. Personal interviews and conversations.

122. *Midence, Hector (2004). Vice President for Meso-America, World Conservation Union (IUCN). Tegucigalpa, Honduras. Personal interview.

123. *Molina, Rene Ayala (2003). Manager for Territorial Development, Vice Minister for Livelihoods and Development. San Salvador, El Salvador. Personal interview.

124. *Moncada, Lourdes (2003). Aquaculture Unit, General Directorate for Fishing and Aquaculture (DIGPESCA), Secretariat of Agriculture and Ranching (SAG). Government of Honduras. Tegucigalpa, Honduras. Personal interview.

125. Moore, Franklin (2004 – Present). Deputy Assistant Administrator – Office of Environment, Bureau of Economic Growth, Agriculture, and Trade (EGAT), USAID. Personal interview, conversations, and correspondence.

126. Moore, William (Bill) (2004). Presentation at Central American Aquaculture Conference, Tegucigalpa, Honduras, 25 August.

127. *Morales, Carlos (2003). Marketing Coordinator, World Wildlife Fund (WWF). Guatemala City, Guatemala. Personal interview.

128. *Mug-Villanueva, Moises (2003). Fisheries Expert, World Wildlife Fund (WWF). San Jose, Costa Rica. Personal interview.

129. *Muñoz, Edas (2002 – 2005). Regional Coordinator for the Gulf of Fonseca, USAID's PROARCA-COSTAS. Tegucigalpa, Honduras. Personal interviews and correspondence.

130. Nygren, Anja (2004 – 2006). Professor, University of Helsinki, Finland. Personal correspondence.

131. *Ochoa, Emilio (2003 – 2005). Executive Director of EcoCostas. Personal interviews and correspondence.

132. *Olivas, Lorenzo (2005). COHDEFOR/PROMANGLE. San Lorenzo, Honduras. Email correspondence.

133. Olive, Steve (2003). USAID Environment Officer. Managua, Nicaragua. Personal interview and field excursion to Puerto Morazan, Nicaragua (Gulf of Fonseca).

134. Olsen, Stephen (2002 – 2007). Director, University of Rhode Island Coastal Resources Center. Personal interviews, correspondence, and numerous conversations regarding the GOF.

135. Olsen, Derrick M. (2004). US Diplomat, Political Officer, US Embassy. Tegucigalpa, Honduras. Personal interview.

136. Olson, Steven (2004). US Diplomat, Counselor for Economic Affairs, US Embassy. Guatemala City, Guatemala. Personal interview.

137. *Ortiz, Max Campos (2003). Executive Secretariat, Central American System for Regional Integration (SICA). San Salvador, El Salvador. Personal interview.

138. *Oyuela, Omar (2005). Formerly with AFE/COHDEFOR, Government of Honduras. Personal interview.

139. *Oveida, Ivonne (2003 and 2005). Protected Areas Specialist, AFE/COHDEFOR. Government of Honduras. Tegucigalpa, Honduras. Personal interviews, conversations, and correspondence.

140. Patterson, Arthur (2002 – 2007). National Ocean Service International Office, Personal interaction, correspondence, and communication.

141. *Penalbo, Ana Cecilia (2003). Nicaraguan Specialist, Inter-American Institute for Cooperation on Agriculture (IICA). Personal interview.

142. Pilz, George (2004 – 2005). Botanist, University of Zamorano. Honduras, Personal conversations.

143. Popp, William (Bill) (2004). US Diplomat, US Embassy. Managua, Nicaragua. Personal interview.

144. Praster, Thomas A. (2004). US Diplomat, Bureau of Oceans, Environment, and Scientific Affairs (OES). Washington D.C. Personal interview.

145. *Quezada, Jorge (2003). Director of Biological Resources, Ministry of Environment and Natural Resources (MARN). San Salvador, El Salvador. Personal interview.

146. Ramirez, Marcela (2002 – 2007). Foreign Service National, US Embassy Regional Environmental Hub Office. San Jose, Costa Rica. (Later a contractor for the US Southern Command and US Army War College). Personal correspondence, interviews, and communication.

147. *Ramon, Juan (2003). Extension Agent, Center for Aquatic Ecosystems Research (CIDEA), University of Central America (UCA). Puerto Morazon, Nicaragua. Personal interview and field exploration.

148. *Ramos, Luis Antonio (2003). Ministry of Environment and Natural Resources (MARN). San Salvador, El Salvador. Personal interview.

149. Rea, Harry (2002 – 2004). USAID, Associate Director, Office of Agriculture and Food Security, Economic Growth, Agriculture, and Trade, USAID. Washington D.C. Personal interviews and conversation.

150. Restrepo, Jorge Ivan (2002 – Present). Director, University of Zamorano, Institute for Biodiversity. Honduras. Personal interviews and conversations.

151. Ricci, Glenn (2001 – Present). Senior Program Manager, University of Rhode Island, Coastal Resources Center. Personal interviews, conversations, and continual communication.

152. Rockeman, Kurt (2003 – 2004). Director for Environmental Programs, USAID Guatemalan Central America Project (GCAP). Guatemala City, Guatemala. Personal interviews.

153. *Rodriguez, Mary (2002 - 2004). International Cooperation Specialist, USAID. San Salvador, El Salvador. Personal interviews and conversations.

154. *Romero, Hernan (2003). Sector Specialist, Inter-American Development Bank (IDB). San Salvador, El Salvador. Personal interview.

155. *Romero, Joaquin (2004 - 2005). Manager of Aquaculture Operations, Grupo Granjas Marinas San Bernardo (GGSMB), Sea Farms International. Choluteca, Honduras (25 August). Subsequent interviews, conversations, and communication.

156. Rueda, Alfredo (2002 – 2004). Regional Coordinator, University of Zamorano. Honduras. Personal interviews.

157. *Ruiz, Rose Marie (2003). International Oceans Institute, National Autonomous University. San Jose, Costa Rica. Personal interview.

158. *Salgado, Gerardo (2004). Vice Minister, Ministry of Environment and Natural Resources (SERNA). Tegucigalpa, Honduras. Personal interview.

159. *Salaverría, Sonia (2003). Technician, Ministry of Agriculture and Ranching (CENDEPESCA). San Salvador, El Salvador. Personal interview.

160. *Samayoa, Jose Luis (2003). Director of Citizen Participation, Ministry of Environment and Natural Resources (MARN). San Salvador, El Salvador. Personal interview.

161. Schwartz, Martin (2002 – 2005). University of Zamorano (formerly with USAID's Regional Office and CARE International). Personal interviews, conversations, and documents provided.

162. Senseney, Robert (Bob) (2002 – 2004). Senior Counselor, US Department of State, Office of Science and Technology Policy. Washington D.C. Personal interviews

163. *Sepulveda, Sergio (2003). Director Rural Sustainable Development, Inter-Institute for Cooperation on Agriculture (IICA). San Jose, Costa Rica, Personal interview.

164. Shaw, Mark (2004). US Diplomat, US Embassy. Tegucigalpa, Honduras. Personal interview.

165. Sherman, Kenneth (2004). Oceanographer, National Marine Fisheries Service (NMFS), Nacional Oceanic and Atmospheric Administration, US Department of Commerce. Rhode Island. Personal interview.

166. Sonzini, Luis (2003). Project Manager, Italian Development Agency (GVC). Managua, Nicaragua. Personal interview.

167. *Sosa, Laura (2002 – Present). Forestry Technician, COHDEFOR PROMANGLE, Government of Honduras. (Personal interviews, correspondence, conversations, and observation).

168. *Sosa, Lidia Erika (2003). Foreign Service National, Commercial Assistant, US Embassy. San Salvador, El Salvador. Personal interview and conversation.

169. Squillante, Leslie (2002 – 2007). Assistant Director, University of Rhode Island Coastal Resources Center. Personal conversations.

170. Stanley, Denise (2003 – 2007). Professor, University of California Fullerton. Personal correspondence and telephone conversation.

171. Stolberg, Harold (2003). Programme Coordinator Latin America, National Science Foundation. Arlington, VA. Personal interview.

172. Sublett, Jennifer (2004). Economic, Science, and Technology Officer (EST), US Embassy. Managua, Nicaragua. Personal interview.

173. *Talavera, Mario Jiminez (2002 and 2004). Minister of Agriculture. Tegucigalpa, Honduras. Personal interviews.

174. Tenney, Anne (2003). Division of Ocean Sciences, National Science Foundation (NSF). Arlington, VA. Personal interview.

175. *Tesak, Ildiko de (2003). President, ADESGOLFO. San Salvador, El Salvador. Personal interview.

176. Thomas, Barbara (2003). US Embassy. San Salvador, El Salvador. Personal interview in Honduras.

177. Tinkman, Stetson H., (2003). Fishery Officer, Office of Marine Conservation, US Department of State. Washington D.C. Personal interview.

178. Tobey, J. (2002 – 2007). Collaborated with Dr. James Tobey, Senior Program Manager, University of Rhode Island/Coastal Resources Center. Interviews, correspondence, and conversations.

179. *Tojeira, Jose Maria (2003). Director of the University of Central America (UCA). San Salvador, El Salvador. Personal interview.

180. *Torres, Helga de (2003). Communications, ADESGOLFO. San Salvador, El Salvador. Personal interview.

181. *Turcia Rosaly, Luis (2003). PROGOLFO. Tegucigalpa, Honduras. Personal interview.

182. Urdaneta, Nicholle (2004). Economic Affairs Officer, US Embassy. Tegucigalpa, Honduras. Personal interview.

183. *Varela, Jorge (2002 – Present). Executive Director, Committee of the Defense of the Flora and Fauna of the Gulf of Fonseca (CODDEFFAGOLF), Tegucigalpa, Honduras, Personal interviews (specific dates below) and correspondence.

184. *Varela, J. (2005). Personal interview. 22 July.

185. *Varela, J. (2006a). Personal correspondence. 31 May.

186. *Varela, J. (2006b). Personal interview. 7 September.

187. *Vélez, Luis (2002). Coordinator for Agricultural Business, University of Zamorano. Honduras. Personal interview.

188. *Vera, Amanda Solis (2003). Vice Rector, University of Central América. Manágua, Nicarágua. Personal interview.

189. *Villanueva, Guillermo (2005). Representative of the Inter-American Institute for Cooperation Agriculture. Tegucigalpa, Honduras. Personal interview.

190. Villanueva, Roberto Martinez (2003). Strategic Partnerships Specialist, Inter-American Institute for Cooperation on Agriculture (IICA), Washington D.C. Personal interview.

191. Vives, René (2002 – 2004). Research Assistant, University of Central América, Center for Aquatic Ecosystems Research. Manágua, Nicarágua.

192. Volk, Richard (2003 – 2005). Water Resources, USAID's Economic, Growth, Agriculture, and Trade. Washington D.C. and Rhode Island. Personal conversations.

193. *Wagner, Carlos A. Landero (2003). Ministry of Environment and Natural Resources (MARENA). Manágua, Nicarágua. Personal interview.

194. Wilbur, Richard (Dick) (2002 – 2004). Bureau of Oceans, Environment, and International Scientific (OESI), US Department of State. Washington D.C. Personal interviews, conversations, and correspondence.

195. *Windevoxhel, Nestor (2002 – 2005). Director, USAID/Guatemalan Central American Project (GCAP) and then Regional Director, The Nature Conservancy. Personal interviews, correspondence, and interactions.

196. Wise, Michael (2003). US Peace Corps Country Director. San Salvador, El Salvador. Personal interview.

197. *Zelaya, Alberto (2003 and 2005). General Manager, Grupo Granjas Marinas San Bernardo. Choluteca, Honduras. Personal interviews.

198. *Zepeda, Ernesto Lopez (2003). Director of Natural Patrimony, Ministry of Environment and Natural Resources. San Salvador, El Salvador. Personal interview.

199. *Zetino, Ana Martha (2003 and 2004). University of El Salvador, Department of Biology. San Salvador. Personal Interview. February 2003 and August 2004.

200. Zweig, Ronald (2003). Senior Aquaculturist, World Bank. Washington D.C. Personal interview.

Bibliography

Newspapers Consulted:

El Cronista 1970 - 1980
El Heraldo 1970 - Present
El Tiempo 1970 - Present
La Tribuna 1970 - Present
Financial Times
Honduras This Week
Inter-Press News Service
La Prensa
New Internationalist
The Economist
Washington Post

Books, Journal Articles, Newsmagazines, Internet Sources:

(1993). Environment: WWF Award Goes to Indian and Honduran Groups. *Inter Press Service* San Jose, Costa Rica. 30 November, 1993 (Retrieved via LexisNexis).

(1994). Honduras-Environment: Thousands Protest Destruction of Mangroves. *Inter Press Service* Tegucigalpa, Honduras. 12 September, 1994 (Retrieved via LexisNexis).

(1996). Greens Target Shrimp Farming. *Latin American Weekly Report* 31 October, 1996 (Retrieved via LexisNexis).

(1996). Shrimp Boycott. *Caribbean & Central America Report*, Latin American Newsletters, Ltd. 3 October, 1996 (Retrieved via LexisNexis).

(1999). Awards for Environmental Heroes. *Bangkok Post* (FT Asia Intelligence Wire). Bangkok, Thailand. 20 April, 1999 (Retrieved via LexisNexis).

(2002). Honduras: International Relations and Defense. *The Economist Intelligence Unit Ltd.* London. (Retrieved via LexisNexis)

(2003). Promise of a Blue Revolution: How aquaculture might meet most of the world's demand for fish without ruining the environment. 7 August, 2003. *The Economist.*

(2004). Farming the Sea, Costing the Earth: Why We Must Green the Blue Revolution. London, Environmental Justice Foundation: 77.

Alphabetical listing of Authors:

Adams, H. and L. Searle (1986). *Critical Theory Since 1965.* Tallahassee, FL, Florida State University Press.

Adger, N. W., T. A. Benjaminsen, K. Brown, and H. Svarstad. (2001). Advancing a Political Ecology of Global Environmental Discourses. *Development and Change* 32(4): 681-715.

Agnew, J. and S. Corbridge (1995). *Mastering Space: Hegemony, Territory and International Political Economy.* London, Routledge.

Agrawal, A. (2005). *Environmentality: Technologies of Government and the Making of Subjects.* Raleigh/Dirham, Duke University Press.

Alexis, S. A. (1999). *Evaluacion de los Manglares del Golfo de Fonseca Mediante un Analisis Multitemporal de Imagenes del Satelite LANDSAT-TM Entre los Anos (1989-1995), (1995-1998), y (1989-1998).* Tegucigalpa, Honduras, Administracion Forestal del Estado Corporacion Hondurena de Desarrollo Forestal (AFE-COHDEFOR): 45.

Amin, S. (1993). Social Movements at the Periphery, in P. Wignaraja (ed.), *New Social Movements in the South Empowering the People.* London, Zed Books: 95.

Arendt, H. (1958). *The Human Condition.* Chicago, IL, The University of Chicago Press.

Azul Grupo Saenz, A. (2000). *Servicios Ambientales de los Manglares: Que Perdemos Cuando los Transformamos?* Ecuador and Honduras, Red Manglar: 53.

Baechler, G. (2006). Environmental Security: A Geographic Information System Analysis Approach – The Case of Kenya. *Environmental Management* 37(2)

Baker, J. L. (2000). *Directions in Development: Evaluating the Impact of Development Projects on Poverty: A Handbook for Practitioners.* Washington D.C., The International Bank for Reconstruction and Development (World Bank).

Barnes, R. S. K. and R. N. Hughes (1999). *An Introduction to Marine Ecology.* Oxford, Blackwell Science.

Barrett, P. and C. Cohen, (eds). (1991). *Funk & Wagnalls Standard Dictionary.* New York, NY, HarperCollins Publishers.

Barry, T. and K. Norsworthy (1990). *Honduras: A Country Guide.* Albuquerque, New Mexico, The Inter-Hemispheric Education Resource Center.

Bayliss-Smith, T. and S. Owens (1994). The Environmental Challenge, in D. Gregory, R. Martin and G. Smith (eds). *Human Geography: Society, Space, and Social Science.* London, The Macmillan Press Ltd.: 113

Benford, R. D. and S. A. Hunt (1992). Dramaturgy and Social Movements: The Social Construction and Communication of Power. *Sociological Inquiry* 62: 36-55.

Berg, B. (2001). *Qualitative Research Methods for the Social Sciences*. Boston, MA, Allyn and Bacon.

Blaikie, P. and H. Brookfield (1987). *Land Degradation and Society*. London, Methuen.

Blomley, N. K. (1994). *Law, Space, and the Geographies of Power*. London, Guilford Press.

Blount, B. G., T.L. (1999). *Ethnoecology: Knowledge, Resources, and Rights*. Athens, GA, University of Georgia Press.

Blumer, H. (1954). Social Structure and Power Conflict, in A. W. Kornhauser (ed.), *Industrial Conflict*. New York, McGraw-Hill: 232-239.

Blumer, H. (1962). Society and Symbolic Interaction, in A. M. Rose (ed.), *Human Behavior and Social Processes: An Interactionist Approach*. Boston, MA, Houghton Mifflin Company: 179-192.

Borman, K. M., M. D. LeCompte, and J.P Goetz. (1986). Ethnographic and Qualitative Research Design and Why it Doesn't Work. *The American Behavioral Scientist* 30(1): 42-57.

Bower, R. T. and P. de Gasparis (1978). *Ethics in Social Research: Protecting the Interests of Human Subjects*. New York, Praeger Publishers.

Boychuck, Rick. (1992). *Saving the Sea, The Blue Revolution*. New Internationalist. Issue 234, August.

Boyd, C. E. (1999). *Codes of Practice for Responsible Shrimp Farming*. St. Louis, MO, Global Aquaculture Alliance: 42.

Boyd, C. E. (2001). *Environmental Issues in Shrimp Farming*. IV Simposio Centroamericano de Acuacultura, Honduras.

Boyd, C. E. and M. C. Haws (1999). *Good Management Practices (GMPs) to Reduce Environmental Impacts and Improve Efficiency of Shrimp Aquaculture in Latin America*. V Central American Symposium on Aquaculture, San Pedro Sula, Honduras, Asociacion Nacional de Acuicultores de Honduras, Latin American Chapter of the World Aquaculture Society, and Pond Dynamics/Aquaculture Collaborative Research Support Program.

Boyd, C. E., and Bartholomew W. Green (2002). 'Shrimp Farming and the Environment: Coastal Water Quality Monitoring in Shrimp Farming Areas, An Example from Honduras', Report prepared under the World Bank, NACA, WWF and FAO Consortium Program on Shrimp Farming and the Environment. Work in Progress for Public Discussion. Published by the Consortium.

Boyd, C. E., J.A. Hargreaves and J.W. Clay (2002). 'Shrimp Farming and the Environment: Codes of Practice and Conduct for Marine Shrimp Aquaculture', Report prepared under the World Bank, NACA, WWF and FAO Consortium Program on Shrimp Farming and the Environment. Work in Progress for Public Discussion. Published by the Consortium.

Boyer, J. and A. Pell (1999). Mitch in Honduras: A Disaster Waiting to Happen. North American Congress on Latin America, *Report on the Americas* 33(2): 36-41.

Braun, B. and N. Castree, (eds.) (1998). *Remaking Reality: Nature at the Millenium*. London, Routledge.

Briggs, C. L., (ed.) (1996). *Disorderly Discourse: Narrative, Conflict and Inequality. Oxford Studies in Anthropological Linguistics*. New York, NY, Oxford University Press, Inc.

British Broadcasting Corporation (BBC). (2004). *Price of Prawns*. Jeremy Bristow, Senior Producer, BBC Natural History Unit. Broadcast December 2004.

Bryant, R. L. and S. Bailey (1997). *Third World Political Ecology*. London, Routledge.

Bulmer-Thomas, V. (1994). *The Economic History of Latin America Since Independence*. Cambridge, Cambridge University Press.

Cahoon, D. *et al.* (2002). 'Hurricane Mitch: Impacts on Mangrove Sediment Elevation Dynamics and Long-Term Mangrove Sustainability', United States Geological Survey. Lafeyette, LA

Calderon, F., A. Piscitelli, and J. L. Reyna. (1992). Social Movements: Actors, Theories, Expectations, in A. Escobar and S. E. Alvarez (eds), *The Making of Social Movements in Latin America: Identity, Strategy, and Democracy*. Boulder, CO, Westview Press: 19-36.

Callon, M. (1986). Some Elements of a Sociology of Translation: Domestication of the Scallops and the Fishermen of St Brieuc Bay, J. Law (ed.) *Power, Action and Belief: A New Sociology of Knowledge*. London, Routledge: 196-233.

Central Intelligence Agency (2002). *CIA – The World Factbook 2002*, Langley, VA, Central Intelligence Agency.

Central Intelligence Agency (CIA) (2007). *The World Factbook*. Online: https://www.cia.gov/cia/publications/factbook/geos/ho.html Updated April 2007

Chaparro, R. *et al.*, (2002). "Providing Educational Support to Mitigate Future Disasters by Developing an Informed Citizenry", University of Puerto Rico Sea Grant, Report to USAID on Post-Mitch Reconstruction efforts. 25 February.

Cloke, P., C. Philo, and D. Sadler. (1991). *Approaching Human Geography: An Introduction to Contemporary Theoretical Debates*. London, Paul Chapman Publishing Ltd.

CODDEFFAGOLF (1990a). *Boletin Informativo* No. 1

CODDEFFAGOLF (1990b). *Boletin Informativo* No. 2.

CODDEFFAGOLF (1991). *Boletin Informativo* No. 3

CODDEFFAGOLF (1991a). *Boletin Informativo* No. 4.

CODDEFFAGOLF (1991b). *Boletin Informativo* No. 5.

CODDEFFAGOLF (1991c). *Boletin Informativo* No. 6.

CODDEFFAGOLF (1998). *Boletin Informativo* No. 46.

CODDEFFAGOLF (1999). *Boletin Informativo* No. 47.

CODDEFFAGOLF (1999). *Boletin Informativo* No. 51.

CODDEFFAGOLF (2001). *Boletin Informativo* No. 57.

COHDEFOR/CONGESA, 1999: Forestry Inventory of 1965 and Annual Statistics of COHDEFOR, Tegucigalpa, Honduras: 136.

Colindres, Ibis; Allison, G; and Belaunde, Luisa Elvira (2000). Estudio Participativo: Uso de Especies Forestales por los Pobladores del Bosque Seco de la Zona Sur (Choluteca, Valle, El Paraiso) de Honduras. Siguatepeque, Honduras, COHDEFOR-ODA-ESNACIFOR Administracion Forestal del Estado: 97.

Collinson, H., (ed.) (1997). *Green Guerrillas: Environmental Conflicts and Initiatives in Latin America and the Caribbean.* New York, Black Rose Books.

CONGESA (2001). Valoracion Economica de los Manglares del Golfo de Fonseca, Honduras. Choluteca, Honduras, Administracion Forestal del Estado Corporacion Hondurena de Desarrollo Forestal (AFE-COHDEFOR), Organizacion Internacional de las Maderas Tropicales (OIMT) Manejo y Conservacion de los Manglares del Golfo de Fonseca, Honduras (PROMANGLE) Consultores en Gestion Ambiental (CONGESA): 136.

Conklin, H. C. (1954). An Ethnoecological Approach to Shifting Agriculture. *New York Academy of Sciences: Transactions of the New York Academy of Sciences* 17: 133-142.

Corbridge, S. (1986). *Capitalist World Development: A Critique of Radical Development Geography.* London, Macmillan Press.

Crane, G. T. and A. Amawi (1997). *The Theoretical Evolution of International Political Economy: A Reader.* New York, NY, Oxford University Press.

Cray, C. (2001). Dubious Development: The World Bank's Foray Into Private Sector Investment. *Multinational Monitor* 22(9): 20.

Cruz, G. A. and S. L. Thorn (1993). Plan de Emergencia para el Manejo de los Recursos Costeros del Golfo de Fonseca. Tegucigalpa, Honduras, Comision Nacional del Ambiente (CONAMA): 111.

Currie, D. J. (1995). Honduras: Ordenacion y Desarrollo del Cultivo del Camaron. Tegucigalpa, Honduras, Programa Regional de Apoyo al Desarrollo de la Pesca en el Istmo Centroamericano: PRADEPESCA-Union Europea-OLDEPESCA: 48.

Currie, D. J. (1995). Ordenamiento y Uso Racional del Recurso Disponible para el Cultivo de Camaron Marino en Centroamerica. Tegucigalpa, Honduras, Programa Regional de Apoyo al Desarrollo de la Pesca en el Istmo Centroamericano (PRADEPESCA): 88.

Davidson, D. (2001). *Essays on Actions and Events.* Oxford, Clarendon Press.

Davila Sampson Hilario, M. (1997). Comparacion Fisico-Quimica del Agua de Dos Esteros del Sur de Honduras en Epoca Seca. Departmento de Recursos Naturales y Conservacion Biologica. Valle Zamorano, Escuela Agricola Panamericana: 28.

Davis, D. E. (1999). The Power of Distance: Re-theorizing Social Movements in Latin America. *Theory and Society* 28: 585-638.

De Ferranti, D.; Masood, A; Loser, C; and Seade, J. (2000). Honduras Interim Poverty Reduction Strategy Paper. Tegucigalpa, Honduras, International Monetary Fund and the International Development Association.

De Vries, P. (1992). A Research Journey: On Actors, Concepts and the Text, in N. Long and A. Long (eds), *The Battlefields of Knowledge: The Interlocking of Theory and Practice in Social Research and Development.* London, Routledge: 47-84.

Deere, C. D. and M. Leon (2000). Neo-liberal Agrarian Legislation, Gender Equality, and Indigenous Rights: the Impact of New Social Movements, in A. Zoomers and G. van der Haar (eds), *Current Land Policy in Latin America: Regulating Land Tenure Under Neo-liberalism.* Amsterdam, Netherlands, Royal Tropical Institute: 75 - 92.

Demmers, J. (2004). Global Neoliberalisation and Violent Conflict, in J. Demmers et al. (eds), *Good Governance in the Era of Global Neoliberalism: Conflict and Depolitisation in Latin America, Eastern Europe, Asia, and Africa.* London, Routledge: 331 – 341.

Denzin, N. K. and Y. S. Lincoln (1994). *Handbook of Qualitative Research.* Thousand Oaks, Sage Publications.

Dove, M. R. (2005). Knowledge and Power in Pakistani Forestry: The Politics of Everyday Knowledge. In Paulson, S. and L. Gezon (eds.) *Political Ecology Across Spaces, Scales and Social Groups.* New Brunswick, NJ, Rutgers University Press: 217-238.

Duncan, J., N. C. Johnson, and R.H. Schein. (2004). *A Companion to Cultural Geography.* Oxford, Blackwell Publishing Ltd.

Dunford, M. and D. Perrons (1983). *The Arena of Capital.* London, The Macmillan Press Ltd.

Dunkerley, J. (1988). *Power in the Isthmus: A Political History of Modern Central America.* London and New York, Verso.

Duran, J. R. (1990a). Environment: Honduran Shrimp Industry Draws Criticism. *Inter Press Service* Tegucigalpa, Honduras. (Retrieved via LexisNexis).

Duran, J. R. (1991). Honduras: Groups Clash Over Environmental Defense. *Inter Press Service* (LexisNexis). Tegucigalpa, Honduras.

Duran, J. R. (1992). Honduras: Environmental Group Denounces Assassination of Activist. *Inter Press Service*. Tegucigalpa, Honduras. (Retrieved via LexisNexis)

Duran, J. R. (1993). Honduras: Southern Region Pulls Out of Economic Slump. *Inter Press Service*, Tegucigalpa, Honduras. (Retrieved via LexisNexis)

Duron, G. and C. Aburto R (2000). *Population, Consumption and Environmental Degradation Spiral: The Case of Southern Honduras*. Tegucigalpa, Honduras, International Center for Research on Women: 51.

Eckstein, S. (1989). *Power and Popular Protest: Latin American Social Movements*. Berkeley, CA, University of California Press.

Edkins, J. (1999). *Poststructuralism and International Relations: Bringing the Political Back In*. Boulder, CO and London, Lynne Rienner Publishers, Inc.

El-Ashry, M. L. (2002). The Challenge of Sustainability: An Action Agenda for the Global Environment. Washington D.C., Global Environment Facility: 124.

Ellison, A. M. and E. J. Farnsworth (2001). *Marine Community Ecology*. Sunderland, MA, Sinauer Associates, Inc.

Escobar, A. (1992a). Culture, Practice and Politics: Anthropology and the Study of Social Movements. *Critique of Anthropology* 12(4): 395-432.

Escobar, A. (1992b). Culture, Economics, and Politics in Latin American Social Movements Theory and Research, in A. Escobar and S. E. Alvarez (eds), *The Making of Social Movements in Latin America: Identity, Strategy, and Democracy*. Boulder, CO, Westview Press: pp. 62-85.

Escobar, A. (1998). Whose Knowledge, Whose Nature? Biodiversity, Conservation and the Political Ecology of Social Movements. *Political Ecology* 5: 53-82.

Escobar, A. and S. E. Alvarez, (eds). (1992). *The Making of Social Movements in Latin America: Identity, Strategy, and Democracy. Series in Political Economy and Economic Development in Latin America*. Boulder, CO, Westview Press.

Espinal, M. (2004). Dotacion de las Herramientas Tecnicas para el Manejo y Desarrollo del Area de Manejo Habitat/Especies Bahia de Chismuyo. Tegucigalpa, Honduras, Programa Ambiental Regional para Centroamerica (PROARCA/APM) Administracion Forestal del Estado Corporacion Hondurena de Desarrollo Forestal (AFE-COHDEFOR) The Nature Conservancy: 107.

Estrada Acevedo Arturo, J. (1995). Monitoreo de Pesticidas en Agua de Esteros del Golfo de Fonseca, Honduras. Ingeniero Agronomo. Valle Zamorano, Escuela Agricola Panamericana: 48.

European Union (2002). European Commission Regional Strategy Paper for Central America 2002-2006, European Commission: 31.

Evers, T. (1985). Identity: The Hidden Side of New Social Movements in Latin America, in D. Slater (ed.), *New Social Movements and the State in Latin America*: 43-71.

Fetterman, D. M. (1989). *Ethnography: Step by Step*. London, Sage Publications.

Forsyth, T. (2003). *Critical Political Ecology: The Politics of Environmental Science*. London, Routledge.

Foucault, M. (1980). *Power/Knowledge: Selected Interviews and Other Writings 1972-1977*. New York, NY, Pantheon Books.

Foucault, M. (1984). *The Foucault Reader*. New York, NY, Pantheon Books.

Foweraker, J. (1995). *Theorizing Social Movements in Latin America*. London, Pluto.

Foweraker, J. (1998). Social Movements and Citizenships Rights in Latin America, in M. Vellings (ed.), *The Changing Role of the State in Latin America*. Boulder, CO, Westview Press: 271-296.

Fox, S. (2000). Communities of Practice, Foucault and Actor-Network Theory. *Journal of Management Studies* 37(6): 853-867.

Friends of the Earth (1990). Whale-Sized Fight Over Shrimp, *Newsmagazine of Friends of the Earth* 20(2): 12-13.

Gammage, S. (1997). Estimating the Returns to Mangrove Conversion: Sustainable Management or Short Term Gain? International Institute for Environment and Development: Environmental Economics Programme. London: 73.
Gammage, S. et al. (2002). An Entitlement Approach to the Challenges of Mangrove Management in El Salvador. *Ambio: Royal Swedish Academy of Sciences* 31(4): 285-294.

Gammage, S. et al. (2000). *A Platform for Action: For the Sustainable Management of Mangroves in the Gulf of Fonseca*. Washington D.C., International Center for Research on Women.

Garcia, M. P. (1992). The Venezuelan Ecology Movement: Symbolic Effectiveness, Social Practices, and Political Strategies in A. Escobar and S. E. Alvarez (eds), *The Making of Social Movements in Latin America: Identity, Strategy, and Democracy*. Boulder, CO, Westview Press: 150-170.

Garcia, Rosa del Carmen; Duarte, Rosa M; Gomez, Rosamaria; Reanos, Olga Sofia; Gonzalez, Jose Rigoberto (1994). Plan de Ordenacion y Desarrollo Pesquero y Acuicola de Honduras (Resumen Ejecutivo). Tegucigalpa, Honduras, Secretaria de Recursos Naturales (SERNA) Direccion General de Pesca y Acuicultura (DIGEPESCA) Secretaria de Planificacion Coordinacion y Presupuesto: 16.

Gass, V. (2002). *Democratizing Development: Lessons from Hurricane Mitch Reconstruction*. Washington D.C., Washington Office on Latin America.

Gee, J. P. (1999). *An Introduction to Discourse Analysis: Theory and Method*. London and New York, Routledge.

Gezon, L. L. (1999). Of Shrimps and Spirit Possession: Toward a Politcal Ecology of Resource Management in Northern Madagascar. *American Anthropologist* 101(1): 58-67.

Giron, Sergio; Vasquez, Cristobal; Gonzalez, Rigoberto; Marin, Mirna; (1989). Evaluacion del Potencial de Tierras Nacionales Aptas para la Acuicultura del Camaron Marino y de los Procedimientos Administrativos en la Adjudicacion de las mismas. Tegucigalpa, Honduras, Secretaria de Planificacion y Presupuesto Coordinacion Direccion General de Recursos Naturales Renovables RR.NN. Coordinacion Tecnica: 64.

Gispert, C., (ed.) (2001). *Atlas Geografico Universal y de Honduras*. Barcelona, Spain, OCEANO Grupo Editorial, S.A.

Gluckman, M. (1960). *Custom and Conflict in Africa*. Oxford, Basil Blackwell.

Gonzalez, M. A. and N. Melendez (1999). Legislacion Ambiental de Centroamerica. San Salvador, El Salvador, Comision Centroamericana de Ambiente y Desarrollo: 113-134.

Government of the Republic of Honduras. (2001). *Interim Poverty Reduction Strategy Paper*. Tegucigalpa, Honduras, Government of the Republic of Honduras.

Green, D. (1995). *Silent Revolution: The Rise of Market Economics in Latin America*. London, Cassell.

Greenpeace (1996). Greenpeace Warning. Latin American NewsLetters, Ltd.; Caribbean & Central America Report (LexisNexis): 4.

Greenwire (1993). Awards Roundup: 103-Year-Old Glades Defender Gets Top Honor. Greenwire, Environment and Energy, LLC (LexisNexis).

Gregory, D. (1994). *Geographical Imaginations*. Oxford, Blackwell.

Gregory, D., R. Martin and G. Smith (1994). *Human Geography: Society, Space, and Social Science*. London, The Macmillan Press Ltd.

Grupo Granjas Marinas San Bernardo (2007). www.seafarmsgroup.com

Guevara, R. (1991). Ponencia Sobre la Acuicultura en Honduras, con Enfasis en el Cultivo de Camaron. Ponencia, Tegucigalpa, Honduras, Secretaria de Recursos Naturales (SERNA) Direccion General de Pesca y Acuicultura (DIGEPESCA).

Gutting, G. (2001). *French Philosophy in the Twentieth Century*. Cambridge, Cambridge University Press.

Haack, S. (1978). *Philosophy of Logics*. Cambridge, Cambridge University Press.

Habermas, J. (1992). *The Structural Transformation of the Public Sphere: An Inquiry into a Category of Bourgeois Society*. Boston, MIT Press.

Hajer, M. A. (1995). *The Politics of Environmental Discourse*. Oxford, Clarendon Press.

Hall, D. (2003). The International Political Ecology of Industrial Shrimp Aquaculture and Industrial Plantation Forestry in Southeast Asia. *Journal of Southeast Asian Studies* 34(2): 251-264.

Hall, P. M. (1972). A Symbolic Interactionist Analysis of Politics. *Sociological Inquiry* 42(3-4): 35-76.

Handley, L. R., *et al.* (2002). *Mapping Coastal Habitats*, United States Geological Survey: 86.

Harvey, D. (1996). *Explanation in Geography*. New York, St. Martins Press.

Harvey, D. (2005). *A Brief History of Neoliberalism*. Oxford, Oxford University Press.

Heerin, J. (2004). History of Aquaculture. Personal Document.

Heise, D. R. (1979). *Understanding Events: Affect and the Construction of Social Action*. London, Cambridge University Press.

Held, D. (1980). *Introduction to Critical Theory*. Berkeley, CA, University of California Press.

Hellin, J. (1999). Land Degradation in Honduras: A Challenge to an Ecological Marxism? *Capitalism, Nature, Socialism* 3(39): 105-125.

Hellman, J. A. (1995). The Riddle of New Social Movements: Who They Are and What They Do. Capital, Power and Inequality, in S. H. Halebsky (ed.), Latin America. Boulder, CO, Westview Press: 165-183.

Hensel, P. and D. Cahoon (2002). *Hurricane Mitch: A Regional Perspective on Mangrove Damage, Recovery, and Sustainability*, Lafayette, LA. United States Geological Survey.

Hensel, P. and C. E. Proffitt (2002). *Hurricane Mitch: Acute Impacts on Mangrove Forest Structure and an Evaluation of Recovery Trajectories*. Executive Summary. U.S. Department of the Interior and the U.S. Geological Survey.

Hernandez, G., *et al.* (1999). *MesoAmerican Wetlands: Ramsar Sites in Central America and Mexico. San Jose, Costa Rica*, The World Conservation Union (IUCN) MesoAmerica Wetlands and Coastal Zones Program.

Holden, R. H. (1981). Central America Is Growing More Beef and Eating Less - As the Hamburger Connection Widens and they're doing it all for you. *Agribusiness Journal* 2(10).

Honduras, Secretariat of Industry and Commerce (2002). *Geographic Atlas of Honduras* (10th edition). Tegucigalpa.

Homer-Dixon, T. (1993). Physical Dimensions of Global Change, in Choucri (ed.) *Global Accord: Environmental Challenges and International Responses*. Cambridge, MA: MIT Press.

Honduras, ANDAH (1992). Historia y Desarrollo de la Camaricultura en Honduras. Choluteca, Honduras, Asociacion Nacional de Acuicultores de Honduras (ANDAH).

Honduras, Central Bank of Honduras (1987). Banco Central de Honduras: Memoria 1987. Tegucigalpa, Honduras, Banco Central de Honduras: 21.

Honduras, Central Bank of Honduras (2002). Banco Central de Honduras: Memoria 2002. Tegucigalpa, Honduras, Banco Central de Honduras.

Horkheimer, M. and T. W. Adorno (1944). *Dialectic of Enlightment*. New York, Continuum.

Horwich, P. (1990). *Truth*. Oxford, Oxford University Press

Howarth *et al.* (2000). *Discourse Theory and Political Análisis: Identities, Hegemonies, and Social Change*. Manchester, Manchester University Press.

Hunt, S., *et al.* (1994). Identity Fields: Framing Processes and the Social Construction of Movement Identities, in E. Laraña, H. Johnston and J. R. Gusfield (eds.), *New Social Movements: from Ideology to Identity*. Philadelphia, PA, Temple University Press, 185-208.

Hutchings and Saenger (1987) in Vindel Calix Felipie, L. (1997). Sobreexplotacion de la Vegetacion Manglar y Sus Efectos al Ecosistema de La Bahia de Chismuyo, Honduras: Comparacion Estructural del Bosque Manglar Segun el Tipo de Explotacion Economica. Guatemala City, Guatemala, The Nature Conservancy, World Wildlife Fund, and University of Rhode Island Coastal Resources Center (PROARCA/Costas).

International Union for the Conservation of Nature (IUCN)(1992). Conservacion de los Ecosistemas Costeros del Golfo de Fonseca. Tegucigalpa, Honduras, Agencia Danesa para el Desarrollo Internacional (DANIDA) and the International Union for the Conservation of Nature: 47.

Jeffrey, P. (1999). Rhetoric and Reconstruction in Post-Mitch Honduras. North American Congress on Latin America, *Report on the Americas* 33(2): 28-35.

Japanese International Cooperation Aagency (JICA) (2002). Country Profile on Environment: Honduras. Japan, Planning and Evaluation Department Japan International Cooperation Agency.

Johannessen, C. L. (1963). *Savannas of Interior Honduras*. Berkeley, CA, University of California Press.

Johnston, R. J. et al., (1986). *Dictionary of Human Geography* (2nd edition). Oxford., England, Blackwell Publishing.

Johnston, R. J. et al., (2000). *Dictionary of Human Geography* (4th edition) Oxford., England, Blackwell Publishing.

Jukofsky, D. (2000). Central American Tribunal Weighs Water Contamination Cases. *Environment News Service, Global News Wire, Financial Times Information* (LexisNexis). San Jose, Costa Rica.

Keil, R., D. Bell, P. Penz and L. Fawcett. (1998). *Political Ecology: Global and Local*. London and New York, Routledge.

Kempton, W. (2001). Cognitive Anthropology and the Environment, in C. Crumley (ed.), *New Directions in Anthropology and Environment: Intersections*. Walnut Creek, AltaMira Press: 49-71.

Kuntz, L. I. (2001). A Coastal Balancing Act; *Planet*. UNESCO Courier.

Lacayo Bustillo, G. (2002). *Golfo de Fonseca: Region Clave en Centroamerica*. Tegucigalpa, Honduras, Guaymuras.

Lacerda, L. D. and Y. Schaeffer-Novelli (1999). Mangroves of Latin America: The Need for Conservation and Sustainable Utilization. Ecosistemas de Manglar en America Tropical. in A. Yanez-Arancibia and A. L. Lara-Dominguez. Instituto de Ecologia; Xalapa, Mexico, UICN/ORMA Costa Rica; NOAA/NMFS Silver Spring, MD USA: 5-8.

Lara-Dominguez, A. L. e. a. (2002). *Sustainable Management of Mangroves in Central America*. Vera Cruz, Mexico, Instituto de Ecologia A.C. Programa de Recursos Costeros in collaboration with the Department of Oceanography and Coastal Sciences and Coastal Ecology, Institute School of the Coast and Environment, Louisiana State University.

Larios Vallejo, M., et al. (2003). Honduras: Frente al Cambio Climatico. Tegucigalpa, Food and Agricultural Organization (FAO) Comision Centroamericana de Ambiente y Desarollo (CCAD): 59.

Lash, S. (1991). *Post-Structuralist and Post-Modernist Sociology*. Cambridge, Edward Elgar Publishing Limited.

Law, J. (1992). Notes on the Theory of the Actor-Network: Ordering, Strategy, and Heterogenity. *Systems Practice* 5(4): 379-393.

Leach, M. and R. Mearns (eds). (1996). *The Lie of the Land: Challenging Received Wisdom on the African Environment*. Portsmouth, NH, Heinemann.

Leonard, H. J. (1987). *Natural Resources and Economic Development in Central America: A Regional Environmental Profile*. Washington, D.C., International Institute for Environment and Development.

Little, W., H. W. Fowler and J. Coulson; revised and edited by C. T. Onions (1985). *The Shorter Oxford English Dictionary*. Guild Publishing, p. 1634.

Long, A. (1992). Goods, Knowledge, and Beer: the Methodological Significance of Situational Analysis and Discourse, in N. Long and A. Long (eds), *The Battlefields of Knowledge: The Interlocking of Theory and Practice in Social Research and Development*. London, Routledge.

Long, N. and A. Long, (eds). (1992). *The Battlefields of Knowledge: The Interlocking of Theory and Practice in Social Research and Development*. London, Routledge.

Lowe (1995), in T. Honderich (ed.), *The Oxford Companion to Philosophy*, pp. 881-882.

Lucas, A. (1998). The Mangroves are Declining, the Fishermen are Lamenting: The Political Ecology of Mangrove Degradation in South Sulawesi, in K. P. Robinson (ed.), *Living Through Histories: Culture, History and Social Life in South Sulawesi*. Canberra, Australian National University: 196 - 228.

Lustig, N. (2000). *Crisis and the Poor: Socially Responsible Macroeconomics*. Washington D.C., Inter-American Development Bank: 31.

Lutz, G. C. (2001). *Strategic Planning for Development of Honduran Aquaculture*. Baton Rouge, LA, Louisiana State University Agricultural Center: Aquaculture Research Station, Alianza Louisiana Presidential Investment Program of Honduras: 22.

Luxner, L. (1992). Central American Entrepreneurs Diving Into Aquaculture Business. *Orlando Sentinel*, 24 May 1992. Orlando, FL: Sentinel Communications Co.

Luxner, L. (1993). Honduras Cultivates Export Diversity. *Journal of Commerce*: Section: Foreign Trade: 5A. (LexisNexis).

Luxner, L. (1995). Honduras; Coffee Industry; Central America; Industry Overview. *Tea and Coffee Trade Journal* 167(1): 20.

Macintosh, D. J. and M. Phillips (2002). *Annexes to the: Thematic Review on Coastal Wetland Habitats and Shrimp Aquaculture. Case studies 7-13*. Washington D.C., Report prepared under the World Bank, NACA, WWF and FAO Consortium Program on Shrimp Farming and the Environment. Work in Progress for Public Discussion. Published by the Consortium.

Maguire, A. and J. Welsh-Brown (1986). *Bordering on Trouble: Resources and Politics in Latin America*. Bethesda, MD, Adler & Adler Publishers, Inc.

Maheu L. (ed.), *Social Movements and Social Classes; the Future of Collective Action*. London, Sage: 107-122.

Malesevic, S. and I. Mackenzie (2002). *Ideology After Poststructuralism*. London, Pluto Press.

Marschatz, Astrid; Argenal, Alejandro; Imendia, Carlos A (2002). Convergencia Economica de Centroamerica: Evidencia Empirica Entre 1920-2000 Consideraciones Preliminares para Definir Politicas Estructurales de Cohesion. San Salvador, El Salvador, Banco Centroamericano de Integracion Economica: Departmento de Planificacion y Presupuesto (BCIE): 36.

Martinez, J. C. (1999). Honduras Post Mitch: Problemas y Oportunidades para el Desarrollo de la Economia Rural: Documento de Trabajo. Washington D.C., Banco Interamericano de Desarrollo: Departamento Regional de Operaciones II: 36.

Martinez-Alier, J. (2002). The Environmentalism of the Poor: A Study of Ecological Conflicts and Valuation. North Hampton, MA, Edward Elgar Publishing Inc.

McCay, B. J. and L. Fortmann (1996). Voices from the Commons: Evolving Relations of Property and Management. Cultural Survival Quarterly (CSQ) 20(1): 24.

McKee, K. L. and T. E. McGinnis II (2002). Hurricane Mitch: Effects on Mangrove Soil Characteristics and Root Contributions to Soil Stabilization, Lafeyette, LA. United States Geological Survey.

Mejia, T. (1996). Environment-Central America: Greens Attack Shrimp Industry. Inter Press Service. San Lorenzo, Honduras. (LexisNexis).

Melucci, A. (1995) "A Strange Kind of Newness: What's 'New' in New Social Movements?" in Laraña, Johnston and Gusfield (eds.) New Social Movements.

Mejia, T. (1996). Environment: Forum Calls for Moratorium on Shrimp Farming. Inter Press Service. Choluteca, Honduras. (LexisNexis).

Mesoamericana (1999). Mesoamericana: Boletin Oficial de la Sociedad Mesoamericana para la Biologia y la Conservacion. 4: 99-104.

Meyer, D. E. (1996). Marine Shrimp Culture Development in southern Honduras. Acta Hydrobiol 37(suppl. 1): 111-120.

Meyer, Daniel E.; Luna, Nohemy; Reyes, Gerardo (1989). Memoria Encuentro Nacional de Acuicultura COINDAH. CEIBA: A Scientific Journal Issued by the Escuela Agricola Panamericana 30(2): 81.

Miller, K., E. Chang, N. Johnson. (2001). Defining Common Ground for the MesoAmerican Biological Corridor. Washington D.C., World Resources Institute.

Milton, K. (1996). Environmentalism and Cultural Theory: Exploring the Role of Anthropology in Environmental Discourse. London and New York, Routledge.

Mitsch, W. J. and J. G. Gosselink (1986). Wetlands. New York, Van Nostrand Reinhold Company Inc.

Montaner, J. B. (1995). An Economic Analysis of Land Titling in Honduras. Agricultural Economics. Oxford, UK, Oxford University: 221.

Murdoch, J. (1998). The Spaces of Actor-Network Theory. *Geoforum* 29(4): 357-374.

Nations, J. and H. J. Leonard (1986). Grounds of Conflict in Central America, in A. Maguire and J. Welsh-Brown (eds), *Bordering on Trouble: Resources and Politics in Latin America*. Bethesda, MD, Adler & Adler Publishers, Inc: 55-98.

Naylor, B. (1992). Vanishing Homelands: Honduran Shrimp. Interview transcript: All Things Considered. National Public Radio (NPR). 18 April1992.

Naylor, R. L., R.J Goldburg, J. H. Primavera, N. Kautsky, M. Beveridge, J. Clay, C. Folke, J. Lubchenco, H. Mooney, and M. Troell. (2000). Effect of Aquaculture on World Fish Supplies. *Nature* 405(June): 1017-1024.

Nazarea, V. D. (1999). *Ethnoecology: Situated Knowledge/Located Lives*. Tucson, AZ, The University of Arizona Press.

Nichols, R. (1998). Shrimp Demand Takes Toll. *The Record* (Special from *The Philadelphia Inquirer*). Bergen County, New Jersey: F05.

Nuijten, M. (1992). Local Organization as Organizing Practices: Rethinking Rural Institutions, in N. Long and A. Long (eds), *The Battlefields of Knowledge: The Interlocking of Theory and Practice in Social Research and Development*. London, Routledge.

Nygren, A. (1995). Deforestation in Costa Rica an Examination of Social and Historical Factors. *Forest and Conservation History* 39(1): 27-35.

Nygren, A. (1998). Environment as Discourse: Searching for Sustainable Development in Costa Rica. *Environmental Values* 7(2): 201-222.

Nygren, A. (1999). Local Knowledge in the Environment-Development Discourse: From Dichotomies to Situated Knowledges. *Critique of Anthropology* 19(3): 267-288.

Nygren, A. (2000). Development Discourses and Peasant-Forest Relations: Natural Resource Utilization as Social Process. *Development and Change* 31(1): 11-34.

Ochoa, E., S. Olsen, and N. Windevohel. (2001). *Avances del Manejo Costero Integrado en PROARCA/Costas*. Guayaquil, Ecuador, University of Rhode Island Coastal Resources Center.

OLDEPESCA El Cultivo del Camaron en el Istmo Centroamericano: Intervencion de la Secretaria Tecnica. Panama City, Panama, Organizacion Latinoamericana de Desarrollo Pesquero: Programa Regional de Apoyo al Desarrollo de la Pesca en el Istmo Centroamericana, Secretaria Tecnica: 111-118.

Otis, J. (1993). Government Weighs Protection Plan. *United Press International*. Tegucigalpa, Honduras. (LexisNexis).

Oyuela, O. (1994). Los Manglares del Golfo de Fonseca – Honduras. El Ecosistema de Manglar en America Latina y La Cuenca del Caribe: Su Manejo y Conservacion. D. O. Suman. Miami and New York City, Rosentiel School of Marine and Atmospheric Science, University of Miami and The Tinker Foundation: 144.

Pacheco, R. E. M. (1999). Captura de Post-Larvas de Camaron y Fauna de Acompanamiento en Dos Esteros en el Sur de Honduras. Departmento de Zootecnia. Zamorano-Honduras, Escuela Agricola Panamericana: 14.

Page, B. (2003). The Political Ecology of *Pruns africana* in Cameroon. *Area* 35(4): 357-370.

Page, K. D. and M. J. Schwartz (1996). USAID Capacity Building in the Environment: A Case of the Central American Commission for Environment and Development, Environment and Natural Resource Information Centre with Central American Commission for Environment and Development: 15.

Painter, J. (1995). *Politics, Geography & 'Political Geography': A Critical Perspective*. London, Arnold.

Park, R. E. (1915). *The Principles of Human Behavior*. Chicago, The Zalaz Corporation.

Paulson, S., L. L. Gezon, and M. Watts. (2003). Locating the Political in Political Ecology: An Introduction. *Human Organization* 62(3): 205-216.

Peet, R. (2002). Ideology, Discourse, and the Geography of Hegemony: From Socialist to Neoliberal Development in Postapartheid South Africa. *Antipode*: 54-84.

Peet, R. and M. Watts (eds.) (1996). *Liberation Ecologies: Environment, Development, Social Movements*. London and New York, Routledge.

Peluso, N. and M. Watts (eds.) (2001). *Violent Environments*. Ithaca and London, Cornell University Press.

Perez-Brignoli, H. (1989). *A Brief History of Central America*. Berkeley, CA, University of California Press.

Perreault, T. and P. Martin (2005). Geographies of Neoliberalism in Latin America. *Environment and Planning Part A; International Journal of Urban and Regional Research*, 37(5): 191-201.

Peters, M. A. (2001). *Poststructuralism, Marxism, and Neoliberalism: Between Theory and Practice*. Lanham, MD, Rowman and Littlefield Publishers, Inc.

Phillips, J. (2002). *Activities in Honduras In Support of the Hurricane Mitch Reconstruction Program: A Final Report Submitted to The U.S. Agency for International Development*, Reston, VA. United States Geological Survey: 73.

Phillips, M., J. Clay, R.Zweig, C.G. Lundin, R. Subasinghe (2001). Shrimp Farming and the Environment. *InterCoast* (Spring): (1 – 2)

Polletta, F. (1997). Culture and its Discontents: Recent Theorizing on the Cultural Dimensions of Protest. *Sociological Inquiry* 67(4): 431-450.

Portes, A. (1997). Neoliberalism and the Sociology of Development: Emerging Trends and Unanticipated Facts. *Population and Development Review* 23: 229-259.

Poster, M. (1989). *Critical Theory and Postructuralism: In Search of a Context.* Ithaca and London, Cornell University Press.

PROARCA/COSTAS (1996). *Golfo de Fonseca: Perfil Preliminar.* Guatemala City, Guatemala, The Nature Conservancy World Wildlife Fund Coastal Resources Center PROARCA/Costas and USAID: 23.

PROARCA/COSTAS (2001). *Final Report for PROARCA/COSTAS* (The Coastal Zone Management Component of PROARCA). Guatemala City, Guatemala, The Nature Conservancy (TNC); World Wildlife Fund (WWF); University of Rhode Island Coastal Resources Center (URI/CRC): 53.

PROGOLFO (1997). *Borrador del Documento del Proyecto: Conservacion de los Ecosistemas Costeros en el Golfo de Fonseca.* Nicaragua, Honduras, and El Salvador, Ministerio de Relaciones Exteriores and DANIDA: 70.

PROGOLFO (1998a). *Diagnostico del Estado de los Recursos Biofisicos, Socioeconomicos e Institutcionales.* Honduras, Nicaragua, El Salvador, Comision Centroamericana de Ambiente y Desarrollo (CCAD) Secretaria de Recursos Naturales y Ambiente de Honduras (SERNA) Ministerio de Medioambiente y Recursos Naturales de El Salvador (MARN) Ministerio de Ambiente y Recursos Naturales de Nicaragua (MARENA) Proyecto Regional Ordenamiento Ecologico de los Ecosistemas Costeros del Golfo de Fonseca (PROGOLFO) Union Mundial para la Naturaleza (UICN) and DANIDA: 20.

PROGOLFO (1999). Proyecto Conservacion de los Ecosistemas Costeros del Golfo de Fonseca: Resumen Ejecutivo del Documento de Implementacion del Proyecto Periodo Octubre 1998 a Septiembre 1999 Honduras. Honduras, Secretaria de Recursos Naturales y Ambiente (SERNA) Comision Centroamericana de Ambiente y Desarrollo (CCAD) and DANIDA: 15.

PROGOLFO (2000). *Plan Operativo Anual 2001.* Choluteca, Honduras, Proyecto Conservacion de los Ecosistemas Costeros en el Golfo de Fonseca: 62.

PROMANGLE-AFE-COHDEFOR (1997). *Proyecto Manejo y Conservacion de los Manglares en el Golfo de Fonseca, Honduras: Tercera Revision del Documento de Proyecto PD 44/95 (F) Para La Implementacion del Proyecto por Fases.* Santa Cruz de la Sierra, Bolivia, International Tropical Timber Organization: Comite de Repoblacion y Ordenacion Forestales: 97.

PROMANGLE-AFE-COHDEFOR (1999). *Analisis de los Resultados de La Encuesta Socioeconomica del Proyecto Manejo y Conservacion de los Manglares del Golfo de Fonseca, Honduras (PROMANGLE).* La Lujosa, Marcovia; Honduras, Administracion Forestal del Estado Corporacion Hondurena de Desarollo Forestal (AFE-COHDEFOR)

Organizacion Internacional de Las Maderas Tropicales (OIMT) Proyecto Manejo y Conservacion de Los Manglares del Golfo de Fonseca, Honduras (PROMANGLE).

Rabinow, P. (ed.) (1984). *The Foucault Reader*, New York, NY, Pantheon Books.

Radcliffe, S. (2001a). Popular and State Discourses of Power, in T. B. S. Hanson and F. Raleigh, *States of Imagination: Ethnographic Explorations of the Post-Colonial State.* Duke University Press: 219-242.

Radcliffe, S. A. (2001b). Development, the State, and Transnational Political Connections: State and Subject Formations in Latin America. *Global Networks: A Journal of Transnational Affairs* 1(1): 19-36.

Rafael del Cid, J., I. Walker, and H. Cardenas. (1988). *Sociedad y Ambiente: Los Desafios para el Desarrollo Sostenible de Honduras.* Central America, Tegucigalpa, Honduras. Fundacion Centroamericana para el Desarrollo Humano: 78.

REMIDE (2002). Plan de Emergencia Local: Comunidad de Pueblo Nuevo, Municipio de Marcovia. Choluteca, Honduras, Programa Regional Para la Mitigacion de Desastres.

Rhoades, R. E. and J. Harlan (1999). Epilogue: Quo Vadis? The Promise of Ethnoecology, in V. D. Nazarea, *Ethnoecology: Situated Knowledge/Located Lives.* Tuscon, AZ, The University of Arizona Press.

Rivera-Monroy, V. H. *et al.* (2002). Hurricane Mitch: Integrative Management and Rehabilitation of Mangrove Resources to Develop Sustainable Shrimp Mariculture in the Gulf of Fonseca, Honduras, Lafyette, LA. United States Geological Survey.

Robbins, P. (2003). Political Ecology in Political Geography. *Political Geography* 22: 641-645.

Roberts, K. M. (1997). Beyond Romanticism: Social Movements and the Study of Political Change. *Latin American Research Review* 32(2): 137-151.

Robinson, G. M. (1990). *Conflict and Change in the Countryside: Rural Society, Economy and Planning in the Developed World.* London, Belhaven Press.

Robinson, W.I. (2003). *Transnational Conflicts: Central America, Social Change, and Globalization.* London, Verso Press.

Rodriguez, C. M. (2002). Central American Policy on the Conservation and Wise Use of Wetlands. San Jose, Costa Rica, Central American Commission on Environment and Development (CCAD).

Rodriguez, J. (1998). *State of Environment and Natural Resources in Central America.* San Jose, Costa Rica, Central American Commission on Environment and Development.

Roe, E. (1994). *Narrative Policy Analysis: Theory and Practice.* Durham, NC and London, Duke University Press.

Rojas, M. H. *et al.* (2001). Gender and Community Conservation. *Contemporary Women's Issues, Gender Matters Quarterly.* United States Agency for International Development, No. 3: 1.

Rollet, B. (1981). *Bibliography on Mangrove Research 1600-1975.* Paris, United Nations Educational, Scientific, and Cultural Organization.

Rosenberg, M. and R. H. Turner, (eds). (1981). *Social Psychology: Sociological Perspectives.* New York, Basic Books, Inc., Publishers.

Rosenberry, Bob (2004). Interview with Jim Heerin, Chairman of Sea Farms International, *Shrimp News International,* June 30.

Rowlands, J. M. (1995). *Empowerment Examined: An Exploration of the Concept and Practice of Women's Empowerment in Honduras.* Durham, UK, Department of Geography, Durham University: 246.

RPI, Research Planning Incorporated (1994a). Propuesta de Inventario de Recursos Costeros y Mapas de Sensibilidad de Habitats Costeros para el Golfo de Fonseca. Guatemala City, Guatemala, Proposal to the Comision Centroamericana de Ambiente y Desarrollo (CCAD): 14.

Saad-Filho, A. and D. Johnston (eds.) (2005). *Neoliberalism: A Critical Reader.* London, Pluto Press.

Salman, T. and W. Assies (2000). Revisioning Cultures of Politics: An Essay on Social Movements, Citizenship, and Culture. *Critique of Anthropology* 20(3): 289-307.

Sampson, E. E. (1981). Cognitive Psychology as Ideology. *The American Psychologist* 36(7): 730-7-43.

Sanchez, A. A. (1999). *Evaluacion de los Manglares del Golfo de Fonseca Mediante un Analisis Multitemporal de Imagenes del Satelite LANDSAT-TM Entre los Anos (1989-1995), (1995-1998), y (1989-1998).* Tegucigalpa, Honduras, Administracion Forestal del Estado Corporacion Hondurena de Desarrollo Forestal (AFE-COHDEFOR): 45.

Sanchez-Paez, Heliodoro; Ulloa-Delgado; Giovanni Andres; Alvarez-Leon, Ricardo (eds). (2000). *Hacia la Recuperacion de los Manglares del Caribe de Colombia.* Bogota, Colombia, Ministerio del Medio Ambiente Asociacion Colombiana de Reforestadores - ACOFORE Organizacion Internacional de Maderas Tropicales - OIMT.

Sandstrom, K. L., D. D. Martin, and G.A. Fine. (2001). Symbolic Interactionism at the End of the Century in G. Ritzer and B. Smart (eds) *Handbook of Social Theory.* London, Thousand Oaks, New Delhi, SAGE: 217-231.

Sarup, M. (1988). *An Introductory Guide to Post-Stucturalism and Postmodernism.* Athens, The University of Georgia Press.

Schellenberg, J. A. (1996). *Conflict Resolution: Theory, Research, and Practice.* Albany, NY, State University of New York Press.

Schleifstein, M. (1999). *The Long Recovery*. The Times-Picayune Publishing Co. New Orleans, LA: A1.

Schuurman, F. J. (1993). Modernity, Post-Modernity and the New Social Movements. In F. Schuurrman (ed.), *Beyond the Impasse: New Directions in Development Theory.* London, Zed Press: 187-206.

Sea Farms International (2007). http://www.seafarmsgroup.com/

Sequeira, M. (1996). Environment: World Conservation Congress Dominated by North. *Inter Press Service*. Montreal, Canada. (LexisNexis).

Sequeira, M. (1999). Central America-Environment: Defending Water Rights. *Inter Press Service*. San Jose, Costa Rica. (LexisNexis).

Sequeira, M. (2000). Trade-Central America: Three Way Agreement Criticised. *Inter Press Service*. San Jose, Costa Rica. (LexisNexis).

SERNA (1988). *Estudio de Perfectibilidad Tecnica y Economica para el Proyecto de Desarrollo de La Pesca Artesanal en Honduras*. Tegucigalpa, Honduras, Secretaria de Recursos Naturales (SERNA) Direccion General de Recursos Naturales Renovables Deparmento de Planificacion: 33.

SERNA (1994). *Plan de Ordenacion y Desarrollo Pesquero y Acuicola de Honduras.* Tegucigalpa, Honduras, Secretaria de Recursos Naturales Direccion General de Pesca y Acuicultura Secretaria de Planificacion Coordinacion y Presupuesto: 56.

SERNA (1998). *Estudio Sobre Diversidad Biologica de la Republica de Honduras.* Tegucigalpa, Honduras, Republica de Honduras, Secretaria de Estado en los Despachos de Recursos Naturales y Ambiente, Direccion General de Biodiversidad: 159.

Seur, H. (1992). The Engagement of Researcher and Local Actors in the Construction of Case Studies and Research Themes: Exploring Methods of Restudy, in N. Long and A. Long (eds), *The Battlefields of Knowledge: The Interlocking of Theory and Practice in Social Research and Development.* London, Routledge.

Sheffner, J. (1995). Moving in the Wrong Direction in Social Movement Theory. *Theory and Society* 24: 595-612.

Sherman, K. and Q. Tang (1999). *Large Marine Ecosystems of the Pacific Rim: Assessment, Sustainability, and Management.* Gland, IUCN.

Skidmore, T. E. and P. H. Smith (2001). *Modern Latin America.* New York, Oxford University Press.

Snedaker, S. C. and J. G. Snedaker, (eds). (1984). *The Mangrove Ecosystem: Research Methods.* Paris, United Nations Educational, Scientific, and Cultural Organization (UNESCO), Scientific Committee on Oceanic Research (SCOR), Working Group 60 on Mangrove Ecology.

Stanley, D. (1994). *Politicas Estatales e Impactos Economicos y Ecologicos de las Nuevas Exportaciones: El Caso de Camaron NTX en Honduras*. Tegucigalpa, Honduras, Universidad Nacional Autonoma de Honduras (UNAH): 1-15.

Stanley, D. (1996). Labor Market Outcomes in a Natural Resource Boom: The Case of Mariculture Exports in Honduras. Doctoral Dissertation, Department of Agricultural Economics, University of Wisconsin-Madison: Madison, WI

Stanley, D. (1998). Explaining Persistent Conflict Among Resource Users: The Case of Honduran Mariculture. *Society and Natural Resources* 11: 267-278.

Stanley, D. (1999a). Labor Market Structure, New Export Crops, and Inequality: The Case of Mariculture in Honduras. *Economic Development and Cultural Change*: 71-89.

Stanley, D. (1999b). Understanding Conflict in Lowland Forest Zones: Mangrove Access and Deforestation Debates in Southern Honduras, in L. U. Hatch and M. E. Swishern (eds.), *Managed Ecosystems: The Mesoamerican Experience*. New York and Oxford, Oxford University Press: pp. 231-236.

Stanley, D. and C. Alduvin (2002). *Science and Society in the Gulf of Fonseca – The Changing History of Mariculture in Honduras*. Washington D.C., Report prepared under the World Bank, NACA, WWF and FAO Consortium Program on Shrimp Farming and the Environment. Work in Progress for Public Discussion. Published by the Consortium.

Stanley, D. and S. Bunnag (2001). A New Look at the Benefits of Diversification: Lessons from Central America. *Applied Economics* 33: 1369-1383.

Stevenson, M. (1995). Mexican Shrimp Farm Plan Under Fire. *United Press International*. Mexico City. (LexisNexis).

Stevis, D. and V. J. Assetto (2001). *The International Political Economy of the Environment: Critical Perspectives*. Boulder, CO and London, Lynne Rienner Publishers, Inc.

Stonich, S. (1991). The Promotion of Non-traditional Agricultural Exports in Honduras: Issues of Equity, Enviroment and Natural Resource Management. *Development and Change*. London, Sage. 22: 725-755.

Stonich, S. C. (1992). Struggling With Honduran Poverty: The Environmental Consequences of Natural Resource-Based Development and Rural Transformations. *World Development* 20(3): 385-399.

Stonich, S. C. (1993). *"I Am Destroying the Land!" The Political Ecology of Poverty and Environmental Destruction in Honduras*. Boulder, CO, Westview Press.

Stonich, S. C. (1996). Reclaiming the Commons: Grassroots Resistance and Retaliation in Honduras. *Cultural Survival Quarterly* (CSQ) 20(1): 31.

Stonich, S. C. and P. Vandergeest (1999). Violence, Environment, and Industrial Shrimp Farming, Toronto, Unpublished Manuscript.

Stonich, S. C. and C. Bailey (2000). Resisting the Blue Revolution: Contending Coalitions Surrounding Industrial Shrimp Farming. *Human Organization* 59(1): 23-36.
Stonich, S. C. and P. Vandergeest (2001). Violence, Environment, and Industrial Shrimp Farming, in M. Watts and N. Peluso (eds.), *Violent Environments*. Ithaca and London, Cornell University Press.

Stromgaard, Peter; Krantz, Lasse; Casse, Thorkil; Nohr, Henning (1993). *Draft Appraisal Report Conservation of Coastal Ecosystems in the Gulf of Fonseca: Nicaragua, Honduras, El Salvador*. San Jose, Costa Rica, Ministry of Foreign Affairs and DANIDA (Danish Development Agency) Ornis Consult Ltd.: 77.

Stryker, S. (1981). Symbolic Interactionism: Themes and Variations, in M. Rosenberg and R. H. Turner (eds.), *Social Psychology: Sociological Perspectives*. New York, Basic Books, Inc., Publishers: 1-29.

Tarrow, S. (1992). *Mentalities, Political Cultures, and Collective Action Frames: Constructing Meaning through Action. Frontiers in Social Movement Theory*. New Haven, Connecticut, Yale University Press: 174-202.

Tarrow, S. (1996). Social Movements in Contentious Politics: A Review Article. *The American Political Science Review* 90(4): 874-883.

Thayer, H. S. (1981). *Meaning and Action: A Critical History of Pragmatism*. Indianapolis, IN, Hackett Publishing Company.

The World Conservation Union (1996). Conservation of Coastal Ecosystems in the Gulf of Fonseca: Technical Project Document, The World Conservation Union, IUCN.

Thorpe, A. (2002). *Agrarian Modernisation in Honduras*. Lewiston, NY, The Edwin Mellen Press.

Tilly, C. (1993-1994). Social Movements as Historically Specific Clusters of Political Performances. *Berkeley Journal of Sociology* 38: 1-30.

Tobey, J., J. Clay, P. Vergne. (1998). *Maintaining a Balance: The Economic, Environmental, and Social Impacts of Shrimp Farming in Latin America*. Narragansett, RI, University of Rhode Island, Coastal Resources Center.

Travis, S. E. (2002). *Hurricane Mitch: Shrimp Population Assessments in the Gulf of Fonseca, Honduras*. United States Geological Survey.

Turner, B. (ed.) (2006). *The Cambridge Dictionary of Sociology*. Cambridge University Press.

United Nations (1960). *Analysis y Proyecciones del Desarrollo Economico XI: El Desarrollo Economico de Honduras*. Mexico City, Mexico, Naciones Unidas Departmento de Asuntos Economicos y Sociales: 174.

United Nations (2004). *Honduras: Evolucion Economica Durante 2003 y Perspectivas para 2004.* Mexico City, Mexico, Naciones Unidas Comision Economica para America Latina y El Caribe – CEPAL: 45.

United Nations Development Program (UNDP), and United Nations Food and Agricultural Organization (FAO) (1976). Datos Economicos de los Principales Cultivos en la Zona Sur de Honduras. Choluteca, Honduras: 48.

United Nations Education, Scientific, and Cultural Organisation (UNESCO) (1979a). The Mangrove Ecosystem: Scientific Aspects and Human Impact. Cali, Colombia, United Nations Educational, Scientific, and Cultural Organization: 46.

United Nations Education, Scientific, and Cultural Organisation (UNESCO) (1979b). The Mangrove Ecosystem: Human Uses and Management Implications. Dacca, Bangladesh, United Nations Educational, Scientific, and Cultural Organization: 19.

United States Agency for International Development (USAID) (1989). Perfil Ambiental de Honduras 1989. Tegucigalpa, Honduras, United States Agency for International Development (USAID): 346.

United States Department of Commerce (DOC) (2001). Honduras Country Commercial Guide 2002. Washington D.C., United States Department of Commerce, Commercial Service: 102.

United States Department of Commerce (DOC) (2001). NOAA Ocean Service, Office of Response and Restoration, 'Hurricane Mitch Reconstruction: Gulf of Fonseca Contaminant Survey and Assessment'. Seattle, WA.

United States Department of State (DOS) (2002). Remarks by Otto Reich, Assistant Secretary of State for Western Hemisphere Affairs, at the Center for Strategic and International Studies: US Foreign Policy in the Western Hemisphere. *Federal News Service, Inc.* (Retrieved via LexisNexis). Washington D.C.

United States Department of State (DOS) (2002). Organizations Receive $3 Million in Grants for Bird Conservation; Migratory birds Considered Vital to Western Hemisphere's Ecosystem. *Federal Information and News Dispatch*, Inc., United States Department of State Office of International Information Programs. Washington D.C.

Valderrama, D. and C. Engle (2002). Economics of Better Management Practices (BMP) for Semi-Intensive Shrimp Farms in Honduras and Shrimp Cooperatives in Nicaragua. Washington D.C., Report prepared under the World Bank, NACA, WWF and FAO *Consortium Program on Shrimp Farming and the Environment.* Work in Progress for Public Discussion. Published by the Consortium.

Van der Borgh, C. (1995). A Comparison of Four Development Models in Latin America. *The European Journal of Development Research* 7(2): 276-296.

Vandergeest, P., M. Flaherty, P. Miller (1999). A Political Ecology of Shrimp Aquaculture in Thailand. *Rural Sociology* 64(4): 573-596.

Vannucci, M. (1986). *Why Mangroves? The Mangroves: Proceedings of National Symposium on Biology, Utilization and Conservation of Mangroves*, Kolhapur; India, Shivaji University.

Varela, J. (2002). The Human Rights Consequences of Inequitable Trade and Development Expansion: The Abuse of Law and Community Rights in the Gulf of Fonseca, Honduras, http://www.earthisland.org/map/ineq-trd.htm. 2004.

Varela Marquez, J., K. Cissna, and C. Stonich. (2001). Artisanal Fisherfolk of the Gulf of Fonseca, in S. C. Stonach (ed.), *Endangered Peoples of Latin America: Struggles to Survive and Thrive*. Santa Barbara (eds), Greenwood Press: 53-70.

Vasquez Cristobal, J. and F. Wainwright W. (2002). Zonificacion de los Bosques de Mangle del Golfo de Fonseca, Honduras, *C.A.* Choluteca, Honduras, Administracion Forestal del Estado Corporacion Hondurena de Desarrollo Forestal (AFE-COHDEFOR) Organizacion Internacional de las Maderas Tropicales (OIMT) Manejo y Conservacion de los Manglares del Golfo de Fonseca, Honduras (PROMANGLE): 78.

Vayda, A. P. and B. B. Walters (1999). Against Political Ecology. *Human Ecology* 27(1): 167-175.

Vayda, A. P., Walters, G; and Setyawati, I; (2004). Doing and Knowing: Questions about Studies of Local Knowledge, in A. Bicker, P. Sillitoe and J. Pottier (eds), *Investigating Local Knowledge: New Directions, New Approaches*. London, Ashgate Publishers.

Velásquez Mazariegos, S. (1998). Informe de Consultoria Clasificacion Digital de Imagenes de Satellite y Elaboracion de la Base de Datos del Area de PROGOLFO. San Jose, Costa Rica, Proyecto Conservacion de los Ecosistemas Costeros del Golfo de Fonseca (PROGOLFO), Unión Mundial para la Naturaleza, UICN.

Vergne, P., Hardin, M. and Dewalt, B. (1993). Environmental Study of the Gulf of Fonseca. Washington D.C., Tropical Research and Development, Inc. for USAID.

Vindel Calix Felipie, L. (1997). Sobreexplotacion de la Vegetacion Manglar y Sus Efectos al Ecosistema de La Bahia de Chismuyo, Honduras: Comparacion Estructural del Bosque Manglar Segun el Tipo de Explotacion Economica. Guatemala City, Guatemala, The Nature Conservancy, World Wildlife Fund, and University of Rhode Island Coastal Resources Center (PROARCA/Costas): 108.

Walters, B. B. (2003). People and Mangroves in the Philippines: Fifty Years of Coastal Environmental Change. *Environmental Conservation* 30(2): 293-303.

Walters, B. B. (2004). Local Management of Mangrove Forests in the Philippines: Successful Conservation or Efficient Resource Exploitation? *Human Ecology* 32(2): 177-195.

Walters, B. B., A. Cadelina, A. Cardano, and E. Visitacion. (1999). Community History and Rural Development: Why Some Farmers Participate More Readily Than Others. *Agricultural Systems* 59: 193-214.

Walton, J. (1989). Debt, Protest, and the State in Latin America. Power and Popular Protest: Latin American Social Movements. S. Eckstein. Berkeley, University of California Press: 299-328.

Warren, D. S. (2000). Honduras-FHIS V. Washington D.C., World Bank.

Watts, M. and N. Pelosi (2001). *Violent Environments*, Ithaca and London, Cornell University Press.

Whatmore, S. (1999). Hybrid Geographies: Rethinking the 'Human' in Human Geography, in D. Massey, J. Allen and P. Sarre (eds.), *Human Geography Today*. Cambridge, Polity Press: 22-39.

Weinberg, B. (1991). *War on the Land: Ecology and Politics in Central America*. London, Zed Books Ltd.

Wignaraja, P. (1993). *New Social Movements in the South: Empowering the People*. London, Zed Books Ltd.

Wilburn, S. M. (2005). Socioeconomic Data, Southern Honduran Coastal Communities. Original Research.

Windevoxhel, N. J. (1997). PROARCA/SIGMA Strategy Document. Guatemala City, Guatemala, El Programa Ambiental Regional para CentroamÈrica (PROARCA).

Windevoxhel, N. J.; Rodriguez, J.J.; and Lahmann, E.J. (1997). Situation of Integrated Coastal Zone Management in Central America; Experiences of the IUCN Wetlands and Coastal Zone Conservation Program. Moravia, Costa Rica, World Union for the Conservation of Nature (IUCN): 31.

Wolf, E. R. (1999). Cognizing "Cognized Models". *American Anthropological Association* 101(1): 19-22.

World Wildlife Fund (1998). World Wildlife Fund Position Statement on Shrimp Aquaculture. Tegucigalpa, Honduras, World Wildlife Fund (WWF).

Wright, L. (1989). *US Policy Toward El Salvador, Honduras, and Nicaragua, 1979-1985*. M.Phil. Thesis, Department of International Relations. London, London School of Economics: 339.

Wunder, S. (1995). *The Gulf of Fonseca Project Proposal an Appraisal of Selected Aspects: Report Prepared for the Royal Danish Embassy*. Managua, Nicaragua, Danish Department of International Development Cooperation (DANIDA): 34.

Wynia, G. W. (1978). *The Politics of Latin American Development*. Cambridge, Cambridge University Press.

Yanez-Arancibia, A. and A. L. Lara-Dominguez, (eds). (1999). *Ecosistemas de Manglar en America Tropical.* Xalapa, Mexico, Instituto de Ecologia; UICN/ORMA Costa Rica; NOAA/NMFS Silver Spring, MD USA.

Zimmerer, K. S. (2000). The Reworking of Conservation Geographies: Nonequilibrium Landscapes and Nature-Society Hybrids. *Annals of the Association of American Geographers* 90(2): 356-369.

Zuleta, A. V. (1993). Modernizacion del Sector Publico Pesquero: Informe Technico Ordenacion de las Principales Pesquerias de Crustaceos (Version Preliminar). Tegucigalpa, Honduras: 132.

LaVergne, TN USA
22 December 2009
167810LV00007B/11/P